The Ecological Management and Sustainable Development of Forests

The Ecological Management and Sustainable Development of Forests

Chao Wang
Fan Zhang
Wei Liu

MDPI

Basel • Beijing • Wuhan • Barcelona • Belgrade • Novi Sad • Cluj • Manchester

Editors

Chao Wang
School of Labor Economics
Capital University of Economics and Business
Beijing
China

Fan Zhang
Institute of Geographic Sciences and Natural Resources Research
Chinese Academy of Sciences
Beijing
China

Wei Liu
School of Geography and Environment
Shandong Normal University
Jinan
China

Editorial Office
MDPI AG
Grosspeteranlage 5
4052 Basel, Switzerland

This is a reprint of articles from the Special Issue published online in the open access journal *Forests* (ISSN 1999-4907) (available at: www.mdpi.com/journal/forests/special_issues/U4Y269205L).

For citation purposes, cite each article independently as indicated on the article page online and as indicated below:

Lastname, A.A.; Lastname, B.B. Article Title. *Journal Name* **Year**, *Volume Number*, Page Range.

ISBN 978-3-7258-2240-9 (Hbk)
ISBN 978-3-7258-2239-3 (PDF)
doi.org/10.3390/books978-3-7258-2239-3

Cover image courtesy of Chao Wang

Contents

Preface

In the context of climate change and rapid urbanization, it is urgent to investigate the ecological conditions of forests, analyze their ecological process, assess their ecological status, and explore effective measurements and policies to ensure their sustainable development. This Reprint focuses on environmental science/Earth science/geography, providing an overview of the most recent advances in the ecological management and sustainable development of forests.

Our intention is to build a platform for communication so that we can discuss our research results in a more targeted way. This Reprint is for anyone interested in environmental science, ecology, forest management, and other related directions, which can provide students, scholars, and managers with a theoretical basis and typical case references. Limited by time and effort, we present only a limited number of studies. In the future, we will continue to make efforts to achieve the sustainable management of forests and strongly communicate with all parties involved.

As the Editors of this Reprint, we thank the Editors of *Forests* for providing us with a platform and their editing and publishing work. We thank all the authors of the 12 papers collated in this Special Issue. All authors provide novel findings and perspectives to the direction of forest ecosystem and management.

This Reprint and Special Issue were funded by the National Natural Science Foundation of China (Grant No. 72304192).

Chao Wang, Fan Zhang, and Wei Liu
Editors

forests

MDPI

Editorial

The Ecological Management and Sustainable Development of Forests

Chao Wang [1,*], Fan Zhang [2] and Wei Liu [3]

[1] School of Labor Economics, Capital University of Economics and Business, Beijing 100070, China
[2] Institute of Geographic Sciences and Natural Resources Research, Chinese Academy of Sciences, Beijing 100101, China; zhangf.ccap@igsnrr.ac.cn
[3] School of Geography and Environment, Shandong Normal University, Jinan 250358, China; liuw@sdnu.edu.cn
* Correspondence: wangc@cueb.edu.cn

Forest ecosystems play a dominant role in regulating climate change and sustainable development. In the context of climate change and rapid urbanization, forest ecosystems exchange energy and materials with other open natural–social–economic systems and face threats of degradation to their structure, function, and services. Climate change alters forest structure due to variations in physical conditions, and social and economic activities result in considerable land-use and land-cover change (LUCC), as well as having complex impacts on forests and other natural systems. In line with the requirements of the Sustainable Development Goals (SDGs), many ecological protection measures and programs have been established to strengthen forest ecosystem function. To summarize, it is urgent to investigate the ecological conditions of forest, analyze their ecological process, assess their ecological status, and explore effective measurements and policies to ensure their ecological management and sustainable development.

This Special Issue, entitled "The Ecological Management and Sustainable Development of Forests", presents 12 high-quality original research papers, including both micro- and macro-scale studies (Figure 1), that are critical to clarifying the mechanisms that alter the structure and function of forest ecosystems, strengthening ecosystem restoration and conservation, and supporting the achievement of the SDGs. This Special Issue gives an overview of the most recent advances in the ecological management and sustainable development of forests.

Citation: Wang, C.; Zhang, F.; Liu, W. The Ecological Management and Sustainable Development of Forests. *Forests* **2024**, *15*, 871. https://doi.org/10.3390/f15050871

Received: 7 May 2024
Accepted: 15 May 2024
Published: 17 May 2024

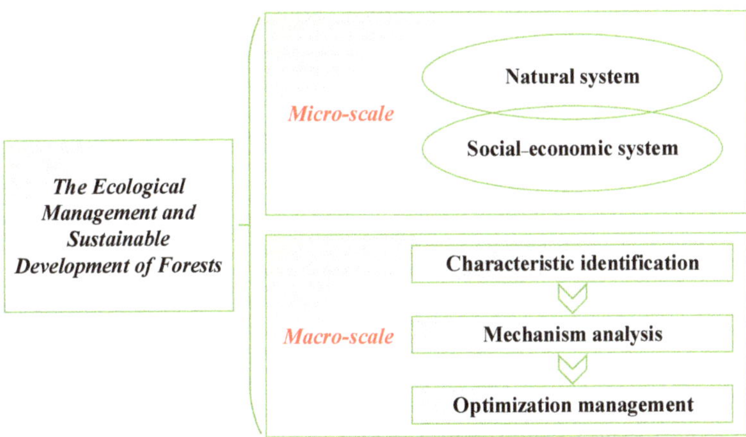

Figure 1. The 12 papers published in this Special Issue.

At the micro-scale, the included authors carried out field experiments and surveys, investigated the characteristics of the forestry economy, and explored the mechanisms and influencing factors in the alteration of trees or forests. Three papers [1–4] were published on the topic of natural systems. Hua et al. (2023) [1] conducted a field experiment in a Eucalyptus plantation to compare the effects of three-year fertilization and five-year dry-season irrigation on height growth rate during the wet and dry seasons. The results revealed that long-term dry-season irrigation significantly increased the height growth rate of *Eucalyptus urophylla* × *E. grandis* by improving the hydraulic conductivity and photosynthetic rate of leaves during the dry season. Moreover, the results revealed that the leaves' photosynthetic capacity contributed more to accelerating height growth than water conduction within the leaves. Oszako et al. (2023) [2] analyzed the responses of young pine seedlings to different types of soil media mixtures: phosphogypsum mixed with organic ash or with sewage sludge. The results showed that the phosphogypsum-based preparations used showed no harmful (toxic) effects on the potted pine seedlings. The loosely prepared preparation made from a mixture of phosphogypsum and organic ash began to positively affect the development of the seedlings' root systems, and it was also easier to mix with the soil surface than phosphogypsum with sewage sludge, which was also sticky. The authors also found that obtaining detailed conclusions regarding heavy metals and their effects on seedling development, and the changes in the microbiome, required observations to be made over a longer period. Shi et al. (2022) [3] compared three models for assessing aboveground biomass based on a sample of 30 trees, introduced a linear seemingly unrelated regression (SUR) approach to determine the best model, and estimated the biomass and carbon storage of *T.* Zhongshanshan stands in the Yangtze River Basin (YRB) in China. The results showed that the total tree biomass values were 53.43, 84.87, 140.67, 192.71, and 156.65 t ha^{-1} in 9-, 11-, 13-, 15-, and 22-year-old *T.* Zhongshanshan. The current *T.* Zhongshanshan stands in the YRB area can store 124.76 to 217.64 t ha^{-1} of carbon. Wang, Zhao et al. (2023) [4] carried out household surveys and found that cooperative membership (herbal medicine planting) leads to considerable improvements in forest farmers' household income and assets. Householders who were migrant workers were more likely to make the decision to participate in cooperatives compared with non-migrant workers. In addition, this paper outlined the problems in the current development of cooperatives and proposed feasible strategies and policy recommendations to guide policy for sustainable forest development.

At the macro-scale, this Special Issue discusses three aspects of forest ecosystem characteristics and management, i.e., characteristic identification [5–8], mechanism analysis [9–11], and optimization management [12].

Identifying the spatio-temporal characteristics of land use, ecosystem services, vegetation cover, etc., can provide basic evidence of impact mechanisms and inform related policy formulation. Xing et al. (2023) [5] focused on the ecological impacts of ecological restoration strategies and projects, and investigated the evolution of the land-use transition (LUT) pathways and ecosystem service value (ESV) in four geomorphological regions of the Beiluo River Basin. The results indicated that forest land increased by 18.27%, and the overall ESV increased by USD 3.209 billion (54.16%). The authors suggested that ecological restoration projects enhanced the main ecological function of individual regions, and conducted a detailed analysis of the impacts of the "Grain for Green" project. Wen et al. (2023) [6] proposed a three-step approach to explore the multi-aspect dynamics of land change, including the differences among land-use categories, spatial characteristics of urban expansion, and forest fragmentation, and explored the characteristics of land-use change in a low-income mountainous city (Enshi City, China). The findings confirmed that recent planning measures were effective in restoring the natural environment, and that the identified key areas can support sustainable forest management during urban growth. Liu et al. (2023) [7] explored the overall change characteristics of vegetation cover and the distribution patterns of different terrains on the complex terrain of arid and semi-arid Xinjiang. The authors integrated the ESTARFM model, the dimidiate pixel model, unary linear regression, and the digital elevation model. The results showed that the overall vegetation cover was high, and serious degradation and improvement phenomena were both identified. Forest land is important in resolving the ecological risks of the lakeside area and building its ecological base. Wang, Wu et al. (2023) [8] explored the spatial and temporal evolutionary features of forest land in the Erhai rim region using bivariate spatial autocorrelation and multi-scale geographical weighted regression (MGWR) models. The results revealed that reasonable forest land expansion can effectively alleviate the growth of landscape ecological risk in the Erhai rim region, whereas the shrinkage of forest land would aggravate this risk.

Clarifying systemic mechanisms is conducive to understanding the relationships between forest ecosystems and other related systems, and more comprehensively and accurately identifying the key aspects of forests. Teng et al. (2023) [9] analyzed the characteristics of vegetation coverage (VC) change based on high-resolution remote sensing data in the Qilian Mountains (QLM) and identified the climatically and anthropogenically driven pattern of VC change. They found that VC presented a remarkable upward trend from 1990 to 2018 and identified a significant positive correlation with precipitation change and annual average temperature, as well as a significant negative correlation with annual average precipitation, current VC status, livestock density, and slope. Dong et al. (2023) [10] aimed to understand the determining factors of the spatial distribution of forest cover, conducting a nuanced case study in Fujian Province, China, in 2020. The paper showed the specific relationships between the spatial distribution of forest cover and natural conditions and socio-economic factors. The results indicated that natural factors could shape the spatial distribution of forest cover, while socio-economic factors could play a more significant role in the spatial distribution of forest cover. Mo et al. (2023) [11] focused on the *Pinus massoniana* (PM) plantations on Guangxi Paiyang Forest Farm and studied the synergistic and trade-off relationships between ecosystem services. The results showed that the ecosystem services maintained significant positive correlations (synergy), with a mutually reinforcing relationship. Their results and conclusions are essential for maintaining the structure, function, and health of plantation forest ecosystems.

Achieving the sustainable management and high-quality development of forest ecosystems is the ultimate goal of this Special Issue, with most of the papers presented proposing specific strategies according to their results and conclusions. Taking Poland as a typical case, Referowska-Chodak and Kornatowska (2023) [12] analyzed the effects of Poland's forest management evolution over the last 75 years on forest biodiversity. The evolution of forest management practice implemented in Poland's forests by the State Forests National Forest Holding led to the restoration of/an increase in biodiversity. Their paper also found some unsolved organizational, political, financial, conceptual, and natural/anthropogenic issues.

Author Contributions: Conceptualization, C.W., F.Z. and W.L.; writing—original draft preparation, C.W.; writing—review and editing, F.Z. and W.L.; project administration, C.W. All authors have read and agreed to the published version of the manuscript.

Funding: This research was funded by the National Natural Science Foundation of China (Grant No. 72304192).

Acknowledgments: We are grateful to the authors who submitted papers to this Special Issue, the editors, and the reviewers for their constructive work.

Conflicts of Interest: The authors declare no conflicts of interest.

References

1. Hua, L.; Chen, P.; Luo, J.; Su, Y.; Li, J.; He, Q.; Yang, H. The Impact of Long-Term Dry-Season Irrigation on Eucalyptus Tree Height Growth: Insights from Leaf Photosynthesis and Water Conduction. *Forests* **2023**, *14*, 2017. [CrossRef]
2. Oszako, T.; Pasławski, T.; Szulc, W.; Rutkowska, B.; Rutkiewicz, A.; Kukina, O.; Bakier, S.; Borowik, P. Short-Term Growth Response of Young Pine (*Pinus silvestris*) Seedlings to the Different Types of Soil Media Mixture with Phosphogypsum Formulations under Poland Forest Environmental Conditions. *Forests* **2023**, *14*, 518. [CrossRef]

3. Shi, Q.; Hua, J.; Creech, D.; Yin, Y. Biomass Estimation and Carbon Storage of Taxodium Hybrid Zhongshanshan Plantations in the Yangtze River Basin. *Forests* **2022**, *13*, 1725. [CrossRef]
4. Wang, J.; Zhao, Z.; Gao, L. Research on the Impact of Cooperative Membership on Forest Farmer Household Income and Assets—Case Study from Liaoning Herbal Medicine Planting Cooperatives, China. *Forests* **2023**, *14*, 1725. [CrossRef]
5. Xing, J.; Zhang, J.; Wang, J.; Li, M.; Nie, S.; Qian, M. Ecological Restoration in the Loess Plateau, China Necessitates Targeted Management Strategy: Evidence from the Beiluo River Basin. *Forests* **2023**, *14*, 1753. [CrossRef]
6. Wen, C.; Wang, L. Landscape Dynamics in a Poverty-Stricken Mountainous City: Land-Use Change, Urban Growth Patterns, and Forest Fragmentation. *Forests* **2022**, *13*, 1756. [CrossRef]
7. Liu, Z.; Chen, D.; Liu, S.; Feng, W.; Lai, F.; Li, H.; Zou, C.; Zhang, N.; Zan, M. Research on Vegetation Cover Changes in Arid and Semi-Arid Region Based on a Spatio-Temporal Fusion Model. *Forests* **2022**, *13*, 2066. [CrossRef]
8. Wang, M.; Wu, Y.; Wang, Y.; Li, C.; Wu, Y.; Gao, B.; Wang, M. Study on the Spatial Heterogeneity of the Impact of Forest Land Change on Landscape Ecological Risk: A Case Study of Erhai Rim Region in China. *Forests* **2023**, *14*, 1427. [CrossRef]
9. Teng, Y.; Wang, C.; Wei, X.; Su, M.; Zhan, J.; Wen, L. Characteristics of Vegetation Change and Its Climatic and Anthropogenic Driven Pattern in the Qilian Mountains. *Forests* **2023**, *14*, 1951. [CrossRef]
10. Dong, J.; Zhou, C.; Liang, W.; Lu, X. Determination Factors for the Spatial Distribution of Forest Cover: A Case Study of China's Fujian Province. *Forests* **2022**, *13*, 2070. [CrossRef]
11. Mo, R.; Wang, Y.; Mo, Y.; Li, L.; Ma, J. The Trade-Offs and Synergies of Ecosystem Services in *Pinus massoniana* Lamb. Plantations in Guangxi, China. *Forests* **2023**, *14*, 581. [CrossRef]
12. Referowska-Chodak, E.; Kornatowska, B. Effects of Forestry Transformation on the Ecosystem Level of Biodiversity in Poland's Forests. *Forests* **2023**, *14*, 1739. [CrossRef]

forests

Article

The Impact of Long-Term Dry-Season Irrigation on *Eucalyptus* Tree Height Growth: Insights from Leaf Photosynthesis and Water Conduction

Lei Hua [1], Penglong Chen [1], Jun Luo [1], Yan Su [2], Jiyue Li [2], Qian He [2,*] and Huizhu Yang [1,3,*]

1 State Environmental Protection Key Laboratory of Urban Ecological Environment Simulation and Protection, South China Institute of Environmental Sciences, Ministry of Ecology and Environment of China, Guangzhou 510655, China; hualei@scies.org (L.H.); chenpenglong@scies.org (P.C.); luojun@scies.org (J.L.)
2 Guangdong Key Laboratory for Innovative Development and Utilization of Forest Plant Germplasm, College of Forestry and Landscape Architecture, South China Agricultural University, Guangzhou 510642, China; suyan@scau.edu.cn (Y.S.); ljyue@scau.edu.cn (J.L.)
3 The Shanghai Key Lab for Urban Ecological Processes and Eco-Restoration, East China Normal University, Shanghai 200241, China
* Correspondence: heqian@scau.edu.cn (Q.H.); yanghuizhu@scies.org (H.Y.)

Abstract: Tree height is a crucial characteristic of plant ecological strategies and plantation productivity. Investigating the influence of dry-season irrigation on the tree height growth in *Eucalyptus* plantations contributes to a deeper understanding of precise improvement and sustainable development in such plantations. We conducted a field experiment in a *Eucalyptus* plantation with three-year fertilization and five-year dry-season irrigation to compare their effects on height growth rate during wet vs. dry seasons. Our findings revealed that long-term dry-season irrigation significantly increased the height growth rate of *Eucalyptus urophylla* × *E. grandis* by improving leaf hydraulic conductivity and photosynthetic rate during the dry season. However, in the wet season, the tree height growth rate in the fertilization treatment outperformed the other treatments significantly. Interestingly, we also found that leaf photosynthetic capacity contributed more to accelerating height growth than water conduction within the leaves. By examining the differences in leaf structural and functional traits, our results shed light on the impact of long-term dry-season irrigation on the height growth of *E. urophylla* × *E. grandis* plantations. Furthermore, this research provides both theoretical and empirical evidence supporting the application of dry-season irrigation and the potential for further enhancing plantation productivity in seasonally arid areas.

Keywords: height growth rate; photosynthetic rate; leaf hydraulic conductivity; *Eucalyptus* plantation

check for updates

Citation: Hua, L.; Chen, P.; Luo, J.; Su, Y.; Li, J.; He, Q.; Yang, H. The Impact of Long-Term Dry-Season Irrigation on *Eucalyptus* Tree Height Growth: Insights from Leaf Photosynthesis and Water Conduction. *Forests* **2023**, *14*, 2017. https://doi.org/10.3390/f14102017

Academic Editor: Rosana López Rodríguez

Received: 9 September 2023
Revised: 24 September 2023
Accepted: 3 October 2023
Published: 8 October 2023

1. Introduction

The *Eucalyptus* tree, widely known for its economic and ecological significance, plays a crucial role in various industries, including timber, paper, and biomass production [1]. *Eucalyptus* plantations are widely cultivated in the southern region of China, which is characterized by a subtropical monsoon climate featuring distinct dry and wet seasons. Specifically, the dry season spans from October to March of the subsequent year, during which the annual precipitation accounts for less than 20% of the total rainfall [2]. Relevant studies have demonstrated that the growth of *Eucalyptus* is adversely affected during the dry season due to water scarcity [3–5]. Hence, delving into the physiological mechanisms by which *Eucalyptus* plantations adapt to dry-season irrigation becomes essential in driving their growth and productivity to new heights.

The growth of trees requires plants to synthesize organic matter through photosynthesis and obtain sufficient water to maintain the normal metabolism of canopy organs. *Eucalyptus* trees are highly dependent on water availability for their physiological processes, including photosynthesis and water conduction [6]. Photosynthesis, the primary process responsible for converting solar energy into chemical energy, directly influences the growth and development of trees [7,8]. Studies have shown that water scarcity significantly reduces photosynthetic activity [9,10], leading to decreased growth rates in *Eucalyptus* trees. Additionally, water conduction, the process through which water is transported from roots to leaves, plays a crucial role in maintaining water balance and facilitating tree growth [11–13]. Previous studies have commonly examined the effects of water availability on *Eucalyptus* growth by analyzing traits associated with photosynthesis and water conduction separately [3,14,15]. However, there is a lack of research examining the extent to which leaf photosynthesis and water transport contribute to the growth of *Eucalyptus* tree height under dry-season irrigation in seasonal arid regions.

Structural traits have significant effects on regulating both leaf photosynthesis and hydraulic functions in plants [16–18]. Among these, leaf thickness and vein density play critical roles in regulating leaf photosynthetic rate and hydraulic conductivity. Leaf water conductivity refers to the ability of leaves to effectively transport water to chloroplasts and has an impact on the accumulation of carbon in leaves [19]. Leaf thickness is a key factor in regulating the diffusion distance for CO_2 to reach the chloroplasts. Furthermore, it can impact the movement of water to the photosynthetic site and its subsequent transpiration from the leaf [20,21]. Additionally, leaf veins serve as the pathway of least resistance for water transport within leaf blades, and their abundance also determines the speed of transport [22]. Increased vein density and reduced leaf thickness can effectively reduce the transportation distance for CO_2 and H_2O, thereby enhancing the photosynthetic rate and leaf hydraulic conductivity [21,23].

Furthermore, fertilization practices are commonly employed in *Eucalyptus* plantations to enhance growth and productivity, but the productivity of *Eucalyptus* plantations in China lags behind that of countries such as Brazil [24]. Therefore, by modifying traditional planting practices, we aim to optimize the productivity of *Eucalyptus* plantations by implementing drip irrigation during the dry season, aiming to alleviate the growth limitations imposed by seasonal drought on plantations. Previous studies have suggested that fertilization can improve photosynthetic efficiency and water uptake capacity in trees [25,26]. However, the impact of long-term dry-season irrigation after fertilization on leaf photosynthesis and water conductivity capacity during both wet and dry seasons still lacks research.

In this study, various dry-season irrigation and fertilization treatments were implemented for the *E. urophylla* × *E. grandis* plantations. Height growth rate, leaf functional traits, and structural traits were determined among the four different treatments. We aim to investigate the following inquiries: (1) How does dry-season irrigation affect the leaf structural and functional traits of *E. urophylla* × *E. grandis* plantations during both wet and dry seasons? (2) How do these alterations affect the plant height growth rate under dry-season irrigation?

2. Materials and Methods

2.1. Study Site and Plant Material

The experimental forest site was situated at the Teaching & Research Base of South China Agricultural University (SCAU) in Zengcheng District, Guangzhou (23°14′48″ N, 113°38′20″ E). This region has a typical subtropical monsoon climate with an average annual precipitation of approximately 1900 mm, with the majority occurring during the wet season from April to September. The average annual temperature is 21.6 °C. Analysis of meteorological data for Guangzhou City reveals that during the dry season from 2017 to 2022, Zengcheng District experienced an average monthly precipitation of only 55.68 mm (Figure 1). Additionally, the average monthly temperature of 18.39 °C during the dry season falls within the temperature range associated with maximum growth of *Eucalyptus* trees, as studies have shown that temperatures above 18 °C are conducive to their growth [27]. The soil is the common red soil in South China. The initial experimental soil conditions were as follows: pH: 4.92, organic matter: 7.03 g/kg, total nitrogen: 0.35 g/kg, total phosphorus: 0.15 g/kg, and total potassium: 8.83 g/kg. The field water holding capacity of the soil was

20.41%, with a bulk density of 1.55 g/cm^3. In April 2017, healthy *E. urophylla* × *E. grandis* seedlings, measuring approximately 20–35 cm in height and free from any signs of disease or mechanical damage, were planted under optimal conditions. The planting density was 1667 plants per hectare.

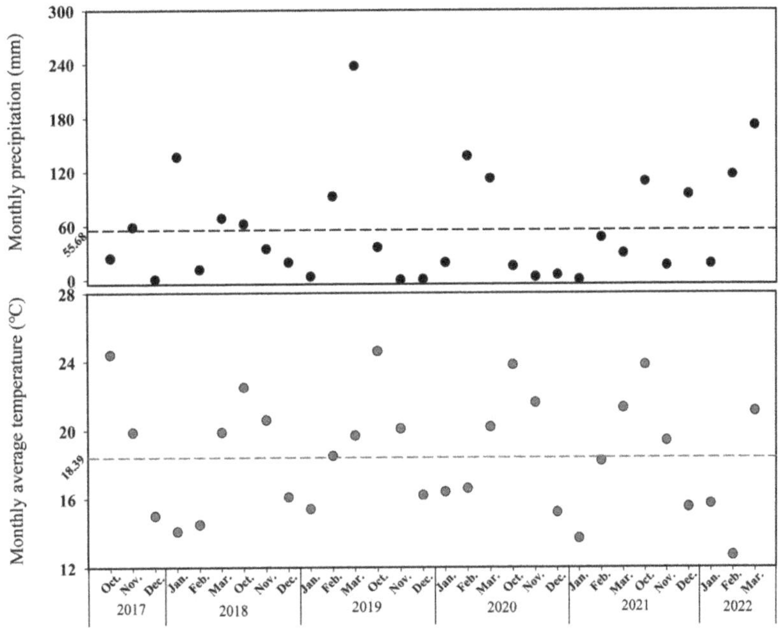

Figure 1. The monthly precipitations and average temperatures in the dry seasons of Zengcheng District, Guangzhou, from 2017 to 2022. Dash lines indicate the average monthly precipitation (55.68 mm) and the average monthly temperature (18.39 °C), respectively.

2.2. Experimental Design

The experiment utilized an orthogonal design, incorporating irrigation and fertilizer treatments [5]. During the dry season of each year, water was provided exclusively through drip irrigation equipment. The forestland was partitioned into five terraces, with each terrace housing four distinct treatments, as illustrated in Figure 2. The plots varied in size from 20 to 92 trees and were arranged randomly. The treatments for each terrace consisted of (1) a blank control group (CK); (2) dry-season irrigation only (W); (3) conventional fertilization only (F); and (4) both dry-season irrigation and fertilization (WF). To minimize interference from neighboring treatments, the measured trees were positioned in the center of each treated plot. Dry-season irrigation involved a 4 h duration, maintaining the soil's relative water content (measured at a depth of 40 cm and 40 cm away from the trees) above 80% for three days following irrigation. Drip irrigation was applied for a total of 8 h per week at a rate of 4 L/h, amounting to 32 L/week per tree. The irrigation period spanned from 1 October 2017 to 31 March 2022, covering five consecutive dry seasons. The collected precipitation in the wet season and some groundwater were used for irrigation. A waterproof and anti-corrosion partition, 50 cm in depth, was buried between each treatment plot. According to Yang et al. [28], although the soil water content fluctuated, the treatments with dry-season irrigation (W and WF) were almost always higher than those without water supply (CK and F). Fertilizer application rates matched those typically used for *Eucalyptus* production in China and were sourced from Guangdong Dayi Agroforestry Ecological Technology Co. The foundation fertilizer was added in March 2017 with 400 g per tree, with a total of 24 g N, 72 g P_2O_5, and 24 g K_2O. For the F and WF treatments, the first superficial fertilizer was provided in July with 300 g per tree and contained 45 g N, 21 g P_2O_5, and 24 g K_2O. The second and third topdressings, with 400 g per tree containing 60 g N, 28 g P_2O_5, and 32 g K_2O, were applied in July 2018 and July 2019, respectively. Fertilization was carried out continuously for three years (F and WF), consistent with prevailing practices in *Eucalyptus* plantations across China. Following five years of treatment, data collection was conducted in August and October of 2021, as well as January of 2022. Each plot was replicated three times, resulting in a total of 15 samples per treatment group.

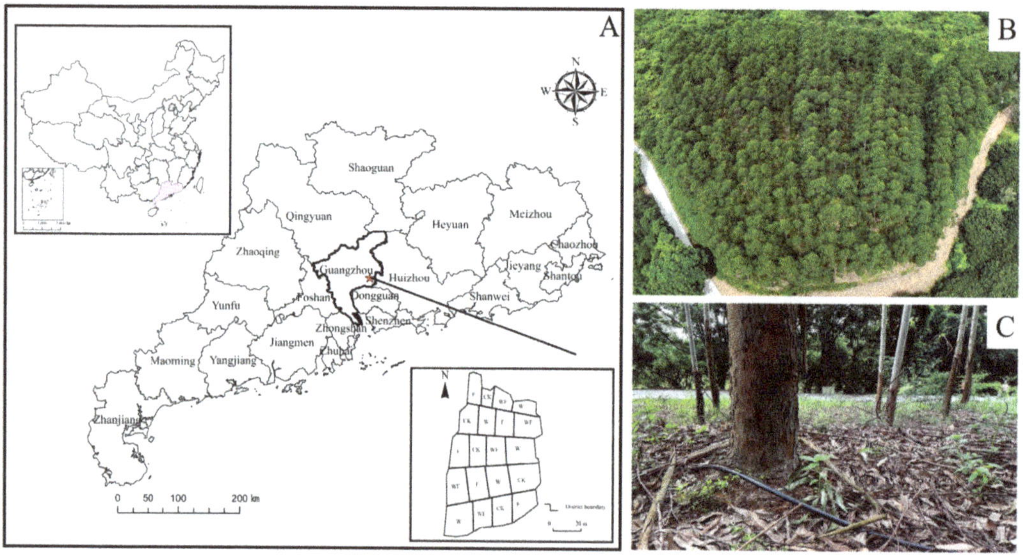

Figure 2. The information on experimental plots (**A**), overview of the site (**B**), and the dry-season irrigation facilities (**C**). CK, W, F, and WF represent control conditions, 5 years of dry-season irrigation, only conventional fertilization (only fertilizer in the first three years), and both dry-season irrigation and fertilization.

2.3. Measurement of Height Growth Rate

The tree height was measured using a comparative method with a special carbon fiber tall tree pruner (net length of 20 m). Tree height measurements were taken for all trees of the four treatments in August and October of 2021 and January of 2022 (a total of 745 trees). By calculating the difference in tree height between two months and dividing it by the number of days elapsed, we determined the height growth rate (H_r) of the wet and dry seasons, respectively, for each treatment.

2.4. Measurement of Leaf Functional Traits

Leaf gas exchange measurements were performed on cloudless days in August 2021 and January 2022, between 9:00 am and 11:00 am, with the Li6800 photosynthetic instrument (Li-Cor, Lincoln, NE, USA). For each treatment, three measurements were carried out, whereby the photosynthetic rate (A_n) was measured in five fully developed leaves per plant. Therefore, a total of 225 values of the photosynthetic rate were measured for each treatment in each season. The photosynthetic photon flux density was set at 1500 mol m^{-2} s^{-1} to ensure consistent light-saturated photosynthetic rates among all treatments. Furthermore, we kept the ambient CO_2 levels at 400 μmol mol^{-1} and temperature at 26 °C throughout all the measurements. To minimize the effects of vapor pressure deficit (VPD), the chamber's relative humidity was controlled within the range of 70% to 90%. Before measurement, the leaves were allowed to stand for 5–10 min under the above conditions to stabilize photosynthetic parameters.

Leaf hydraulic conductance (K_{leaf}) was measured following the methods from Brodribb and Holbrook [29]. Branches with leaves were selected and cut at predawn and quickly put into black bags along with damp towels for 1 h. The initial water potential (ψ_{leaf}) of leaves was measured with the pressure chamber (PMS Instruments, Albany, OR, USA). The adjacent leaves were then cut in water and rehydrated for 10 s. The water potential of the rehydrated leaves was immediately measured. Leaf hydraulic conductance was calculated using the following formula:

$$K_{leaf} = C \times \ln(\psi_0/\psi_f)/t$$

where C represents the leaf capacitance, ψ_0 denotes the initial water potential, ψ_f represents the leaf water potential after rehydration, and t is the duration of rehydration (10 s). The values of C before and after turgor loss were determined by leaf pressure–volume relationships and normalized by leaf area, as follows:

$$C = \Delta RWC/\Delta\psi_1 \times (DM/LA) \times (WM/DM)/M$$

where RWC indicates the relative water content, DM represents leaf dry mass (g), LA means leaf area (m^2), WM represents the mass of leaf water at 100% RWC (g), and M is the molar mass of water (g mol^{-1}). One branch per tree was sampled, and three values were measured for each branch, resulting in a total of 45 measurements for each treatment in each season.

2.5. Measurement of Leaf Structural Traits

Vein density was determined following the method described by Sack and Scoffoni [22]. Five leaves were sampled from each tree, and from each leaf, three rectangles (with an area of at least 1 cm^2) were obtained from the top, middle, and bottom sections. These sections were then soaked in a 5% *w*/*v* $NaOH/H_2O$ solution for a period of five days. Subsequently, the leaves were rinsed with water and immersed in a commercial bleach solution (6% *w*/*v* $NaClO/H_2O$) for approximately 15 min. After a second rinse with water, the leaves were stained with a 1% *w*/*v* safranin/ethanol solution for 10 min. Following this, the leaves were kept in 100% ethanol for 20 min before being transferred back to water for imaging. Photos of each section were taken using an optical microscope equipped with a digital camera (Leica ICC50 W). The length of all veins in each image was measured using ImageJ software (version 1.53t) [30]. Leaf vein density (*VD*) was calculated as the sum of vein length divided by the leaf area.

To obtain cross-sections, the top, middle, and bottom of each leaf were carefully cut. These sections were then photographed using the aforementioned light microscope. Furthermore, the leaf thickness (*LT*) was measured using ImageJ software (version 1.53t). For *VD* and *LT*, a total of 225 values were measured for each treatment in each season, respectively.

2.6. Statistical Analysis

The ANOVA analysis was employed to compare trait values across the four treatments. Associations between measured traits were analyzed separately for the wet and dry seasons using Pearson correlation in SPSS (Chicago, IL, USA). To analyze all data together, standard major axis (SMA) regression was conducted using the SMATR software (version 2.0) [31]. Additionally, the "varpart" in the R package vegan v.2.5–6 [32] was utilized to estimate the unique contributions of K_{leaf}, A_n, and leaf structural traits (including *LT* and *VD*) to H_r.

3. Results

3.1. Response of Functional Traits to Five-Year Dry-Season Irrigation

The height growth rate (H_r) and the leaf functional traits differed among treatments in the wet and dry seasons (Figure 3). For leaf hydraulic conductivity (K_{leaf}), during the wet season, the trees subjected to fertilization (F and WF) exhibited a significantly higher

K_{leaf} compared to the non-fertilized trees (CK and W); however, during the dry season, the K_{leaf} of the dry-season irrigation treatments (W and WF) was significantly higher than that of the non-irrigated treatment (CK and F) (Figure 3A). Regarding the photosynthetic rate (A_n), during the wet season, there was no significant difference among the CK, F, and WF treatments, but the A_n of W was significantly lower than that of the F treatment. During the dry season, the A_n of the dry-season irrigation treatments (W and WF) was significantly higher than that of the non-irrigated treatment (CK and F) (Figure 3B). As for the height growth rate (H_r), during the rainy season, there was no significant difference among the CK, W, and WF treatments, but they were all significantly lower than the F treatment. During the dry season, the H_r of the dry-season irrigation treatments (W and WF) was also significantly higher than that of the non-irrigated treatment (CK and F) (Figure 3C).

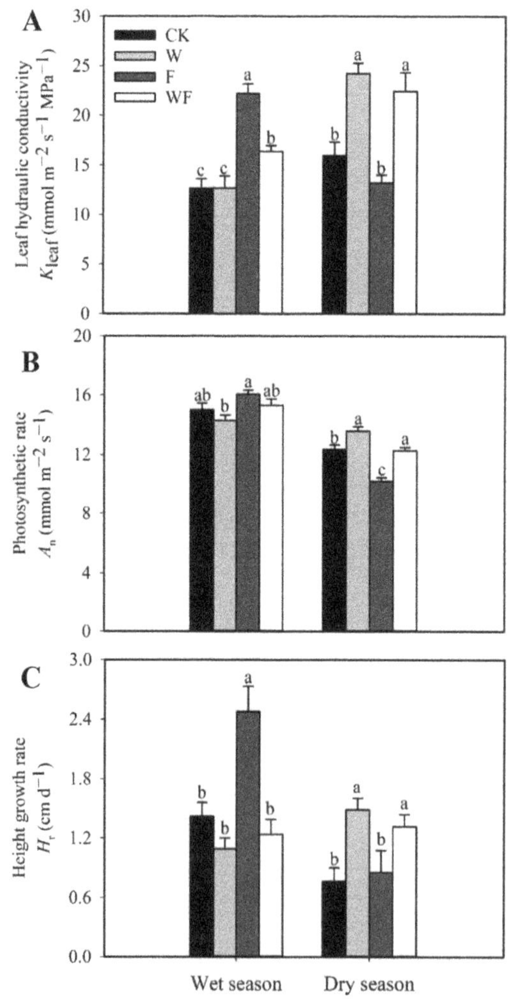

Figure 3. Leaf hydraulic conductivity (**A**), photosynthetic rate (**B**), and height growth rate (**C**) of *Eucalyptus urophylla* × *grandis* under control conditions (CK), dry-season irrigation (W), fertilization (F), and both dry-season irrigation and fertilization (WF) in wet season and dry season. Lowercase letters indicate significant differences ($p < 0.05$). The total values for analysis were 360 (**A**), 1800 (**B**), and 1490 (**C**), respectively.

3.2. Response of Structural Traits to Five-Year Dry-Season Irrigation

Different treatments did not have a significant effect on leaf thickness (*LT*), but they significantly affected leaf vein density (Figures 4 and S1). During the wet season, the F treatment exhibited significantly higher leaf vein density (*VD*) compared to the other treatments, while during the dry season, the F treatment had the lowest *VD*. The *VD* of the dry-season irrigation treatments (W and WF) was significantly higher than that of the non-irrigation treatments (CK and F) during the dry season (Figure 4).

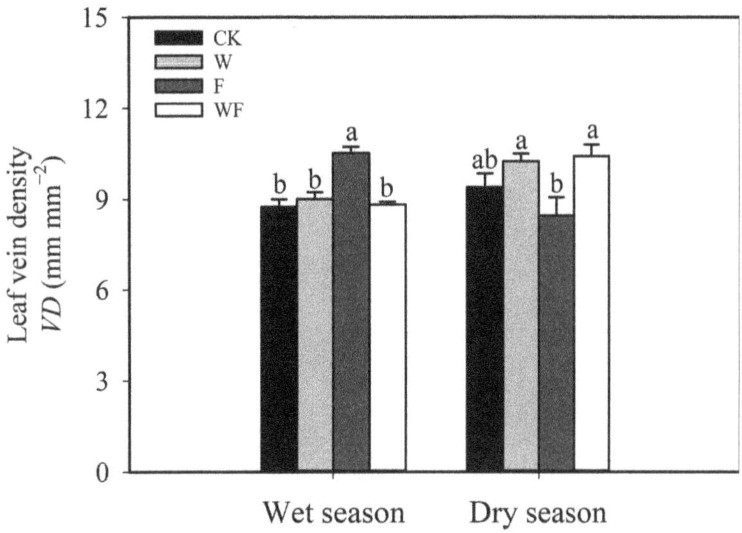

Figure 4. Leaf vein density of *Eucalyptus urophylla* × *grandis* under control conditions (CK), dry-season irrigation (W), fertilization (F), and both dry-season irrigation and fertilization (WF) in wet season and dry season. Lowercase letters represent significant differences ($p < 0.05$). The total value for analysis was 1800.

3.3. Associations between Functional Traits and Structural Traits

Overall, the *LT* showed a negative correlation with both A_n and K_{leaf} (Figure 5A,B). However, their relationships differed in the wet and dry seasons: in the dry season, the *LT* was significantly negatively correlated with A_n, but this relationship was not significant in the wet season. While the *LT* did not have a significant relationship with K_{leaf} in both the wet and dry seasons, the thickness of the leaf lower epidermis cuticle layer ($T_{c\text{-}L}$) was significantly negatively correlated with K_{leaf}. Moreover, regardless of the season, there was a significantly positive correlation between *VD* and K_{leaf} (Figure 5C).

Both A_n and K_{leaf} had an impact on the H_r (Figure 6). The H_r and K_{leaf} were significantly correlated in both wet and dry seasons. Additionally, the H_r of different treatments was higher in the wet season than in the dry season (Figure 6A). The A_n was also significantly positively correlated with H_r, but their relationships were not significant in both wet and dry seasons (Figure 6B).

In order to determine the leaf traits that strongly influenced the height growth rate of *E. urophylla* × *grandis*, we assessed the individual impacts of A_n, K_{leaf}, and leaf structural traits (*LT* and *VD*) on H_r. The unique effect of A_n and leaf structural traits on H_r were 67% and 14%, respectively, whereas that of K_{leaf} was only 5% (Figure 7).

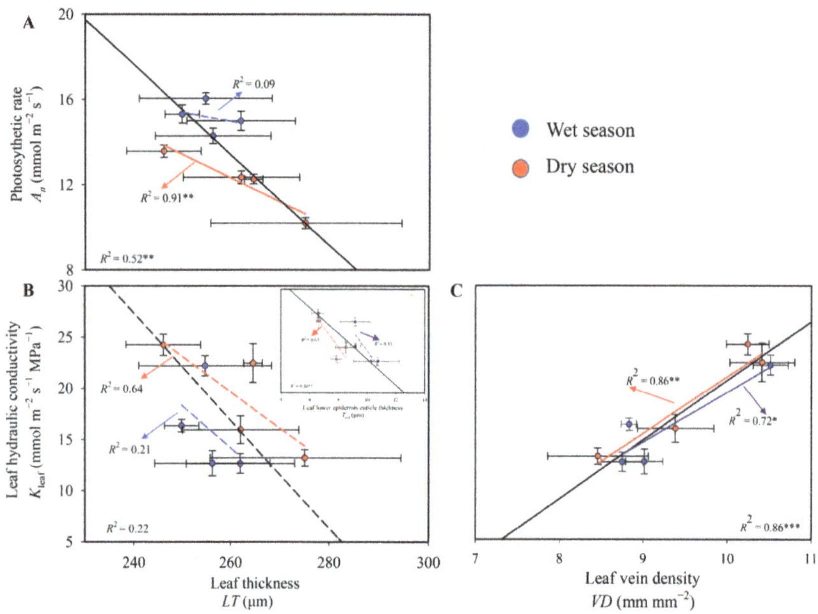

Figure 5. Relationship between leaf structural traits and leaf functional traits. (**A**) Leaf thickness (*LT*) and photosynthetic rate (A_n); (**B**) leaf thickness (*LT*) and leaf hydraulic conductivity (K_{leaf}); (**C**) leaf vein density (*VD*) and leaf hydraulic conductivity (K_{leaf}). Blue and red circles with blue and red lines represent wet season and dry season, respectively. Standardized major axis (SMA) slopes (black lines) are indicated for all data together. Straight lines represent significant correlations: *, $p < 0.1$; **, $p < 0.05$; and ***, $p < 0.01$. Dash lines indicate no significant correlations. The values of A_n, *LT*, and *VD* for each treatment in each season were averaged from 225 measurements, while K_{leaf} was obtained by averaging 45 measurements.

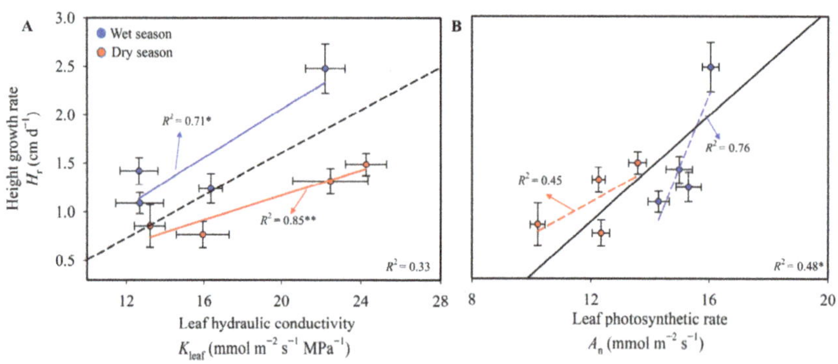

Figure 6. Relationship among functional traits. (**A**) Leaf hydraulic conductivity (K_{leaf}) and height growth rate (H_r); (**B**) photosynthetic rate (A_n) and height growth rate (H_r). Blue and red circles with blue and red lines represent wet season and dry season, respectively. Standardized major axis (SMA) slopes (black lines) are indicated for all data together. Straight lines represent significant correlations: *, $p < 0.1$; **, $p < 0.05$. Dash lines indicate no significant correlations. The values of A_n and K_{leaf} for each treatment in each season were obtained by averaging 225 and 45 measurements, respectively. The values of H_r for each season were obtained by averaging 169 (CK), 212 (W), 178 (F), and 186 (WF) measurements, respectively.

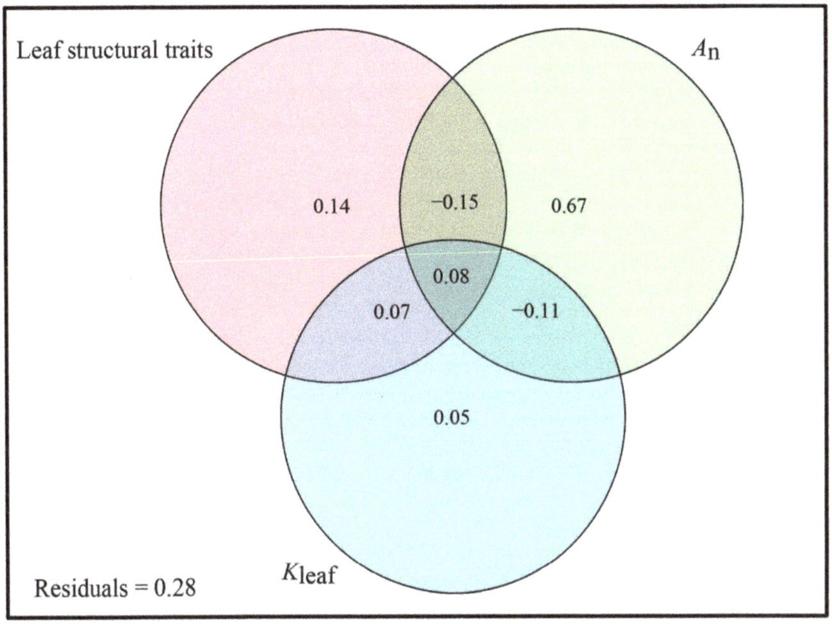

Figure 7. The variance partitioning results of a comprehensive model of height growth rate (H_r), incorporating leaf hydraulic conductivity (K_{leaf}), photosynthetic rate (A_n), and leaf structural traits, including leaf thickness (*LT*) and leaf vein density (*VD*) as explanatory variables. The results are presented as a percentage, representing the explained variance.

4. Discussion

4.1. Effect of Long-Term Dry-Season Irrigation on Plant Height Growth

Consistent with previous studies [33–35], we found that under well-hydrated conditions (during the wet season), the tree height growth rate was primarily influenced by fertilizer treatments. However, there were no significant differences between the WF treatment and the treatments without fertilizer (CK and W). During the dry season, the tree height growth of the W and WF treatments was not limited by drought due to sufficient water supply through drip irrigation, resulting in rapid growth, but there was no significant difference between them. This can be explained by two reasons. First, the external cause for the differential performance of tree height growth rates among different treatments during the wet and dry seasons in this study may be attributed to variations in soil nutrient availability (Table S1). Through five years of water supplementation, the nutrients increased by fertilization in the WF treatment were also utilized more in the previous dry seasons, which was consistent with previous studies where WF's growth was significantly higher than other treatments [4], resulting in no significant difference in tree height growth rates between the WF and W treatments during the dry season in this study. Similarly, in the F treatment, the fertilization nutrients were not fully utilized during the previous five years' dry seasons. Therefore, in well-watered conditions during the wet season, the F treatment exhibited significantly higher nutrient availability, leading to remarkable tree height growth. Second, a larger leaf vein density of the F treatment in the wet season and the W and WF treatments in the dry season can reduce leaf hydraulic resistance, thereby enhancing leaf water transport efficiency and promoting photosynthetic rates [19,21,36], ultimately promoting tree height growth. Moreover, although there were no significant differences in functional traits (K_{leaf}, A_n, and H_r) between the WF and W treatments during the dry season, the functional traits of the W treatment were consistently higher than those of the WF treatment. This may be due to the initial rapid growth of the

WF treatment, facilitated by sufficient water and nutrient conditions [4]. As tree height increases, the resistance to water transport to the canopy leaves increases [37], and the pathway for downward transport of organic matter becomes longer, requiring more energy consumption [38]. Therefore, compared to the shorter trees of the W treatment, the trees of the WF treatment may experience greater growth resistance, resulting in smaller functional traits during the dry season despite both treatments receiving supplemental irrigation.

Somewhat unexpectedly, the leaf thickness of all treatments did not change significantly during the wet or dry season. One possible explanation is that trees subjected to drought stress tend to increase leaf thickness, thereby elongating the water transport distance within the leaves and improving water use efficiency [21,39]. Additionally, taller trees tend to develop thicker and smaller leaves to reduce moisture evaporation [7,40]. Following three years of fertilization and five years of dry-season irrigation, the trees of the W and WF treatments displayed greater tree height. However, the trees of the F and CK treatments experienced more severe drought stress during the dry season. Therefore, all four treatments showed an increase in leaf thickness, resulting in no significant differences in leaf thickness among the treatments during the dry season. During the wet season, leaf thickness variations may be influenced more by genetic factors, as the differences in leaf thickness among the treatments were relatively small, given that they belong to the same tree species.

4.2. Effect of Leaf Structural and Functional Traits on Plant Height Growth

In all four treatments during both wet and dry seasons, the height growth rate was significantly affected by leaf hydraulic conductivity and photosynthetic rate. This illustrates that the adaptions of H_2O transport capacity and carbon assimilation capacity due to dry-season irrigation can enhance plant growth rate. Leaf resistance contributes to approximately 80% of the total resistance in plant water transport [41–44]. Research by Nardini and Salleo [45] revealed that *Laurus nobilis* had 92% of its water transport resistance located in its leaves. Leaves account for a significant proportion of the entire plant hydraulic pathway and are usually more susceptible to water stress compared to other plant organs [41,46], thus affecting leaf photosynthetic carbon assimilation [36]. Therefore, leaves may play a disproportionately important role in plant adaptation to drought [47]. In this study, the hydraulic conductivity of *E. urophylla* × *grandis* leaves significantly influenced tree height growth rates. *E. urophylla* × *grandis* trees that received dry-season irrigation maintained higher leaf hydraulic conductivity, allowing their height growth to remain unrestricted.

Notably, our findings revealed that carbon assimilation capacity had a significantly greater and distinct impact on the increase in height growth rate under prolonged dry-season irrigation when compared to leaf hydraulic conductivity and leaf structural traits. In Ochnaceae species, Schneider et al. [48] reported that enhancing stomatal anatomy to reduce the distance of CO_2 transport proved to be a more efficient strategy, as compared to modifying water transport structures such as increasing leaf vein density. In light of our discoveries, a plausible explanation could be that for plants, modifying water transport-related traits is a more resource-intensive process compared to altering CO_2 transport-related traits. This could lead to inadequate adjustments in leaf hydraulic conductivity and structural traits (*LT* and *VD*). Therefore, under dry-season irrigation, leaves of *E. urophylla* × *grandis* are more inclined to invest in other, more "cost-effective" traits to enhance photosynthetic carbon assimilation efficiency, thereby increasing tree height growth rates. We also found that leaf traits in our study cannot explain the 28% variation in tree height growth rates. As previously indicated by studies [49–52], other traits, such as root system characteristics and wood traits, may have additional effects on tree height growth rates. Therefore, future research should consider including more plant traits as variables in order to further investigate their impact.

However, our results also come with limitations. Our findings are based on *Eucalyptus* plantations in Guangzhou City and may not fully represent all seasonally dry regions, as differences in geographical location, climate, and soil can also influence *Eucalyptus* growth.

Furthermore, this experiment did not include additional treatments regarding the frequency and intensity of irrigation and the amount and type of fertilizer, which makes it difficult to accurately determine the optimal strategies for dry-season irrigation and fertilization.

5. Conclusions

We conducted a field experiment in a *Eucalyptus* plantation with three-year fertilization and five-year dry-season irrigation to compare their effects on height growth rate. Our results indicate that long-term dry-season irrigation has contrasting effects on the tree height growth rate of an *E. urophylla* × *grandis* plantation in southern China. During the dry season, long-term dry-season irrigation significantly enhanced the tree height growth rate of *E. urophylla* × *grandis* by improving leaf water transport efficiency and photosynthetic carbon assimilation efficiency. After 5 years of dry-season irrigation, the height growth rate of *Eucalyptus* trees was increased by 73% compared to non-irrigated conditions in the dry season. However, during the wet season, only the trees with fertilization treatment exhibited a higher height growth rate. Furthermore, our findings indicate that leaf photosynthetic capacity plays a more prominent role in accelerating the height growth rate compared to water conduction in leaves. These results provide valuable insights into the divergent impacts of long-term dry-season irrigation on tree height growth rate and also offer empirical evidence supporting the implementation of dry-season irrigation and the potential for enhancing plantation productivity in regions with seasonal aridity.

Supplementary Materials: The following supporting information can be downloaded at: https://www.mdpi.com/article/10.3390/f14102017/s1, Figure S1: Leaf thickness of *Eucalyptus urophylla* × *grandis* under control conditions (CK), dry-season irrigation (W), fertilization (F) and both dry-season irrigation and fertilization (WF) in wet season and dry season; Table S1: The total nitrogen, total phosphorus, total potassium of the soil under control conditions (CK), dry-season irrigation (W), fertilization (F) and both dry-season irrigation and fertilization (WF).

Author Contributions: Conceptualization, L.H. and H.Y.; formal analysis, L.H.; funding acquisition, L.H. and H.Y.; investigation, L.H. and P.C.; methodology, L.H., Y.S., J.L. (Jiyue Li) and Q.H.; project administration, Y.S.; resources, J.L. (Jun Luo) and Y.S.; supervision, Q.H. and H.Y.; writing—original draft, L.H.; writing—review and editing, L.H., P.C., J.L. (Jun Luo), Y.S., J.L. (Jiyue Li), Q.H. and H.Y. All authors have read and agreed to the published version of the manuscript.

Funding: This research was funded by the National Natural Science Foundation of China, 32101505, the Fundamental Research Funds for the Central Public Welfare Research Institutes of China, PM-zx703-202305-181, and the Shanghai Key Lab for Urban Ecological Processes and Eco-Restoration, SHUES2022A11.

Data Availability Statement: Data will be made available on Figshare upon paper acceptance.

Acknowledgments: The authors wish to thank Quan Qiu for the laboratory assistance and Wenting Yin for the sample site supervision.

Conflicts of Interest: The authors declare no conflict of interest.

References

1. Zhou, X.; Zhu, H.; Wen, Y.; Goodale, U.M.; Li, X.; You, Y.; Ye, D.; Liang, H. Effects of understory management on trade-offs and synergies between biomass carbon stock, plant diversity and timber production in *eucalyptus* plantations. *For. Ecol. Manag.* **2018**, *410*, 164–173. [CrossRef]
2. Hua, L.; He, P.; Goldstein, G.; Liu, H.; Yin, D.; Zhu, S.; Ye, Q.; Riederer, M. Linking vein properties to leaf biomechanics across 58 woody species from a subtropical forest. *Plant Biol.* **2019**, *22*, 212–220. [CrossRef] [PubMed]
3. Hua, L.; Yu, F.; Qiu, Q.; He, Q.; Su, Y.; Liu, X.; Li, J. Relationships between diurnal and seasonal variation of photosynthetic characteristics of *Eucalyptus* plantation and environmental factors under dry-season irrigation with fertilization. *Agric. Water Manag.* **2021**, *248*, 106737. [CrossRef]
4. Hua, L.; Yu, F.; Qiu, Q.; He, Q.; Su, Y.; Liu, X.; Li, J. Dry-season irrigation further promotes the growth of *Eucalyptus urophylla* × *E. grandis* plantations under the conventional fertilization. *New For.* **2023**, *54*, 1085–1102. [CrossRef]
5. Yu, F.; Truong, T.V.; He, Q.; Hua, L.; Su, Y.; Li, J. Dry Season Irrigation Promotes Leaf Growth in *Eucalyptus urophylla* × *E. grandis* under Fertilization. *Forests* **2019**, *10*, 67. [CrossRef]

6. Whitehead, D.; Beadle, C.L. Physiological regulation of productivity and water use in *Eucalyptus*: A review. *For. Ecol. Manag.* **2004**, *193*, 113–140. [CrossRef]
7. Kenzo, T.; Ichie, T.; Watanabe, Y.; Yoneda, R.; Ninomiya, I.; Koike, T. Changes in photosynthesis and leaf characteristics with tree height in five dipterocarp species in a tropical rain forest. *Tree Physiol.* **2006**, *26*, 865–873. [CrossRef]
8. Kitao, M.; Harayama, H.; Yazaki, K.; Tobita, H.; Agathokleous, E.; Furuya, N.; Hashimoto, T. Photosynthetic and Growth Responses in a Pioneer Tree (Japanese White Birch) and Competitive Perennial Weeds (*Eupatorium* sp.) Grown under Different Regimes with Limited Water Supply to Waterlogging. *Front. Plant Sci.* **2022**, *13*, 835068. [CrossRef]
9. Chaves, M.M.; Flexas, J.; Pinheiro, C. Photosynthesis under drought and salt stress: Regulation mechanisms from whole plant to cell. *Ann. Bot.* **2009**, *103*, 551–560. [CrossRef]
10. Zhang, A.; Liu, M.; Gu, W.; Chen, Z.; Gu, Y.; Pei, L.; Tian, R. Effect of drought on photosynthesis, total antioxidant capacity, bioactive component accumulation, and the transcriptome of *Atractylodes lancea*. *BMC Plant Biol.* **2021**, *21*, 293. [CrossRef]
11. Sperry, J.S.; Meinzer, F.C.; McCulloh, K.A. Safety and efficiency conflicts in hydraulic architecture: Scaling from tissues to trees. *Plant Cell Environ.* **2008**, *31*, 632–645. [CrossRef] [PubMed]
12. Wu, G.; Chen, D.; Zhou, Z.; Ye, Q.; Wu, J. Canopy nitrogen addition enhance the photosynthetic rate of canopy species by improving leaf hydraulic conductivity in a subtropical forest. *Front. Plant Sci.* **2022**, *13*, 942851. [CrossRef] [PubMed]
13. Levionnois, S.; Salmon, C.; Alméras, T.; Clair, B.; Ziegler, C.; Coste, S.; Stahl, C.; González-Melo, A.; Heinz, C.; Heuret, P. Anatomies, vascular architectures, and mechanics underlying the leaf size-stem size spectrum in 42 Neotropical tree species. *J. Exp. Bot.* **2021**, *72*, 7957–7969. [CrossRef] [PubMed]
14. Pereira, J.S.; Chaves, M.M.; Fonseca, F.; Araújo, M.C.; Torres, F. Photosynthetic capacity of leaves of *Eucalyptus globulus* (Labill.) growing in the field with different nutrient and water supplies. *Tree Physiol.* **1992**, *11*, 381–389.
15. Câmara, A.P.; Vidaurre, G.B.; Oliveira, J.C.L.; de Toledo Picoli, E.A.; Almeida, M.N.F.; Roque, R.M.; Tomazello Filho, M.; Souza, H.J.P.; Oliveira, T.R.; Campoe, O.C. Changes in hydraulic architecture across a water availability gradient for two contrasting commercial *Eucalyptus* clones. *For. Ecol. Manag.* **2020**, *474*, 118380. [CrossRef]
16. Yin, Q.; Wang, L.; Lei, M.; Dang, H.; Quan, J.; Tian, T.; Chai, Y.; Yue, M. The relationships between leaf economics and hydraulic traits of woody plants depend on water availability. *Sci. Total Environ.* **2018**, *621*, 245–252. [CrossRef]
17. Sack, L.; Frole, K. Leaf Structural Diversity is Related to Hydraulic Capacity in Tropical Rain Forest Trees. *Ecology* **2006**, *87*, 483–491. [CrossRef]
18. Poorter, L.; Bongers, F. Leaf Traits are Good Predictors of Plant Performance across 53 Rain Forest Species. *Ecology* **2006**, *87*, 1733–1743. [CrossRef]
19. Brodribb, T.J.; Feild, T.S.; Jordan, G.J. Leaf Maximum Photosynthetic Rate and Venation Are Linked by Hydraulics. *Plant Physiol.* **2007**, *144*, 1890–1898. [CrossRef]
20. Noblin, X.; Mahadevan, L.; Coomaraswamy, I.A.; Weitz, D.A.; Holbrook, N.M.; Zwieniecki, M.A. Optimal vein density in artificial and real leaves. *Proc. Natl. Acad. Sci. USA* **2008**, *105*, 9140–9144. [CrossRef]
21. Drake, P.L.; Boer, H.J.; Schymanski, S.J.; Veneklaas, E.J. Two sides to every leaf: Water and CO_2 transport in hypostomatous and amphistomatous leaves. *New Phytol.* **2019**, *222*, 1179–1187. [CrossRef] [PubMed]
22. Sack, L.; Scoffoni, C. Leaf venation: Structure, function, development, evolution, ecology and applications in the past, present and future. *New Phytol.* **2013**, *198*, 983–1000. [CrossRef]
23. de Boer, H.J.; Eppinga, M.B.; Wassen, M.J.; Dekker, S.C. A critical transition in leaf evolution facilitated the Cretaceous angiosperm revolution. *Nat. Commun.* **2012**, *3*, 1221. [CrossRef] [PubMed]
24. Chen, W.; Zou, Y.; Dang, Y.; Sakai, T. Spatial distribution and dynamic change monitoring of *Eucalyptus* plantations in China during 1994–2013. *Trees* **2021**, *36*, 405–414. [CrossRef]
25. Asensio, V.; Domec, J.-C.; Nouvellon, Y.; Laclau, J.-P.; Bouillet, J.-P.; Jordan-Meille, L.; Lavres, J.; Rojas, J.D.; Guillemot, J.; Abreu-Junior, C.H. Potassium fertilization increases hydraulic redistribution and water use efficiency for stemwood production in *Eucalyptus grandis* plantations. *Environ. Exp. Bot.* **2020**, *176*, 104085. [CrossRef]
26. Lovelock, C.E.; Feller, I.C.; McKee, K.L.; Engelbrecht, B.M.J.; Ball, M.C. The effect of nutrient enrichment on growth, photosynthesis and hydraulic conductance of dwarf mangroves in Panama. *Funct. Ecol.* **2004**, *18*, 25–33. [CrossRef]
27. Queiroz, T.B.; Campoe, O.C.; Montes, C.R.; Alvares, C.A.; Cuartas, M.Z.; Guerrini, I.A. Temperature thresholds for *Eucalyptus* genotypes growth across tropical and subtropical ranges in South America. *For. Ecol. Manag.* **2020**, *472*, 118248. [CrossRef]
28. Yang, L.; Lin, Y.; Kong, J.; Yu, Y.; He, Q.; Su, Y.; Li, J.; Qiu, Q. Effects of fertilization and dry-season irrigation on the timber production and carbon storage in subtropical *Eucalyptus* plantations. *Ind. Crops Prod.* **2023**, *192*, 116143. [CrossRef]
29. Brodribb, T.J.; Holbrook, N.M. Stomatal Closure during Leaf Dehydration, Correlation with Other Leaf Physiological Traits. *Plant Physiol.* **2003**, *132*, 2166–2173. [CrossRef]
30. Schneider, C.A.; Rasband, W.S.; Eliceiri, K.W. NIH Image to ImageJ: 25 years of image analysis. *Nat. Methods* **2012**, *9*, 671–675. [CrossRef] [PubMed]
31. Warton, D.I.; Wright, I.J.; Falster, D.S.; Westoby, M. Bivariate line-fitting methods for allometry. *Biol. Rev.* **2006**, *81*, 259–291. [CrossRef] [PubMed]
32. Oksanen, J.; Blanchet, F.G.; Friendly, M.; Kindt, R.; Legendre, P.; McGlinn, D.; Minchin, P.R.; O'hara, R.; Simpson, G.L.; Solymos, P.; et al. Package 'vegan'. Community Ecol Package Version 2. Available online: http://mirror.bjtu.edu.cn/cran/web/packages/vegan/vegan.pdf (accessed on 16 February 2016).

33. Prescott, C.E.; Brown, S.M. Five-year growth response of western red cedar, western hemlock, and amabilis fir to chemical and organic fertilizers. *Can. J. For. Res.* **1998**, *28*, 1328–1334. [CrossRef]
34. Xu, F.; Chu, C.; Xu, Z. Effects of different fertilizer formulas on the growth of loquat rootstocks and stem lignification. *Sci. Rep.* **2020**, *10*, 1033. [CrossRef]
35. Walters, S.J.; Harris, R.J.; Daws, M.I.; Gillett, M.J.; Richardson, C.G.; Tibbett, M.; Grigg, A.H. The benefits of fertiliser application on tree growth are transient in restored jarrah forest. *Trees For. People* **2021**, *5*, 100112. [CrossRef]
36. Zhu, S.-D.; Song, J.-J.; Li, R.-H.; Ye, Q. Plant hydraulics and photosynthesis of 34 woody species from different successional stages of subtropical forests. *Plant Cell Environ.* **2013**, *36*, 879–891. [CrossRef] [PubMed]
37. Koch, G.W.; Sillett, S.C.; Jennings, G.M.; Davis, S.D. The limits to tree height. *Nature* **2004**, *428*, 851–854. [CrossRef]
38. Johnson, R.J.; Canny, M.J. Phloem translocation of organic compounds: A possible mechanism to assist osmotically-generated pressure flow in tall trees. *Water* **2013**, *4*, 112–128. [CrossRef]
39. McGregor, I.R.; Helcoski, R.; Kunert, N.; Tepley, A.J.; Gonzalez-Akre, E.B.; Herrmann, V.; Zailaa, J.; Stovall, A.E.L.; Bourg, N.A.; McShea, W.J.; et al. Tree height and leaf drought tolerance traits shape growth responses across droughts in a temperate broadleaf forest. *New Phytol.* **2020**, *231*, 601–616. [CrossRef]
40. England, J.R.; Attiwill, P.M. Changes in leaf morphology and anatomy with tree age and height in the broadleaved evergreen species, *Eucalyptus regnans* F. Muell. *Trees* **2006**, *20*, 79–90. [CrossRef]
41. Sack, L.; Cowan, P.; Jaikumar, N.; Holbrook, N.J.P. The 'hydrology'of leaves: Co-ordination of structure and function in temperate woody species. *Plant Cell Environ.* **2003**, *26*, 1343–1356. [CrossRef]
42. Sack, L.; Holbrook, N.M. Leaf hydraulics. *Annu. Rev. Plant Biol.* **2006**, *57*, 361–381. [CrossRef] [PubMed]
43. Tyree, M.; Cruiziat, P.; Benis, M.; LoGullo, M.; Salleo, S.J.P. The kinetics of rehydration of detached sunflower leaves from different initial water deficits. *Plant Cell Environ.* **1981**, *4*, 309–317. [CrossRef]
44. Yang, S.; Tyree, M.T. Hydraulic architecture of Acer saccharum and A. rubrum: Comparison of branches to whole trees and the contribution of leaves to hydraulic resistance. *J. Exp. Bot.* **1994**, *45*, 179–186. [CrossRef]
45. Nardini, A.; Salleo, S.J.T. Limitation of stomatal conductance by hydraulic traits: Sensing or preventing xylem cavitation? *Trees* **2000**, *15*, 14–24. [CrossRef]
46. Blackman, C.J.; Brodribb, T.J.; Jordan, G.J. Leaf hydraulic vulnerability influences species' bioclimatic limits in a diverse group of woody angiosperms. *Oecologia* **2012**, *168*, 1–10. [CrossRef] [PubMed]
47. Hao, G.-Y.; Hoffmann, W.A.; Scholz, F.G.; Bucci, S.J.; Meinzer, F.C.; Franco, A.C.; Cao, K.-F.; Goldstein, G.J.O. Stem and leaf hydraulics of congeneric tree species from adjacent tropical savanna and forest ecosystems. *Oecologia* **2008**, *155*, 405–415. [CrossRef]
48. Schneider, J.V.; Habersetzer, J.; Rabenstein, R.; Wesenberg, J.; Wesche, K.; Zizka, G. Water supply and demand remain coordinated during breakdown of the global scaling relationship between leaf size and major vein density. *New Phytol.* **2017**, *214*, 473–486. [CrossRef]
49. Comas, L.H.; Eissenstat, D.M. Linking Fine Root Traits to Maximum Potential Growth Rate among 11 Mature Temperate Tree Species. *Funct. Ecol.* **2004**, *18*, 388–397. [CrossRef]
50. Lihui, M.; Xiaoli, L.; Jie, C.; Youke, W.; Jingui, Y. Effects of Slope Aspect and Rainfall on Belowground Deep Fine Root Traits and Aboveground Tree Height. *Front. Plant Sci.* **2021**, *12*, 684468. [CrossRef]
51. Becker, P.; Meinzer, F.C.; Wullschleger, S.D. Hydraulic Limitation of Tree Height: A Critique. *Funct. Ecol.* **2000**, *14*, 4–11. [CrossRef]
52. Petit, G.; Mencuccini, M.; Carrer, M.; Prendin, A.L.; Hölttä, T. Axial conduit widening, tree height and height growth rate set the hydraulic transition of sapwood into heartwood. *J. Exp. Bot.* **2023**, *74*, 5072–5087. [CrossRef] [PubMed]

forests

MDPI

Article

Short-Term Growth Response of Young Pine (*Pinus silvestris*) Seedlings to the Different Types of Soil Media Mixture with Phosphogypsum Formulations under Poland Forest Environmental Conditions

Tomasz Oszako [1,*], Tomasz Pasławski [2], Wieslaw Szulc [3], Beata Rutkowska [3], Artur Rutkiewicz [1], Olga Kukina [4], Sławomir Bakier [2] and Piotr Borowik [5]

1 Forest Research Institute, Forest Protection Department, ul. Braci Leśnej 3, 05-090 Sękocin Stary, Poland
2 Institute of Forest Sciences, Faculty of Civil Engineering and Environmental Sciences, Bialystok University of Technology, Wiejska 45E, 15-351 Bialystok, Poland
3 Institute of Agriculture, Warsaw University of Life Sciences, Nowoursynowska 159, 02-787 Warsaw, Poland
4 Ukrainian Research Institute of Forestry and Forest Melioration Named after G. M. Vysotsky, 61024 Kharkiv, Ukraine
5 Faculty of Physics, Warsaw University of Technology, ul. Koszykowa 75, 00-662 Warszawa, Poland
* Correspondence: t.oszako@ibles.waw.pl

Abstract: The production of phosphoric acid produces "waste heaps" that have not yet been tapped, but which have the character of weak fertilizers and can perhaps be reintroduced into the elemental cycle in the forests. Two variants of mixing with organic ash and with sewage sludge were carried out. One-year-old pine seedlings (*Pinus sylvestris* L.) from the Trzebież forest district (northern Poland) were planted in pots with soil that also came from the same field. Preparations containing phosphogypsum were applied topically to the soil in four doses (1, 2, 3 and 5 t/ha). The trial, which lasted one growing season, was conducted in four replicates. At the end of the trial, the height of the above-ground parts and root length, needle and root area, root neck diameter and photosynthetic performance were measured. The phosphogypsum-based preparations used showed no harmful (toxic) effects on the potted pine seedlings during the six-month trial period. The loosely prepared preparation made from a mixture of phosphogypsum and organic ash began to have a positive effect on the development of the seedlings' root system, and it was also easier to mix with the soil surface than phosphogypsum with sewage sludge, which also contained a sticky form. The photosynthetic performance of one-year-old pine seedlings decreased after one growing season following the application of phosphogypsum preparations and most of the growth parameters tested did not differ from the control, so observations over a longer period (at least two to three growing seasons) are required. However, dosages of 1 and 2 t/ha seem to be the most promising, and these lower dosages are more economical to manage in nurseries or plantations, especially on poor sites. Formulations should be tested for heavy metals and their effects on seedling development. Testing should also be continued to monitor changes in the microbiome.

Keywords: phosphogypsum; sewage sludge; Scots pine; seedlings; soil revitalization

Citation: Oszako, T.; Pasławski, T.; Szulc, W.; Rutkowska, B.; Rutkiewicz, A.; Kukina, O.; Bakier, S.; Borowik, P. Short-Term Growth Response of Young Pine (*Pinus silvestris*) Seedlings to the Different Types of Soil Media Mixture with Phosphogypsum Formulations under Poland Forest Environmental Conditions. *Forests* **2023**, *14*, 518. https://doi.org/10.3390/f14030518

Academic Editors: Cate Macinnis-Ng, Chao Wang, Fan Zhang and Wei Liu

Received: 27 January 2023
Revised: 16 February 2023
Accepted: 2 March 2023
Published: 7 March 2023

1. Introduction

Phosphogypsum is considered a waste product produced in the manufacture of extractable phosphoric acid, which is used, among other things, in the production of phosphate fertilizers [1]. For one tonne of orthophosphoric acid produced, 3.5 to 4.5 tonnes of moist phosphogypsum [2] are obtained. Phosphogypsum as a by-product of fertilizer production requires an attempt to solve the problem of its storage and recycling. However, since it cannot be used in agriculture or even forestry before its category (waste code) is changed, our attempt was limited to potted plants. This research step is necessary to proceed with broader experiments in forests. However, it is worth playing a game because this "waste" contains calcium sulphate dihydrate (the basic ingredient of phosphogypsum), which has no negative impact on the environment unless it is mixed with heavy metals, lanthanides or radionuclides [3]. The literature indicates that contamination of phosphogypsum with heavy metals and radioactive elements applies primarily to waste from the processing of phosphorites and only to a lesser extent to waste from apatites [4]. Apatites have a much lower content of such pollutants compared to phosphorites [2]. Phosphogypsum produced from apatite raw material can be considered as waste, practically free from this type of compounds, which should be tried in forest plantations in the future.

The management of phosphogypsum is difficult due to the above-mentioned impurities it contains; its treatment is energy-intensive, and significant amounts of wastewater are produced, so the main method of managing phosphogypsum is storage, although this is highly controversial due to its environmental impact. Indeed, numerous expert studies have shown that phosphogypsum stored in stockpiles can pose a real threat to the balance of the soil and water environment, mainly due to insufficient protection of the stockpiles [5]. Investigations of the phosphogypsum dump in Wiślinka (northern Poland) show that leachates from the dumps have a very low pH (2–4) and elevated concentrations of phosphates, fluorides, sulphates and nitrates, heavy metals/metalloids (mainly Cd) and radionuclides. Pollutants from the phosphogypsum landfill in Wiślinka entering groundwater may contribute to changes in local physico-chemical conditions of surface waters and also pose a risk to groundwater quality. Phosphogypsum contains heavy metals/metalloids in its composition, and leachates from the landfill also pose a significant risk to the quality of the soil environment. Chemical analyses of soil samples from areas adjacent to the landfill have shown that contaminants contained in the waste migrate from the landfill to the adjacent areas [6]. Another risk is the accumulation of heavy metals/metalloids in plants, which can lead to them entering the food chain. As for the risk from radioactive elements in phosphogypsum, no real risk to the environment near the landfill has been demonstrated.

The heaps heaped up for the storage of phosphogypsum (PG) increase in volume every year and look like 'white mountains' because of their colour. According to estimates from 2009, the amount of phosphogypsum produced worldwide is 100 to 280 million Mg per year [6]. In 2006, phosphogypsum plants worldwide generated about 5 trillion tonnes of waste, of which 70–90% was sent to landfills [7]. In Poland, phosphogypsum waste is generated near Szczecin (Police), Gdańsk (Wiślinka) and Silesia (Wizów), among others. The disposal of phosphogypsum is energy-intensive and generates large amounts of wastewater. Therefore, it is necessary to look for alternatives for the management of this waste [2]. To date, its use in construction, road building or agriculture has been studied [8].

Currently, we would like to test whether phosphogypsum can be used to improve the habitat of pines in a fresh boreal forest. Phosphate fertilizers based on phosphogypsum are not used in Polish agriculture because phosphogypsum contains heavy metals but they should not have negative effects when used in forestry. Therefore, it was hypothesized that this could be a good way to provide pine seedlings with sulphur, calcium and phosphorus in their first years of life. To date, no similar studies have been conducted in Poland. Before conducting comprehensive field studies, the addition of phosphogypsum formulations to the soil is currently being investigated in a pot trial, as they are safer for the environment. The main objective of the present study is to investigate the effects of phosphogypsum formulations on annual pines (*P. sylvestris*), including seedling growth parameters and photosynthetic efficiency. We hypothesize that the phosphogypsum-based preparations studied in the variants—mixtures of phosphogypsum with organic ash and mixtures of phosphogypsum with sewage sludge—could be used as soil conditioners to improve seedling growth in forestry. Furthermore, we presume that the work will emphasize the phosphogypsum formulation's effect on the young potted seedlings under outdoor conditions.

2. Materials and Methods

2.1. Plant Material and Soil

One-year-old Scots pine seedlings (*P. sylvestris*), also prepared by a local forest nursery, were planted in the soil of a forest growing area in the neighbouring Trzebiez Forest District, Division 598h, Tytania Forest District, Trzebiez Forest District, RDSF Szczecin, and preparations were added to the soil at different doses (per hectare) to observe their effects on plant growth and development, including photosynthetic efficiency. In the control (C), two formulations were used for the study: a phosphogypsum/ash mixture (A) and a phosphogypsum/sludge mixture (S) in a ratio ensuring a soil pH recommended for pine culture (4–4.5). The soil used for planting pine seedlings came from a not very nutrient-rich, fresh coniferous forest where pines are commonly planted in Tatynia forestry (Trzebież forestry district). It was assumed that, in case of positive results, further field studies should be conducted on the use of these previously unused "wastes" in forests, especially in fresh coniferous forests, which account for about 20% of the area of state forests.

The pine seedlings were also obtained from Tatynia Forest Nursery and planted in 18 cm × 18 cm pots, in 9 variants and 4 replicates, using the control sample as a reference. The trial was conducted in the following variants: control; calcium sulphate with wood ash and calcium sulphate with sewage sludge. For both formulations 4 dosages were used: 1 t/ha, 2 t/ha, 3 t/ha and 5 t/ha. The dolomite base doses of 1 to 2 t/ha used for liming forest soils were used as dosing criteria, which were also increased to 3 and 5 t/ha to observe whether there would be further benefits for the pine seedlings or rather phytotoxicity of the preparations. The idea was to select the most beneficial dosages of the preparations for further field trials as a result of the tests. Four seedlings of each treatment were used in the experiment.

Seedlings from the nursery and soil were taken on 24 April 2021 and planted in 36 pots, one per pot, on 29 April 2021. Each pot was placed in a sunny, open location to create favourable growing conditions. The trial was fenced with a metal grid to protect it from damage by forest animals. On 23 May 2021, both formulations were applied to the soil surface at the following amounts: 3.24 g; 6.48 g; 9.72 g; and 16.2 g, to simulate forest conditions and future application by surface spreading. A small amount of soil was sprinkled with the formulations to prevent wind dispersal. They were maintained daily and watered as needed. In particular, the plants of the control variant were observed, their good appearance indicating normal environmental conditions conducive to plant growth. In this way, any abnormalities in the treated plants, e.g., phytotoxicity of the preparations used, were easily detected.

2.2. Measurements of Growth Parameters

During the 2021 growing season, three measurements (25 April, 19 July and 31 October) of height, diameter at the root collar and length and area of roots and needles (31 October) were taken [9].

All pine seedlings were removed from their pots and freed from soil residues on 31 October 2021. Root length was measured with a tape measure graduated to 1 mm [10].

After measuring and photographing on graph paper, one seedling each was placed in a paper bag and weighed. The plants in the paper bags were dried at room temperature for 3 weeks and then dried again at 105 °C for about 1.5 h until a constant weight was reached.

The seedlings were individually spread out on graph paper to measure the needle and root area. Each cutting was photographed. From the photos, the area was determined using the CSS Video FrameGrabber software.

2.3. Measurement of Photosynthetic Performance

On 7 July 2021, ten individual needles were taken from each seedling, two of which were placed in special clamps and left in the dark for more than 20 min to silence the photosynthetic processes. Subsequently, 360 measurements were taken with a mobile PEA fluorimeter (10 measurements on each seedling), which were transferred to a computer with appropriate software for further analysis.

Soil phosphorus tests were conducted on 8 August 2022; the pots of each experimental variant were cut in half and soil from two depths (0–5 cm and 6–10 cm) was placed in plastic containers. Each container was described and sent to the laboratory of the Warsaw University of Life Sciences for analysis of the total phosphorus content (phosphorus pentoxide) and acidity (pH) of the soil [11,12]. Soil samples were characterized as follows: pH—by potentiometric method after extraction with 1 mol/dm^3 KCl (ISO 10390:2005), available P value by Egner-Riehm method (DL).

Four parameters were included in the analysis and their values were compared with those of the control seedlings.

- Fv/Fm is a normalized ratio formed by dividing the variable fluorescence by the maximum fluorescence. The parameter Fv/Fm describes the maximum yield of the PS II photoperiod. The photosystems PS I and PS II are dye–protein–lipid complexes that determine the course of all photochemical reactions of the light-dependent phase of photosynthesis.
- Another parameter "area" or AM defines the area over the chlorophyll fluorescence induction curve. The value of the parameter AM (SM or Area) is proportional to the size of the electron acceptor pool in PS II. The measurement of AM is of great practical importance, for example in monitoring the penetration of herbicide photosynthesis inhibitors such as diuron (DCMU) into leaves. The unit of AM is bitomilliseconds (bms), i.e., the product of the fluorescence signal measured in bits and the transition time from Fo to Fm expressed in milliseconds. The faster the increase from FL to FM (faster reduction of the acceptor pool in PSII), the smaller the area above the FL induction curve of chlorophyll. When electron transport from the reaction centres to the plastoquinones is blocked (during stress), the value of AM decreases. A parameter similar to AM is its standardized version SM / FM (SM = AM). This parameter also determines the number of unreduced electron acceptors in PSII, but per total content of active chlorophyll in the sample [13,14].
- The time to reach maximum chlorophyll fluorescence FM (from the beginning of the measurement) is determined by the TFM parameter, which is usually 500–800 ms [15]. The TFM measurement is an alternative method to determine the size of the pool of unreduced plastoquinones. This time may be prolonged if the test object has been exposed to stress factors that slow the transport of high-energy electrons from the reaction centres to the plastoquinones [16]. PI (Performance index)—the performance index PS II reflects the efficiency of the photosynthetic process, taking into account all automatically measured parameters.
- However, the best parameter to characterize photosynthetic performance is the total performance index (PI Total), which reflects all the previously mentioned factors and was therefore first compared between the treatments and the control.

2.4. Data Analysis

The collected data were analysed to test the statistical significance of the differences between the experimental variants studied and the trends in the data. Two types of statistical analyses were used in the presented research.

- Analysis of variance (ANOVA) was performed for pairwise multiple comparisons using Tukey's HSD test (honestly significant difference) at $p < 0.05$. For this method comparison between the three main studied treatments (Control—C, Ashes—A, Sewage sludge—S) was performed and in the presented results we indicate, in the presented figures, whether there was found a statistically significant difference and for which compared pair.
- The evolution of the measured variables as a function of the applied dose of the phosphogypsum preparations was analysed with linear regression models. The models with linear and quadratic terms of the dose were trained. The models in which

the regression coefficients were statistically significant at $p < 0.05$ were selected. For this type of analysis, we analysed whether the found trends in data were statistically significant and in the presented results we indicated that significance by plotting the regression line. For these analyses, we have not compared whether there was a significant difference between individual studied levels of the used dose.

Data processing and statistical analysis of the data were performed using SAS 9.4 software (SAS Institute, Cary, NC, USA) via the SAS Enterprise Guide user interface and the SAS/Stat procedures [17] PROC ANOVA and PROC REG.

3. Results

3.1. Measurement of Photosynthetic Performance

The photosynthetic efficiency of the needles of the treated pines (A and S) was lower than that of the control group (C) and these differences were statistically significant. However, the treatments did not differ among themselves in this respect (Figure 1).

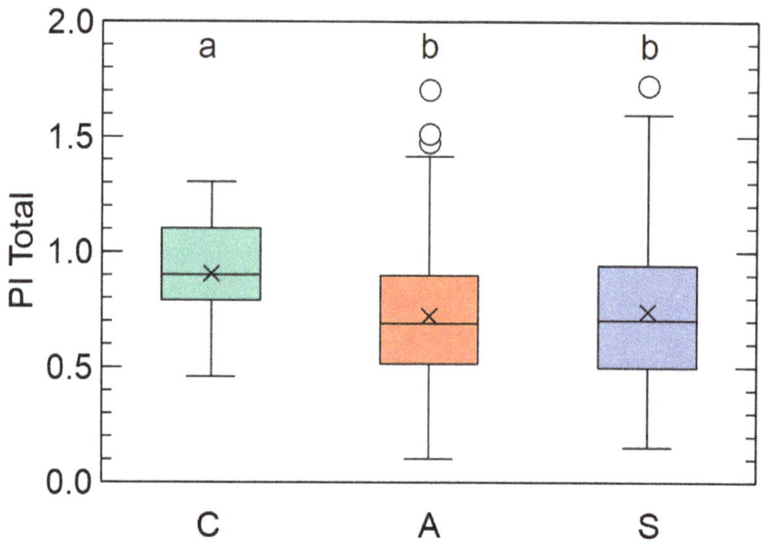

Figure 1. Photosynthetic activity (overall performance index PI total) compared to soil amendments with phosphogypsum mixed with organic ash—A, sewage sludge—S and control—C. The lowercase Latin characters above box plots, if the same, indicate that there is no statistically significant difference between the groups at $p < 0.05$. In the plot, boxes span from 1st to 3rd quantile, with the median indicated by a line, whiskers indicating 1.5 inter-quantile range, the × symbol mean value, and circles outlier observations.

In particular, for the phosphogypsum–ash mixture (from burned organic material), increasing doses of the preparation (3 and 5 t/ha) resulted in a significant decrease in the photosynthetic efficiency of pine needles. For the phosphogypsum–sludge mixture, the lowest photosynthetic efficiency was observed at a dose of 2 t/ha (Figure 2).

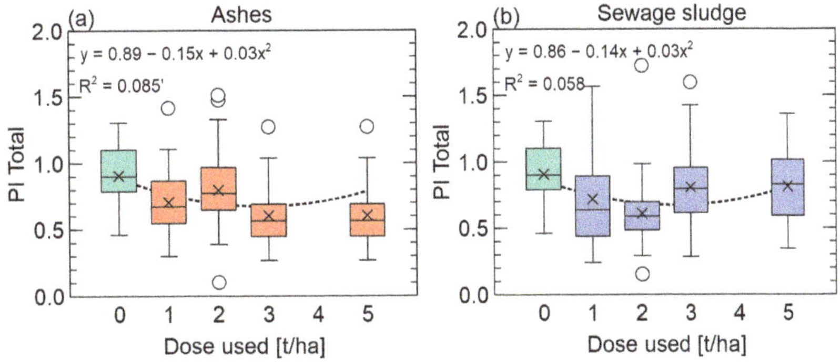

Figure 2. Photosynthetic efficiency (Performance Index total PI) versus used doses of preparations with phosphogypsum. Two studied variants of mixture with ashes (**a**) and sewage sludge (**b**) are presented. The dashed lines represent the linear regression fits of the data. The regression equations with the fitted parameters and the models' R^2 values are printed in the subfigures. The regression parameters are statistically significant at $p < 0.05$. In the plot, boxes span from 1st to 3rd quantile, with the median indicated by a line, whiskers indicating 1.5 inter-quantile range, the × symbol mean value, and circles outlier observations.

3.2. Measurements of the Growth Parameters

The addition of phosphogypsum to the soil resulted in a reduction in height growth in the pines tested and, in the case of S, the difference from the control appeared to be statistically significant (Figure 3). No statistical differences were found in the increase of the thickness of the root necks, although the tendency of the decrease was similar.

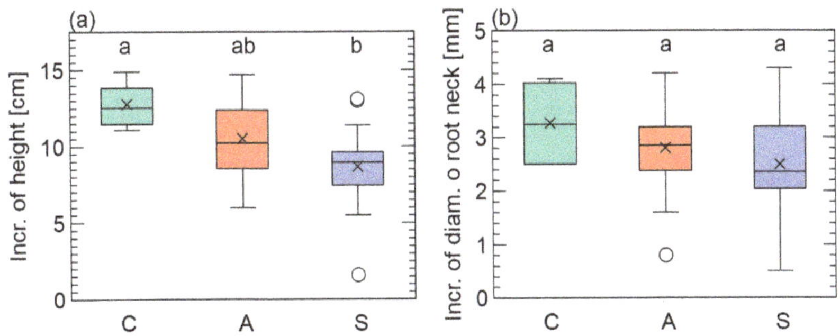

Figure 3. Increase in seedling length (**a**) and root collar diameter (**b**), calculated as the difference between measurements in October and April. Three types of treatments are presented: soil amendments with phosphogypsum mixed with organic ash—A, sewage sludge—S and control—C. The lowercase Latin characters above box plots, if the same, indicate that there is no statistically significant difference between the groups at $p < 0.05$. In the plot, boxes span from 1st to 3rd quantile, with the median indicated by a line, whiskers indicating 1.5 inter-quantile range, the × symbol mean value, and circles outlier observations.

Similarly, increasing dosages of phosphogypsum in mixed preparations in pines resulted in a deterioration of their growth in height and thickness at the root necks (Figure 4).

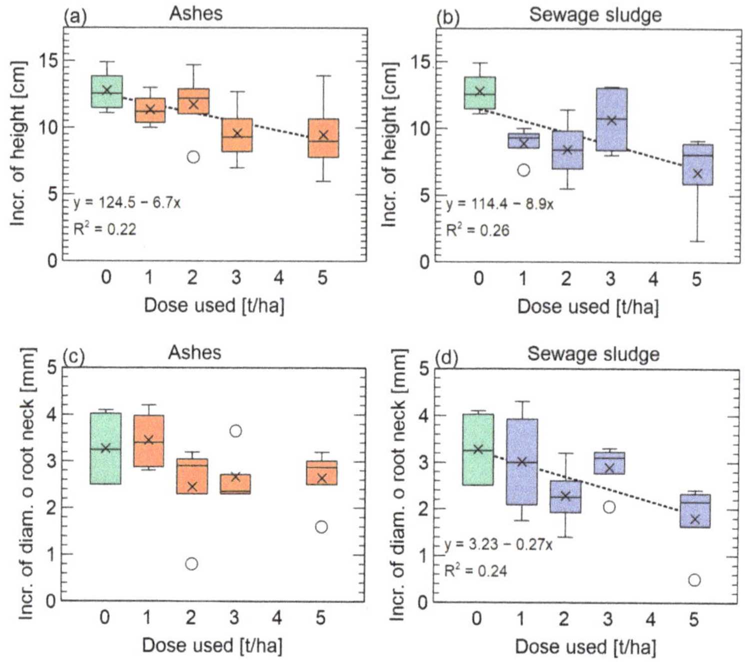

Figure 4. Increase of seedlings' lengths (**a**,**c**) and diameter of root neck (**b**,**d**) versus preparation doses. Two studied variants of mixture with ashes (**a**,**c**) and sewage sludge (**b**,**d**) are presented. The dashed lines represent the linear regression fits of the data. The regression equations with the fitted parameters and the models' R^2 values are printed in subfigures. The results of regression analysis are presented only for the cases when the regression parameters are statistically significant at $p < 0.05$. In the plot, boxes span from 1st to 3rd quantile, with the median indicated by a line, whiskers indicating 1.5 inter-quantile range, the × symbol mean value, and circles outlier observations.

When a mixture of phosphogypsum and sewage sludge was applied to the soil, the surface area of the assimilate (compared to the control and ash) decreased significantly (Figure 5).

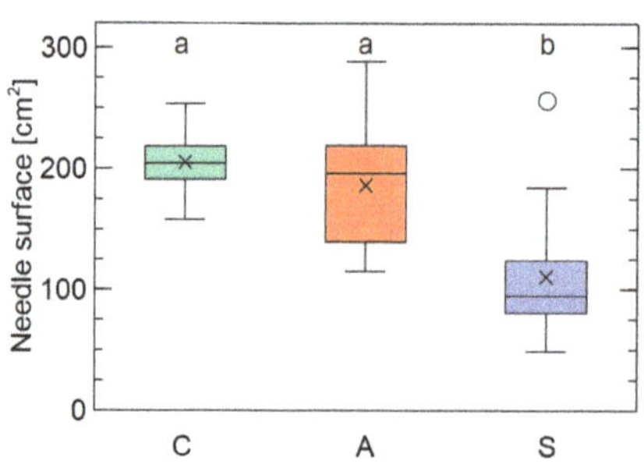

Figure 5. Changes in the needle surface of the Scots pine seedling. Three types of treatments are presented: soil amendments with phosphogypsum mixed with organic ash—A, sewage sludge—S and control—C. The lowercase Latin characters above box plots, if the same, indicate that there is no statistically significant difference between the groups at $p < 0.05$. In the plot, boxes span from 1st to 3rd quantile, with the median indicated by a line, whiskers indicating 1.5 inter-quantile range, the × symbol mean value, and circles outlier observations.

The above phenomenon reduced the needle area at concentrations of 2.3 and 5 t/ha of the sludge–phosphogypsum mixture (Figure 6).

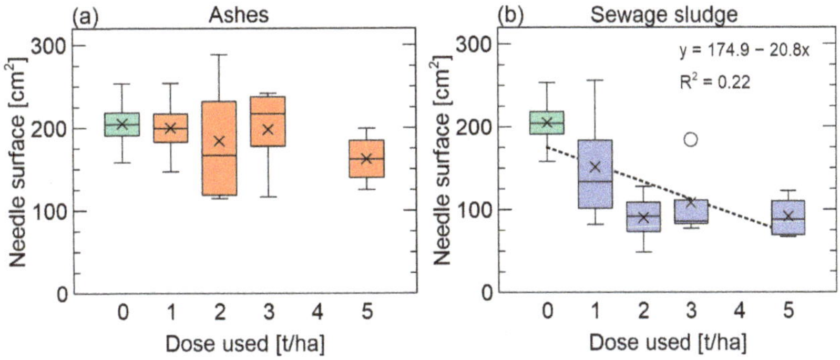

Figure 6. Changes in the needle surface of Scots pine seedlings versus preparation doses. Two studied variants of mixture with ashes (**a**) and sewage sludge (**b**) are presented. The dashed line represents the linear regression fits of the data. The regression equation with the fitted parameters and the model's R^2 value is printed in the subfigure. The results of regression analysis are presented only for the case when the regression parameters are statistically significant at $p < 0.05$. In the plot, boxes span from 1st to 3rd quantile, with the median indicated by a line, whiskers indicating 1.5 inter-quantile range, the × symbol mean value, and circles outlier observations.

A slight improvement in root length (not yet statistically significant) was observed when a mixture of phosphogypsum and ash (A) was added to the soil (Figure 7). However, no statistically significant differences were observed between the treatments and the control at the surface either, although the variant with the addition of sewage sludge (S) performed the worst (Figure 7).

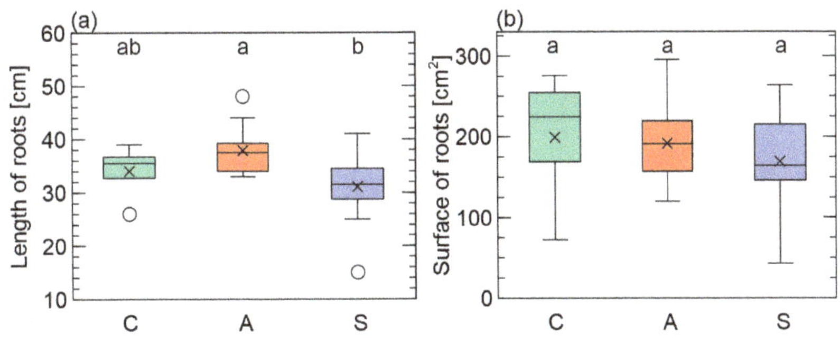

Figure 7. Length (**a**) and surface area (**b**) of roots of Scots pine seedlings. Three types of treatments were used: soil amendments with phosphogypsum mixed with organic ash—A, sewage sludge—S and control—C. The lowercase Latin characters above box plots, if the same, indicate that there is no statistically significant difference between the groups at $p < 0.05$. In the plot, boxes span from 1st to 3rd quantile, with the median indicated by a line, whiskers indicating 1.5 inter-quantile range, the × symbol mean value, and circles outlier observations.

In this case, higher application rates (5 t/ha) appear to have a positive effect on root length when A was added to the soil and a negative effect when S was added (Figure 8). Similar trends of improvement or deterioration in root area were observed for phosphogypsum mixed with A and S (Figure 8).

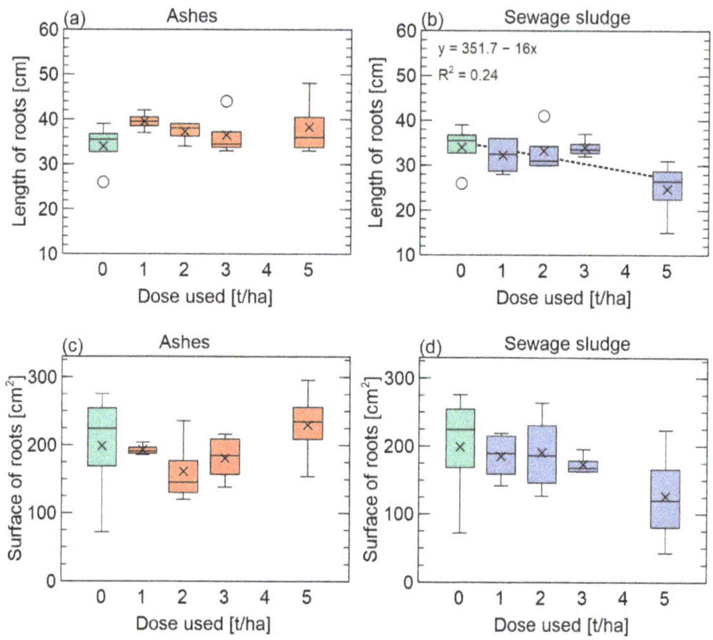

Figure 8. Roots length and surface area of Scots pine seedlings as a function of preparation dosage. Two studied variants of mixing with ash (**a**,**c**) and sewage sludge (**b**,**d**) are shown. The dashed line represents the linear regression fit of the data. The regression equation with the fitted parameters and the R^2 value of the model is shown in the sub-figure. The results of the regression analysis are presented only for the case where the regression parameters are statistically significant at $p < 0.05$. In the plot, boxes span from 1st to 3rd quantile, with the median indicated by a line, whiskers indicating 1.5 inter-quantile range, the × symbol mean value, and circles outlier observations.

Fresh and dry biomass of pine seedlings decreased uniformly with both treatment options, with the greatest decrease occurring after the addition of sewage sludge (Figure 9).

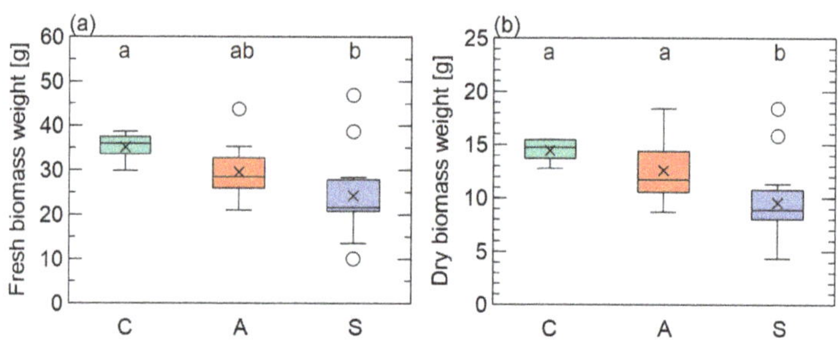

Figure 9. Fresh and dry biomass of Scots pine seedlings. Two studied variants of mixing with ash (**a**) and sewage sludge (**b**) are shown.Three types of treatments were used: Soil amendments with phosphogypsum mixed with organic ash–A, sewage sludge–S and control–C. The small Latin letters above the boxplots, when the same, indicate that there is no statistically significant difference between the groups in $p < 0.05$. In the plot, boxes span from 1st to 3rd quantile, with the median indicated by a line, whiskers indicating 1.5 inter-quantile range, the × symbol mean value, and circles outlier observations.

The higher the treatment doses, the more negative the effects on pine seedling biomass (Figure 10). Phosphogypsum mixed with sewage sludge in an amount of 5 t/ha had a particularly negative effect (Figure 10).

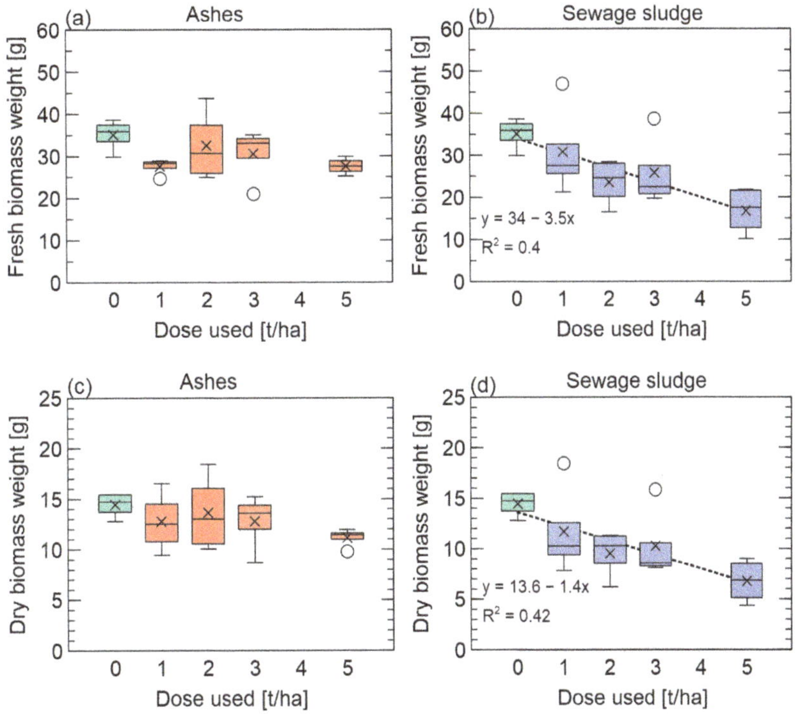

Figure 10. Fresh and dry biomass of Scots pine seedlings compared to preparation doses. Shown are two investigated variants of the mixture with ash (**a**,**c**) and sewage sludge (**b**,**d**). The dashed lines represent the linear regression fits of the data. The regression equations with the fitted parameters and the R^2 values of the models are shown in subfigures. The results of the regression analysis are presented only for the cases where the regression parameters are statistically significant at $p < 0.05$. In the plot, boxes span from 1st to 3rd quantile, with the median indicated by a line, whiskers indicating 1.5 inter-quantile range, the × symbol mean value, and circles outlier observations.

3.3. *Measurements of Soil Parameters*

The acidity of the soil samples studied is generally acidic, which affects the availability of phosphorus to plants (Table 1). In an acidic environment, poorly soluble compounds with aluminium and iron are formed. Although the P2O5 content in some samples was more than twice as high as in the control soils, it was not absorbed by the pines. The addition of ash (A) with a higher calcium Ca content (CaO-oxides) changed the pH of the forest soil (control—pH 3.52), which varied between 3.06 and 4.84 after treatment, the latter value being reached at an addition of 5 t/ha. However, the mean value (A) calculated from all measurements was identical to that of control C (3.52). For the preparation (S), the pH values were lower (from 3.38 to 3.57), due to the acidic nature of the phosphogypsum sludge (average 3.43). In addition, phosphorus tended to be in the upper layers (O) when treated with preparation (A), while it moved deeper into the root zone (1) to some extent after treatment with preparation (S): 0—sample from the soil layer 0–5 cm, 1—sample from the soil layer 5–10 cm.

Table 1. Acidity of the investigated soil samples and their phosphorus content: 0—sample from the soil layer 0–5 cm, 1—sample from the soil layer 5–10 cm.

Sample Type	No of Sample	pH	P [mg/kg]	P_2O_5 [mg/kg]
Ashes	A1/0	3.69	24.53	56.20
	A1/1	3.46	23.03	52.76
	A2/0	3.31	50.56	115.83
	A2/1	3.06	33.21	76.08
	A3/0	3.26	26.20	60.02
	A3/1	3.29	57.73	132.26
	A5/0	4.84	61.74	141.45
	A5/1	3.28	23.53	53.91
Sewage Sludge	S1/0	3.45	29.03	66.51
	S1/1	3.38	36.54	83.71
	S2/0	3.57	59.57	136.47
	S2/1	3.44	77.59	177.76
	S3/0	3.40	43.55	99.77
	S3/1	3.44	62.41	142.98
	S5/0	3.42	45.22	103.60
	S5/1	3.34	16.02	36.70
Control	C	3.52	34.04	77.99

4. Discussion

4.1. *Phosphogypsum—Possibilities for Recycling*

Post-phosphate wastes, including dolomitic wastes and phosphogypsum, have been used as reclamation products in North Carolina [18], which is encouraging for such activities in Poland. In particular, soil acidification is a major obstacle to agricultural and forestry development in Poland, as in many other countries [19]. There are few effective methods to improve soil acidity, e.g., in China and elsewhere, where PG has been shown to increase soil exchangeable calcium (Ca) content. Surface liming of moderately acid subtropical soils, when properly applied, is an effective method of reducing subsoil acidity to −0.60 cm in the first year after application [20]. In a similar trial that has been running since 2002, surface application of limestone and phosphogypsum has been shown to have positive effects on root growth, nutrient supply and crop yields [21]. In tropical soils with low natural fertility, acidity can be rapidly improved by surface application of phosphogypsum in a no-till system (NT). Such improvement can be observed as early as 12 months after application. Phosphogypsum mixtures increase the concentration of K, Ca, Mg, N-NO_3^- and S-SO_4^{2-} in the subsurface layers of the soils [22]. Application of calcium dolomite and phosphogypsum (2100 kg/ha) to rice and bean crops in the highlands (in Brazil), superficially and without mixing into the soil, increases the Ca, Mg and Mn content in the leaves and increases the yields of these crops [23]. As a soil additive and soil conditioner, PG is therefore compared to the liming of soils commonly used in Poland, but PG can be used as a supplement to liming and has a comprehensive effect on improving nutrient uptake and soil acidity [24]. In addition, PG has a solubility almost 150 times higher than lime and can reduce the activity of ions toxic to plants, especially active aluminium ions (Al^{3+}), which are particularly harmful to the fine roots [25]. PG positive effects on the environmental properties of soils include the improvement of their physical properties, especially by improving aggregation and hydraulic conductivity [26]. Gypsum is used to improve soil properties in regions where acidic soils prevail; its good solubility leads to stronger growth and better distribution of the root system of plants, which increases their yields. Numerous studies have confirmed that PG increases the accumulation of carbon (C) in the soil and this effect also has a positive aspect, as carbon sequestration in the soil reduces CO_2 emissions into the atmosphere [27]. In this way, PG also improves the quality of soil organic matter (SOM) in acidic soils, which is important for the sustainability of agricultural ecosystems [28].

4.2. The Growing Problem of Phosphogypsum Deposition in the Environment

The world economy consumes very large quantities of phosphoric acid. The annual share of world production of phosphate rock is approximately: USA—25%; China—18%; Morocco—15%; Russia—9%; Tunisia—8%; Ukraine—6%; Jordan—4%; other countries—15% [29,30]. The only raw materials important to Poland for the production of phosphoric acid are phosphate minerals, which occur naturally as phosphorite and apatite deposits. The largest deposits of phosphate rock are in the USA (Florida, Tennessee, North Carolina, Louisiana); China; Africa (Morocco, Senegal, Togo, Tunisia, Jordan, South Africa, Egypt, Israel, Syria); and Russia (Kola Peninsula). In Poland and elsewhere, the amount of phosphogypsum produced in the manufacture of phosphoric acid is about 5 tonnes per tonne of phosphoric acid extracted [6]. Global production of PG is about 100–280 million tonnes per year. Phosphogypsum is stored in landfills: in so-called ponds in liquid form, in stockpiles in semi-dry form, or submerged in seas and oceans. Phosphogypsum is the most produced of all inorganic chemical wastes and unfortunately still one of the most landfilled.

The storage of PG in large landfills near factories, as in Poland, causes significant environmental damage. PG contains gypsum and a variety of impurities such as phosphates, fluorides and sulphates, and could still be used as fertilizer for forests, although it contains heavy metals and other trace elements. A deficiency of phosphorus (P) has been found in oak stands growing on the Krotoszyn plateau (western Poland) [25]. This deficiency favours damage to the root systems by pathogenic oomycetes of the genus Phytophthora. This can be prevented by an appropriate dosage of phosphogypsum, e.g., mixed with ash. They can improve degraded forest soils that have been used for agriculture for many years and other soils with stands on disaster sites (e.g., after industrial emissions or fires). In clear-cuts, they can be used to compensate for the loss of tree biomass after harvesting [9], while they can affect agricultural products, they have less impact on forest ecosystems and can probably even increase timber production. To date, up to 15% of the world's production of PG is used in the manufacture of building materials, as a soil additive and as a regulating agent in cement production. The increasing scale of mining and the growing demands for environmental protection are leading to the search for new solutions for the management of phosphogypsum [29]. This research is part of this trend and is an attempt to address a global problem at a local level. Many wastes, including PG, have a high fertilizing potential and their release into the environment complies with EU requirements. Legal regulations for the natural use of waste for fertilization purposes vary across the EU. Therefore, in June 2019, the European Parliament adopted a regulation opening the European market for mineral and organic fertilizers and new fertilizer products from waste with soil improver status (Regulation 2019/1009).

4.3. Phosphogypsum—Restrictions on Use

The current hazards posed by the storage of phosphogypsum in open landfills require a constant search for innovative methods of managing phosphogypsum dumps. The main proposed applications are: agriculture, cubic construction, civil engineering, road construction, levelling of pits and reclamation of degraded land [2]. Phosphogypsum can be used in many sectors of the economy, including the production of building materials, road construction and the production of elemental sulphur and sulphuric acid [31]. In some countries, waste phosphate compounds from industry are used in agriculture [20] and their impact on the environment has been positively evaluated [32]. Phosphogypsum contains certain amounts of beneficial plant nutrients (mainly phosphorus) that can have a positive effect on plant growth. The use of phosphogypsum in crops is permitted in Brazil [33]. Phosphogypsum can also be used to rehabilitate damaged soils [34] or as a component of mineral–organic fertilizers [35].

4.4. Phosphogypsum with Sewage Sludge

Phosphogypsum can be mixed with sewage sludge, for example, to increase the effectiveness of phosphites. Microbiological processes with organic acids dissolve the phosphites into plant-available phosphates. Usually, a ratio of organic feedstock to phosphate of about 4:1 is recommended for the production of this type of compost. This direction to increase the efficiency of phosphate meals is also attractive because the phosphates formed by the dissolution of phosphate meals have a protective function against nitrogen by binding the ammonia formed during the decomposition processes of the organic matter of the compost mass to ammonium phosphate [36]. Composting sewage sludge with phosphate meals offers a complex combination of increasing phosphorus solubility and protecting against nitrogen loss [37].

Returning the nutrients accumulated in sewage sludge to the soil is not only necessary from an economic point of view, but also important for maintaining and restoring the ecological balance. The mineral and organic composition of sewage sludge from municipal wastewater treatment plants resembles the organic matter of the soil—humus [37]. This enables their natural use, including forestry and pasture [38]. However, sludge intended for non-industrial use should meet requirements regarding its chemical composition and hygienic condition. Limitations include the content of heavy metals due to their toxic effect on living organisms and their ability to bioaccumulate [39].

4.5. The Fresh Coniferous Forest as a Potential Site for Phosphate Application

Fresh coniferous forests are one of the most widespread forest habitats in Poland. At the end of 2011, these habitats occupied an area of 1,433,387 hectares (20.2% of the area under state forest management). The largest areas are located in the mesoregions, including the Masurian Forest (II.4), the Tuchola Forest (III.1), the Notec Forest (III.17) and the Lower Silesian Forest (V.2) [40].

Fresh coniferous forest colonizes moderately poor sites that are still not very moist but on soils that are already somewhat better than dry coniferous forest. Good podsolic soils with varying degrees of podsolation predominate, but fresh coniferous forest with poorly developed podsolic, podsolic–rustic or rusty soils can also be found. Characteristic of these soils is the swampy overburden humus, which is strongly acidic and has a pH value between 3.5 and 4.

4.6. Special Conditions of the Prepared Experiment

The present study was conducted over a period of six months (April to October 2021), which, based on the analysis of the results, seems to be far too short a period to obtain statistically significant differences between the experimental variants. To obtain more accurate results over a wider range, the study would have to be continued over a longer period, e.g., two or three years. Organic phosphogypsum (PG organic) as a soil conditioner improves crop yields in sandy soils in the long term. Other studies have shown that the use of PG to improve the properties of sandy soils enhances plant growth and development, but this effect is only visible after 2 years [41]. This is due to the creation of a suitable environment that promotes the uptake and assimilation of nutrients by the roots. We would also like to achieve such an effect in the long term with woody plants such as pines.

The main purpose of the experimental design was to determine the dosage to be tested over a longer period of time in plots in pine plantations, provided PG is no longer treated as a waste material. For the same reason, forest soil was taken from the Trzebież forest area near the chemical plant where the phosphogypsum is stored. The pine seedlings were also taken from the same forest area in order to be able to use regeneration material with the same gene pool in the study (currently in pots and later in the forest area).

Although the pine seedlings originated from the First Baltic forest region and were raised in the Fourth Mazovian-Podlasian forest region, the growth period after planting and the ecological conditions are similar in both regions: 200–210 days in the Baltic region and

210 days in the Mazovian-Podlasian region [42]. Annual precipitation is also 550–600 mm per year in country IV and 600–700 mm per year in country I [43,44].

As the trial started at the beginning of the growing season, the tested phosphogypsum mixtures were applied in spring to the soil of the pots in which the annual pines had been planted. Phosphorus is a nutrient with low mobility, as it moves poorly in the soil profile. Therefore, fertilization with this nutrient was done in spring when there is enough time for it to seep into the roots and for the plants to respond. Otherwise, the nutrient may not be available to the roots of the plants. However, it is known that the best time to apply this type of fertilizer, especially in the form of superphosphates, is as early as possible in the autumn, as the heavy metals from the fertilizers can then form poorly soluble compounds in the soil that are inaccessible to the plants [45]. If, on the other hand, spring fertilization is necessary, the use of a multi-component compound fertilizer [46] is recommended. In the future, the tested preparations could be enriched with other nutrients, such as nitrogen (N), and preferably applied in granular form in spring. For the time being, however, the loose form was used, as it was assumed that fertilization would take place before planting the seedlings in the soil preparation phase. Further field trials would need to test the effect of harrowing or disc fertilization to allow the preparation to penetrate deeper into the soil and facilitate phosphorus uptake by the root systems of the seedlings.

The two forms of preparation tested are not yet suitable for use in forestry practise, as the free-flowing preparation is too fine and light, so that it can be blown away by the wind and is very likely to damage the needles and leaves of the seedlings. The other, plastic form, on the other hand, is too dense and compacted and cannot be worked into the soil mechanically (without the addition of dilution liquids). It would be necessary to develop other forms of phosphate fertilizers, e.g., granules, which are also enriched with the nutrients mentioned above.

In the future, preparations from sewage sludge (from phosphogypsum landfills) and organic ash (from boiler plants) should be standardized so that there is no doubt as to whether the content of individual elements in each sample is always the same or different.

4.7. *Growth of Pines after Soil Improvement with Phosphogypsum*

Thus, there was no direct toxic effect of the preparations on pine seedlings, even at high doses (5 t/ha). However, a tendency was observed for both mixture A and mixture S to reduce the development of the seedlings compared to control pine C. After the treatments, the height of the above-ground parts of the seedlings was lower than in the control experiment, although the differences were not statistically significant. For mixture A (irrespective of the dose), there was a tendency to increase root length, in contrast to mixture S. The addition of phosphorus usually has a positive effect on root development [45], which was also confirmed in the present study.

Although the total phosphorus content in the soil increased after the addition of the preparations, it was observed that it remained at the soil surface in the case of mixture A, whereas the phosphorus penetrated into the deeper soil layers in the case of mixture S. It is possible that the phosphorus was better absorbed by the plants in this form. Future analyses of the phosphorus content of plant tissues (needles and roots) could provide an answer to this question. The most favourable effect of the formulations was observed at lower doses of 1 to 2 t/ha, which is encouraging for economic reasons. In view of the results of the present study, further studies should therefore be conducted for the low doses.

In the visual assessment of plant health and growth performance, the phosphogypsum A mixture performed significantly better than the phosphogypsum S mixture in the soil. The unfavourable effect of the preparation S used was evident in the wilting of the needles in the lower part of the seedlings. This phenomenon was probably caused by water droplets falling directly on the phosphogypsum mixture during the rain and spraying the needles in the immediate vicinity.

However, the reduction in needle area in this case could also have been influenced by heavy metals that inhibit plant growth. Therefore, the preparations would have to be examined for cadmium (Cd), since under the influence of cadmium ions the intensity of photosynthesis in wheat cereals is significantly reduced [47]. Presumably, the presence of these ions can have a similar effect on pine seedlings.

Although the heavy metals accumulated in the soil are slowly removed from the soil by the plants, this affects their development [48]. It should also be checked whether heavy metals contained in phosphogypsum are taken up by pine seedlings and, if so, by which plant organs. Are heavy metals transported from the roots to the needles, which then fall off, and what effects could this have on the forest litter? Plants take up heavy metals together with other elements in ionic form. Their toxic effects on plant life processes are mainly due to interactions with functional groups of molecules that make up cells, especially proteins and polynucleotides. The result of these phenomena can be poorer growth and development of the plant or even its death. The harmful effects of heavy metals become apparent at certain concentrations in the plant environment [49,50].

Therefore, it would be equally important to investigate whether fungi that live in symbiosis with trees do not take up heavy metals that could ultimately harm the organisms that consume them. Are heavy metals stored in wood and what impact does this have on the quality of the wood grown?

4.8. Phosphogypsum as an Important Source of Phosphorus for Seedlings and Microorganisms

The presence of phosphorus in the soil is an important limiting factor for the uptake of heavy metals by plants, as higher amounts of easily soluble forms of phosphorus can precipitate poorly soluble phosphates of zinc, cadmium, lead and copper [8]. However, this hypothesis would need to be confirmed by analysing the metal content of soil samples containing the formulations, which will be done later.

The pool of permanently bound phosphorus includes inorganic, poorly soluble compounds and organic compounds that are resistant to mineralization by soil microorganisms. Such compounds can be present in the soil for many years, but are not available to plants and have little impact on soil fertility [45]. About half of the total soil phosphorus reserve is accumulated in soil organic matter. The other half of this reserve is inorganic phosphorus with low mobility, mostly bound to mineral soil particles [45]. A change to a more acidic pH, e.g., strong local soil acidification with high doses of preparations (especially with sewage sludge), could also be the cause of poorer development of pine seedlings, for which the optimal soil pH should be between 4.5 and 5 [51].

Future research should also include the biological properties of the soil (microbiome), whose influence on plant health is difficult to assess, such as fungi (mycorrhizal, saprotrophic, pathogenic) and bacteria. The stockpiles themselves should also be examined for bacteria that could be inadvertently transferred to forest areas.

We have refrained from giving the physical properties of the soil and macronutrients such as N, K, Ca and Mg because we feel that such chemical analysis adds little to the discussion, except for phosphorus (which we have done), and the physical analysis of the soil is useful in field studies (in future tests) and not in this case in a pot experiment. Providing detailed soil properties in our article would not have added anything relevant to the research conducted, as we did not investigate the relationship or impact of soil properties on seedling response to phosphogypsum fertilization. As mentioned earlier, we were mainly interested in two aspects: the absence of phytotoxicity of the doses administered and the directional (positive) response of the seedlings to additional doses of phosphogypsum. In our experiment, we determined the content and translocation of P from the upper layers, where it was applied in the form of phosphogypsum, to the lower layers as a function of the application dose. We believe that, at this stage of the research, we should confirm the validity of the direction of recycling and the use of waste produced in the manufacture of fertilizers, in this case in forestry, to bring about a change in their status from waste to poor fertilizer. It will then be possible to carry out more extensive field studies in forests.

4.9. Photosynthetic Performance of Pine Seedlings

Long-term surface application of phosphogypsum (PG) can enhance plant growth and physiological and biochemical processes [52]. Addition of PG increased root development at greater depths and improved the plant nutrition of maize. These combined effects increased the concentrations of photosynthetic pigments and gas exchange even in low water availability. In addition, the activities of Rubisco, sucrose synthase and antioxidant enzymes were improved, reducing oxidative stress. These improvements in the physiological performance of maize plants resulted in higher grain yield. Overall, the results argue for soil amendments as an important strategy to increase soil fertility and secure crop yields in regions where dry spells occur during the cropping cycle [52].

There are two types of photosystems: PS I (=PS700) and PS II (=PS680). Thus, the chlorophyll in the centre of the reaction PS I has an absorption maximum at a wavelength of 700 nm, while PS II has a maximum at 680 nm. In the chloroplasts, PS II predominates and the stoichiometric ratio of PS II to PS I is about 1.5, but can vary depending on environmental conditions [53]. PS II also contains an oxygen releasing complex (OEC). It is located on the inner surface of the thylakoid membrane. The complex contains the amino acid tyrosine on the D1 protein, four manganese atoms and the outer proteins PsbO (stabilizes the manganese cluster), PsbP and PsbQ. During the light phase of photosynthesis, the water molecules in this complex are split into protons, electrons and oxygen.

In phosphogypsum with sewage sludge these two PI (PIabs and PIinst) had a positive effect at a dose of 5 t/ha; in phosphogypsum with ash the same PI parameters (PIabs and Piinst) had an effect, but the positive effect was at a dose of 2 t/ha. Both preparations also reduce the loss of heat energy in the photosynthetic apparatus of the plants (lower DI/CS).

After analysing all the parameters, we decided to present only the overall parameter PI, as the other parameters show only weak tendencies and contribute nothing to the discussion. The other parameters mentioned and analysed in the methodology were not statistically significantly different from each other.

The weaker photosynthetic efficiency of the treated pines could be due to the presence of heavy metals, especially cadmium, in the phosphogypsum (Figure 1). This aspect should be further investigated in the future.

4.10. Restrictions on the Use of New Preparations Based on Phosphogypsum

However, to date there is little information in the literature on the applicability of PG in forestry [18,54,55] or the assessment of the use of phosphogypsum in a nursery for plant propagation [56]. Much more information is available on the impact of PG on the soil ecosystem [31,32] and on the assessment of alternative and traditional cover systems for phosphogypsum remediation [57,58]. The main problem is the content of heavy metals (HM) which, even if they comply with current standards (of the Member States or the European Community), still pose some environmental risk [8]. When HM enter aquatic ecosystems, there is an exceptionally high bioaccumulation, the concentration of which exceeds the initial uptake of toxins by the last link in the trophic chain (e.g., fish) by tens and hundreds of thousands of times [2]. The peculiarity of the behaviour of HM in the air lies in their wide distribution, which can be tens of kilometres. In soil, due to its different biogeochemical composition under different environmental conditions, different effects must be taken into account: antagonistic, synergistic and sensitizing effects, which have a strong impact on the quality of the environment and consequently on human health.

In forests, litter is formed from accumulated and decaying organic matter. The upper humus accumulation horizon (horizon A1 with a thickness of up to 10 cm) concentrates most absorbent roots, releases phytoncides, absorbs radionuclides, etc.). Heavy metals can be transferred from the roots to the leaves through soil contamination and the leaves can also absorb them directly (through leaf contamination). Raking the leaves is one way to remove heavy metals from the soil. However, removing the foliage reduces the supply of organic material to the soil, which is a strong buffer that binds heavy metals. A more environmentally friendly method is therefore to rake the leaves and place them in grooves

25–30 cm deep, i.e., in the upper soil layer where all biological processes are active. In this case, there is active mineralization of the organic matter, binding of HM and, above all, reduction of their mobile forms, which reduces the risk of exposure to HM. Mixing the preparations into the topsoil is therefore a better solution than simply spreading them and leaving them on the surface.

Signs of plant inhibition and a decrease in the activity of the assimilation apparatus could be the negative effects of increased HM concentrations in woody plants but that was not the aim of this study. However, it is known, for example, that the effect of cadmium nitrate on wheat plants manifests itself in a significant reduction in seed germination and a decrease in biometric indicators. Under the influence of cadmium ions, the intensity of photosynthesis is significantly reduced [59].

The preparations can also be mixed with organic material such as wood waste, pine bark compost or sawdust. The addition of organic material will undoubtedly improve the physico-chemical and biological properties of the soil: improvement of the soil structure; creation of a more airy condition of the upper soil horizons (humus accumulation horizon); activation of biological, including microbiological processes; promotion of a more active growth of the root system and especially the growth of suckering roots. The normal state of the root system under the conditions of a natural increase in the volumetric mass value protects against the penetration of pollutants (HM) into the terrestrial organs of phytochromes (due to the normal, optimal functioning of plant resistance mechanisms).

The present forestry study is a pioneering work, which is why it is hardly mentioned in the literature. It is not known whether there is a possibility of wider management and use of phosphogypsum hills in Poland. The basic research problem was to find an answer to the question whether preparations based on calcium sulphate in two variants—a mixture with organic ash A and with sewage sludge S—can serve as fertilizer and in what quantity per hectare?

5. Conclusions

1. The phosphogypsum-based preparations used showed no harmful (toxic) effects on the pine seedlings grown in pots during the six-month trial period.
2. The preparation made in bulk from a mixture of phosphogypsum and wood ash had a favourable effect on the development of the seedlings' root system and was also easier to mix with the soil than phosphogypsum with sewage sludge, which was in a sticky form.
3. The photosynthetic efficiency of one-year-old pine seedlings decreased after one growing season following the application of phosphogypsum preparations, so that observations over a longer period are necessary.
4. The most favourable dosages of phosphogypsum preparations for the development of pine seedlings (especially in combination with ash) are between 1 and 2 t/ha, although most of the growth parameters tested did not differ from the control. From an economic point of view, lower dosages will be more favourable for use in forestry, but further studies over a longer period of time are needed for this.
5. Formulations should be tested for heavy metals and their effects on seedling development.
6. Testing should continue for at least two to three growing seasons, including monitoring for changes in the microbiome.

Author Contributions: Conceptualization, T.O., P.B. and S.B.; methodology, T.O., P.B. and O.K.; software, T.P., P.B. and O.K.; writing, T.O., P.B. and A.R.; visualization, P.B.; supervision, T.O. and S.B.; resources, T.O. and S.B.; investigation, T.P., W.S. and B.R.; project administration, S.B. and T.O.; funding acquisition, S.B. and T.O. All authors have read and agreed to the published version of the manuscript.

Funding: The study was partly carried out within the framework: WZ/WB-INL/3/2021 and financed from the science funds from Ministry of Science and Higher Education in Poland.

Institutional Review Board Statement: Not applicable.

Informed Consent Statement: Not applicable.

Data Availability Statement: Not applicable.

Conflicts of Interest: The authors declare no conflict of interest.

References

1. Jasiński, J.; Żabiński, M. Quality management and sustainable development in local communes—Evidence from Poland. *Public Organ. Rev.* **2022**, *22*, 763–782. [CrossRef]
2. Saadaoui, E.; Ghazel, N.; Ben Romdhane, C.; Massoudi, N. Phosphogypsum: Potential uses and problems—A review. *Int. J. Environ. Stud.* **2017**, *74*, 558–567. [CrossRef]
3. Cieślik, B.M.; Namieśnik, J.; Konieczka, P. Review of sewage sludge management: Standards, regulations and analytical methods. *J. Clean. Prod.* **2015**, *90*, 1–15. [CrossRef]
4. Cieślik, B.; Konieczka, P. A review of phosphorus recovery methods at various steps of wastewater treatment and sewage sludge management. The concept of "no solid waste generation" and analytical methods. *J. Clean. Prod.* **2017**, *142*, 1728–1740. [CrossRef]
5. Vyshpolsky, F.; Qadir, M.; Karimov, A.; Mukhamedjanov, K.; Bekbaev, U.; Paroda, R.; Aw-Hassan, A.; Karajeh, F. Enhancing the productivity of high-magnesium soil and water resources in Central Asia through the application of phosphogypsum. *Land Degrad. Dev.* **2008**, *19*, 45–56. [CrossRef]
6. Tayibi, H.; Choura, M.; López, F.A.; Alguacil, F.J.; López-Delgado, A. Environmental impact and management of phosphogypsum. *J. Environ. Manag.* **2009**, *90*, 2377–2386. [CrossRef] [PubMed]
7. Hilton, J. Phosphogypsum Management and Opportunities for Use. In Proceedings of the International Fertiliser Society Conference, Cambridge, UK, 14 December 2006.
8. Béjaoui, I.; Kolsi-Benzina, N.; Hadj, Z.B. Cadmium and phosphorus lixiviation in polluted acidic forest soil amended with Tunisian phosphogypsum. *J. New Sci.* **2016**, *29*, 3.
9. Solla, A.; Moreno, G.; Malewski, T.; Jung, T.; Klisz, M.; Tkaczyk, M.; Siebyla, M.; Pérez, A.; Cubera, E.; Hrynyk, H.; et al. Phosphite spray for the control of oak decline induced by Phytophthora in Europe. *For. Ecol. Manag.* **2021**, *485*, 118938. [CrossRef]
10. Borowik, P.; Oszako, T.; Malewski, T.; Zwierzyńska, Z.; Adamowicz, L.; Tarakowski, R.; Ślusarski, S.; Nowakowska, J.A. Advances in the Detection of Emerging Tree Diseases by Measurements of VOCs and HSP s Gene Expression, Application to Ash Dieback Caused by Hymenoscyphus fraxineus. *Pathogens* **2021**, *10*, 1359. [CrossRef]
11. Kabała, C.; Musztyfaga, E.; Gałka, B.; Łabuńska, D.; Mańczyńska, P. Conversion of Soil pH 1:2.5 KCl and 1:2.5 H_2O to 1:5 H_2O: Conclusions for Soil Management, Environmental Monitoring, and International Soil Databases. *Pol. J. Environ. Stud.* **2016**, *25*, 647–653. [CrossRef]
12. Jarosch, K.; Santner, J.; Parvage, M.M.; Gerzabek, M.H.; Zehetner, F.; Kirchmann, H. Four soil phosphorus (P) tests evaluated by plant P uptake and P balancing in the Ultuna long-term field experiment. *Plant Soil Environ.* **2018**, *64*, 441–447. [CrossRef]
13. Schreiber, U.; Bilger, W.; Neubauer, C.; Schulze, E.; Caldwell, M. *Ecophysiology of Photosynthesis*; Schreiber, U., Caldwell, M.M., Eds.; Ecological Studies; Springer Science & Business Media: Berlin/Heidelberg, Germany, 1994; Volume 100, pp. 49–70.
14. Babani, F.; Lichtenthaler, H.K.; Richter, P. Changes of chlorophyll fluorescence signatures during greening of etiolated barley seedlings as measured with the CCD-OMA fluorometer. *J. Plant Physiol.* **1996**, *148*, 471–477. [CrossRef]
15. Lichtenthaler, H.; Schwender, J.; Reindl, A.; Herbers, K. Overexpression of a DNA Sequence Coding for a 1-Desoxy-d-xylulose-5-phosphate Reductoisomerase in Plants. U.S. Patent 6,765,129, 2004.
16. Reigosa Roger, M.J.; Weiss, O. Fluorescence techniques. In *Handbook of Plant Ecophysiology Techniques*; Springer: Berlin/Heidelberg, Germany, 2001; pp. 155–171. Available online: https://www.barnesandnoble.com/w/handbook-of-plant-ecophysiology-techniques-manuel-joaquin-reigosa-roger/1100014338 (accessed on 25 January 2023).
17. SAS/STAT User's Guide. 2018. Available online: https://support.sas.com/documentation/onlinedoc/stat/indexproc.html (accessed on 25 January 2023).
18. Andrews, R.L. Comparison of Bucket-Wheel Spoil and Phosphogypsum/Clay Blend as Substrates for Nonriverine Wet Hardwood Forest Restoration. Ph.D. Thesis, North Carolina State University, Raleigh, NC, USA, 2003.
19. Li, J.Y.; Liu, Z.D.; Zhao, W.Z.; Masud, M.; Xu, R.K. Alkaline slag is more effective than phosphogypsum in the amelioration of subsoil acidity in an Ultisol profile. *Soil Tillage Res.* **2015**, *149*, 21–32. [CrossRef]
20. Fontoura, S.M.V.; de Castro Pias, O.H.; Tiecher, T.; Cherubin, M.R.; de Moraes, R.P.; Bayer, C. Effect of gypsum rates and lime with different reactivity on soil acidity and crop grain yields in a subtropical Oxisol under no-tillage. *Soil Tillage Res.* **2019**, *193*, 27–41. [CrossRef]
21. da Costa, C.H.M.; Crusciol, C.A.C. Long-term effects of lime and phosphogypsum application on tropical no-till soybean–oat–sorghum rotation and soil chemical properties. *Eur. J. Agron.* **2016**, *74*, 119–132. [CrossRef]
22. Crusciol, C.A.; Artigiani, A.C.; Arf, O.; Carmeis Filho, A.C.; Soratto, R.P.; Nascente, A.S.; Alvarez, R.C. Soil fertility, plant nutrition, and grain yield of upland rice affected by surface application of lime, silicate, and phosphogypsum in a tropical no-till system. *Catena* **2016**, *137*, 87–99. [CrossRef]

23. Soratto, R.P.; Crusciol, C.A. Dolomite and phosphogypsum surface application effects on annual crops nutrition and yield. *Agron. J.* **2008**, *100*, 261–270. [CrossRef]
24. Michalovicz, L.; Tormena, C.A.; Müller, M.M.L.; Dick, W.A.; Cervi, E.C. Residual effects of phosphogypsum rates and machinery traffic on soil attributes and common-bean (Phaseolus vulgaris) yield in a no-tillage system. *Soil Tillage Res.* **2021**, *213*, 105152. [CrossRef]
25. Bzdyk, R.M.; Olchowik, J.; Studnicki, M.; Nowakowska, J.A.; Oszako, T.; Urban, A.; Hilszczańska, D. Ectomycorrhizal colonisation in declining oak stands on the Krotoszyn Plateau, Poland. *Forests* **2019**, *10*, 30. [CrossRef]
26. Seidel, E.P.; dos Reis, W.; Mottin, M.C.; Fey, E.; Schneider, A.P.R.; Sustakowski, M.C. Evaluation of aggregate distribution and selected soil physical properties under maizejack bean intercropping and gypsum rates. *Afr. J. Agric. Res.* **2017**, *12*, 1209–1216.
27. Araújo, L.G.; de Sousa, D.M.G.; de Figueiredo, C.C.; Rein, T.A.; de Souza Nunes, R.; dos Santos, J.d.D.G.; Malaquias, J.V. How does gypsum increase the organic carbon stock of an Oxisol profile under sugarcane? *Geoderma* **2019**, *343*, 196–204. [CrossRef]
28. Carmeis Filho, A.C.; Penn, C.J.; Crusciol, C.A.; Calonego, J.C. Lime and phosphogypsum impacts on soil organic matter pools in a tropical Oxisol under long-term no-till conditions. *Agric. Ecosyst. Environ.* **2017**, *241*, 11–23. [CrossRef]
29. Kanter, D.R.; Brownlie, W.J. Joint nitrogen and phosphorus management for sustainable development and climate goals. *Environ. Sci. Policy* **2019**, *92*, 1–8. [CrossRef]
30. Chuan, L.M.; Zheng, H.G.; Zhao, J.J.; Wang, A.L.; Sun, S.F. Phosphogypsum production and utilization in China. *IOP Conf. Ser. Mater. Sci. Eng.* **2018**, *382*, 022099. [CrossRef]
31. Costa, R.F.; Firmano, R.F.; Colzato, M.; Crusciol, C.A.C.; Alleoni, L.R. Sulfur speciation in a tropical soil amended with lime and phosphogypsum under long-term no-tillage system. *Geoderma* **2022**, *406*, 115461. [CrossRef]
32. Plyatsuk, L.; Balintova, M.; Chernysh, Y.; Demcak, S.; Holub, M.; Yakhnenko, E. Influence of phosphogypsum dump on the soil ecosystem in the Sumy region (Ukraine). *Appl. Sci.* **2019**, *9*, 5559. [CrossRef]
33. Knupp, E.A.N.; Silva, C.G.B.; de Oliveira, K.A.P.; de Amorim, F.R.; Siqueira, M.C.; Taddei, M.H.T.; Franco, M.B.; Palmieri, H.E.L.; Jacomino, V.M.F. Radionuclide, metal and non-metal levels in percolated water from soils fertilized with phosphogypsum. *J. Radioanal. Nucl. Chem.* **2014**, *299*, 321–327. [CrossRef]
34. Fernández-Caliani, J.C.; Giráldez, M.; Waken, W.; Del Río, Z.; Córdoba, F. Soil quality changes in an Iberian pyrite mine site 15 years after land reclamation. *Catena* **2021**, *206*, 105538. [CrossRef]
35. Huang, L.; Liu, Y.; Ferreira, J.F.; Wang, M.; Na, J.; Huang, J.; Liang, Z. Long-term combined effects of tillage and rice cultivation with phosphogypsum or farmyard manure on the concentration of salts, minerals, and heavy metals of saline-sodic paddy fields in Northeast China. *Soil Tillage Res.* **2022**, *215*, 105222. [CrossRef]
36. Kosobucki, P.; Chmarzynski, A.; Buszewski, B. Sewage sludge composting. *Pol. J. Environ. Stud.* **2000**, *9*, 243–248.
37. Kominko, H.; Gorazda, K.; Wzorek, Z. The possibility of organo-mineral fertilizer production from sewage sludge. *Waste Biomass Valorization* **2017**, *8*, 1781–1791. [CrossRef]
38. Rigueiro-Rodríguez, A.; Mosquera-Losada, M.R.; Ferreiro-Domínguez, N. Pasture and soil zinc evolution in forest and agriculture soils of Northwest Spain three years after fertilisation with sewage sludge. *Agric. Ecosyst. Environ.* **2012**, *150*, 111–120. [CrossRef]
39. Kijo-Kleczkowska, A.; Otwinowski, H.; Środa, K. Properties and production of sewage sludge in Poland with reference to the methods of neutralizing. *Arch. Gospod. Odpad. Ochr. Środowiska* **2012**, *14*, 59–78.
40. Solon, J.; Borzyszkowski, J.; Bidłasik, M.; Richling, A.; Badora, K.; Balon, J.; Brzezińska-Wójcik, T.; Chabudziński, Ł.; Dobrowolski, R.; Grzegorczyk, I.; et al. Physico-geographical mesoregions of Poland: Verification and adjustment of boundaries on the basis of contemporary spatial data. *Geogr. Pol.* **2018**, *91*, 143–170. [CrossRef]
41. Hanafi, M.M.; Azizi, P.; Vijayanathan, J. Phosphogypsum Organic, a Byproduct from Rare-Earth Metals Processing, Improves Plant and Soil. *Agronomy* **2021**, *11*, 2561. [CrossRef]
42. Tomczyk, A.M.; Szyga-Pluta, K. Variability of thermal and precipitation conditions in the growing season in Poland in the years 1966–2015. *Theor. Appl. Climatol.* **2019**, *135*, 1517–1530. [CrossRef]
43. Woś, A. *Klimat Polski w Drugiej połowie XX Wieku*; Wydawnictwo Naukowe UAM-Uniwersytetu im. Adama Mickiewicza; 2010. Available online: https://press.amu.edu.pl/en/klimat-polski-w-drugiej-polowie-xx-wieku-686.html (accessed on 25 January 2023).
44. Przybylak, R. Poland's climate in the last millennium. In *Oxford Research Encyclopedia of Climate Science*; Oxford University Press: Oxford, UK, 2016.
45. Balemi, T.; Negisho, K. Management of soil phosphorus and plant adaptation mechanisms to phosphorus stress for sustainable crop production: A review. *J. Soil Sci. Plant Nutr.* **2012**, *12*, 547–562. [CrossRef]
46. Delin, S. Fertilizer value of phosphorus in different residues. *Soil Use Manag.* **2016**, *32*, 17–26. [CrossRef]
47. Kuznetsova, S.; Klimachev, D. Effect of cadmium on growth processes and intensity of photosynthesis of plants of wheat (in Russian). *Geogr. Environ. Living Syst.* **2014**, *5*, 20–23.
48. Ahsan, N.; Renaut, J.; Komatsu, S. Recent developments in the application of proteomics to the analysis of plant responses to heavy metals. *Proteomics* **2009**, *9*, 2602–2621. [CrossRef]
49. Dalvi, A.A.; Bhalerao, S.A. Response of plants towards heavy metal toxicity: An overview of avoidance, tolerance and uptake mechanism. *Ann. Plant Sci.* **2013**, *2*, 362–368.
50. Manara, A. Plant responses to heavy metal toxicity. In *Plants and Heavy Metals*; Springer: Berlin/Heidelberg, Germany, 2012; pp. 27–53.

51. Neina, D. The role of soil pH in plant nutrition and soil remediation. *Appl. Environ. Soil Sci.* **2019**, *2019*, 5794869. [CrossRef]
52. Bossolani, J.W.; Crusciol, C.A.C.; Garcia, A.; Moretti, L.G.; Portugal, J.R.; Rodrigues, V.A.; Fonseca, M.d.C.d.; Calonego, J.C.; Caires, E.F.; Amado, T.J.C.; et al. Long-term lime and phosphogypsum amended-soils alleviates the field drought effects on carbon and antioxidative metabolism of maize by improving soil fertility and root growth. *Front. Plant Sci.* **2021**, *12*, 650296. [CrossRef] [PubMed]
53. Taiz, L.; Zeiger, E. Photosynthesis: Physiological and ecological considerations. *Plant Physiol* **2002**, *9*, 172–174.
54. de Sousa, R.N.; da Silva, B.A.; da Costa, V.V.; da Silva Teixeira, R.; Valadares, S.V.; da Silva, I.R.; Venegas, V.H.A.; Vergütz, L. Limestone and phosphogypsum are key drivers of eucalypt production in the highly weathered soils of Brazil. *Plant Soil* **2022**, 1–21. [CrossRef]
55. Kushnir, E.; Nedbaeva, I.; Treschevskaya, E. *The Assessment of Forest Stand and Soil Condition on the Recultivated Area of Human-Made Mineral Formations of Kingisepp Phosphorite Deposit*; Saint Petersburg Forestry Research Institute: St Petersburg, Russia, 2021.
56. Ghazel, N.; Saadaoui, E.; Ben Romdhane, C.; Abbès, N.; Grira, M.; Abdelkebir, S.; Aydi, S.; Abdallah, L.; Mars, M. Assessment of phosphogypsum use in a nursery for plant propagation. *Int. J. Environ. Stud.* **2018**, *75*, 284–293. [CrossRef]
57. Robinson, M. Phosphogypsum Reclamation: Evaluating Alternative and Traditional Cover Systems. 2018. Available online: https://era.library.ualberta.ca/items/3537fbe4-d862-43f5-b222-4c07480543ad (accessed on 25 January 2023).
58. Robinson, M.J.; Dhar, A.; Naeth, M.A.; Nichol, C.K. Phosphogypsum Stack Reclamation Using Soil Amendments and Short-Rotational Woody Species. *Land* **2022**, *11*, 2003. [CrossRef]
59. Abilova, G. Effect of cadmium and lead ions on the growth and content of proline in plants of triticale (Triticosecale Wittm.) (in Russian). *Proc. Karelian Res. Cent. Russ. Acad. Sci.* **2016**, *11*, 27. [CrossRef]

forests

Article

Biomass Estimation and Carbon Storage of *Taxodium* Hybrid Zhongshanshan Plantations in the Yangtze River Basin

Qin Shi [1], Jianfeng Hua [1], David Creech [2] and Yunlong Yin [1,*]

[1] Jiangsu Key Laboratory for the Research and Utilization of Plant Resources, Institute of Botany, Jiangsu Province and Chinese Academy of Sciences (Nanjing Botanical Garden Mem. Sun Yat-Sen), No. 1 Qianhuhou Village, Nanjing 210014, China

[2] Arthur Temple College of Forestry and Agriculture, Stephen F. Austin State University, No. 1936 North Street, Nacogdoches, TX 75962-3000, USA

* Correspondence: yinyl066@sina.com; Tel.: +86-025-8434-7143

Abstract: As a pivotal wetland tree, *Taxodium* hybrid Zhongshanshan has been widely planted in the region of Yangtze River for multipurpose of ecological restoration, field shelter, landscape aesthetics as well as carbon sequestration. However, the carbon allocation patterns across distinct stages of stand development of *T.* Zhongshanshan are poorly documented. Using a sample of 30 trees which were destructively harvested, this study compared 3 models for assessing aboveground biomass. Furthermore, a linear seemingly unrelated regression (SUR) approach was introduced to fit the system of the best selected model that ensured the additivity property. On this basis, biomass and carbon storage of *T.* Zhongshanshan stands in the Yangtze River Basin (YRB) were fairly estimated. Specifically, the study developed height-diameter at breast (*H-DBH*) function. The results showed that the selected 3-parameter polynomial model performed better, and the SUR approach provided more accurate estimates of leaf and stem fractions. The total tree biomass was 53.43, 84.87, 140.67, 192.71 and 156.65 t ha^{-1} in the 9-, 11-, 13-, 15-, and 22-year-old *T.* Zhongshanshan stands, and contributed averagely 94.40% of the ecosystem biomass accumulation. The current *T.* Zhongshanshan stands in the YRB area can store 124.76 to 217.64 t ha^{-1} carbon, of which total tree ranges from 25.32 to 90.89 t ha^{-1}, with 55.19% to 77.66% storing in the soil. The *T.* Zhongshanshan had continuous potential for carbon storage during its growth, particularly in the incipient stages. The findings of this research are firsthand information for forest managers for the sustainable management of *T.* Zhongshanshan in the YRB and similar subtropical areas.

Keywords: biomass estimation; carbon storage; allometric model; *Taxodium* hybrid Zhongshanshan; Yangtze River Basin

check for updates

Citation: Shi, Q.; Hua, J.; Creech, D.; Yin, Y. Biomass Estimation and Carbon Storage of *Taxodium* Hybrid Zhongshanshan Plantations in the Yangtze River Basin. *Forests* **2022**, *13*, 1725. https://doi.org/10.3390/f13101725

Academic Editor: Mark Vanderwel

Received: 27 September 2022
Accepted: 18 October 2022
Published: 19 October 2022

1. Introduction

Increasing global carbon fixation through the expansion of afforestation lands has been proposed as an efficient approach for alleviating elevated concentrations of atmospheric CO_2 [1]. The absolute and relative distribution of carbon storage in plantations is acquainted with a variation of tree species, soil condition and climate. Immediate challenges of forest management are concerned with credible, precise and cost-effective methods to adequately document forest dynamics. As a result of the monetary value being attached to carbon sequestration, there is increased scrutiny of techniques for estimating tree biomass. Burgeoning carbon credit market mechanism such as reducing emission from deforestation and forest degradation necessitates such a need [2]. This requires appropriate methods specific for a given forest type. Traditionally, allometric relationships to aboveground biomass have been used at the experimental scale to estimate aboveground biomass. Generic equations, stratified by ecological zones, for estimating aboveground biomass exist, but they may not accurately reflect the tree biomass in a specific area or region [3]. Small tree individuals not merchantable are often omitted in forest resource investigation.

The allometric model is a statistical formula calculated by regression analysis between tree properties, among which tree diameter at breast height (*DBH*) and height (*H*) have often been used as explanatory variables because of their immediate availability. Developing such relationships of a certain tree species is a time-consuming activity, especially separating biomass components including the leaves and branches. A good database for developing regression equations should contain an age sequence because trees of different diameters distinguish from each other in the component proportion of the aboveground biomass. When there were several tree components in the biomass data, the additivity of models for accessing tree total, sub-total, and separate biomass fractions should be considered, owing to the inherent correlations among the biomass components measured on the same sample trees [4]. Traditional models often ignored such inherent relationships. The seemingly unrelated regression (SUR) model, proposed by Parresol et al. [5], has been adopted by many reports to facilitate biomass models construction since it ensures the additivity among components and total biomass predictions [6,7]. Conventionally, forest inventories measure the *DBH* of all trees in each plot but often few are randomly selected and measured for *H*. Accurate measurement of tree *H* is more difficult than *DBH* measurement [8], which implies that the forest productivity appraisal, in practice, requires *H-DBH* models for *H* estimation. Moreover, in the cases where the actual measurements of height growth are not available, *H-DBH* functions can be used to indirectly predict height growth.

Taxodium hybrid Zhongshanshan (*T. distichum* × *T. mucronatum*), a superior inter-species hybrid, was successfully planted in southeastern China. Being one of the most important tree species, *T.* Zhongshanshan produces excellent quality timber, with a tower-shaped morphological structure, high resistance of bending and cracking, and impressive water tolerance traits. These phenotypic characteristics enhance wind resistance and permit better performance in hostile coastal environments. Owing to its high commercial and ecological value, the artificially established area of *T.* Zhongshanshan in the Yangtze River Basin (YRB) is around 70,000 ha, with quantities over 50 million, and the demand for this conifer tree in landscape plantations and ecological restoration surpasses supply. Despite the fact that an increasing number of studies have been conducted from many perspectives, including crossbreeding, water resistance mechanism and photosynthetic traits [9–12], little information is available on carbon pools with stand ages in *T.* hybrid Zhongshanshan plantations since models to assist management of this species are in most cases lacking. An assessment of carbon storage in *T.* Zhongshanshan plantations is crucial for regional-scale evaluation of carbon dynamics and ameliorating these estimates requires plentiful field studies. The aims of this study were (1) to establish the allometric biomass equations for *T.* Zhongshanshan and its individual component biomass with consideration of stand age; and (2) to estimate the carbon storage of the *T.* Zhongshanshan ecosystem.

2. Materials and Methods

2.1. Study Sites Description

The forests both natural and artificial are rigorously conserved in the YRB, and *T.* Zhongshanshan plants should be sampled without felling or seriously damaging them. Given this, the destructive tree sampling for allometric models was conducted at the experimental base of Institute of Botany, Jiangsu Province and Chinese Academy of Sciences, which is located in the Qixia and Liuhe Districts, Nanjing and Tinghu District, Yancheng (Figure 1). These areas experience a typical subtropical monsoon climate. The mean annual temperature is 15.8 °C and the mean annual precipitation is 1085 mm, of which 65% falls during June-September. The study areas of biomass and carbon storage are located along the Yangtze River, respectively, in the cities of Nanjing, Jingzhou, Chongqing and Kunming (Figure 1). The elevation ranges from 5 m to 1823 m above sea level. Characterized by a subtropical monsoon climate, the annual average temperature ranges from 12.6 to 18.0 °C and mean annual precipitation is about 1076 mm. The areas of the *T.* Zhongshanshan stands cover from 42 to 201 ha and are managed by the local forestry departments. Understory vegetation was dominated by common herbaceous species included: *Erigeron annuus*, *Alternanthera philoxeroides*, *Achyranthes bidentata*, *Chenopodium album*, *Rubus hirsutus*, *Angelica sieboldin*, *Mazus japonicus*, *Plantago depressa*, *Setaria viridis* and *Solidago canadensis*, with sparsely scattered woods (mainly *Morus alba* and *Ulmus pumila*).

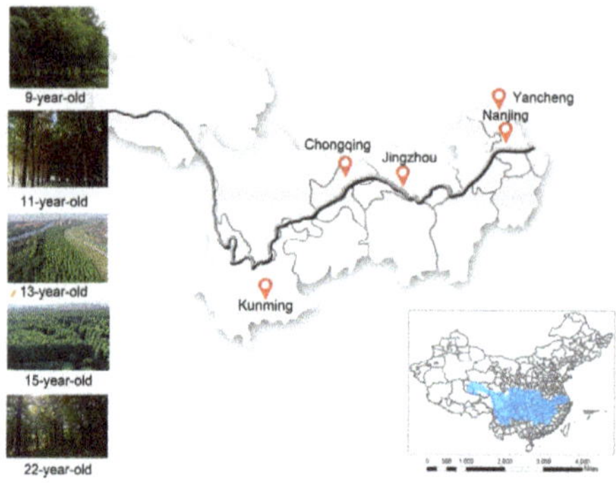

Figure 1. Location of the *T.* Zhongshanshan stands.

2.2. Destructive Tree Sampling

From April 2021 to June 2022, 30 trees aged 8, 12, 18 and 22 years (6–8 trees per age group) in the experimental base were used to establish the allometric equation for *T.* Zhongshanshan using the segmenting method. The tree was cut by an electric saw from the bottom. After *H* and *DBH* measurement, they were cut at 1 m intervals from the tree base. Each section was separated into stem, branch and leaf. The fresh weights of all components were measured in situ, and samples of every component in each standard tree were collected for water content and carbon concentration analysis. Allometric regression equations relating tree *DBH* and *H* were developed across this chronosequence, which was used to calculate the biomass of each tree and the total biomass of each *T.* Zhongshanshan stand.

2.3. Forest Inventory and Measurements

From May to June 2022, preliminary forest inventories were carried out in 9-, 11-, 13-, 15- and 22-year-old *T.* Zhongshanshan stands. We established a size of 1 ha covering minor local heterogeneity in soil and plant conditions, which allowed us to differentiate 4 fixed plots of 25 × 25 m in each stand and were spaced 30 m from each other. After measuring the distance of 25 m in a straight line with a 100-m tape, we inserted horticultural tallies into the 4 corners of each plot and connected them with strings to mark the boundaries. At each sampling stand, coordinate was recorded using a GPS device (A8, Zuolin Technology Co., LTD, Guangdong, China). Within each plot, *H* and *DBH* were recorded by a height measuring device (CGQ-1, Harbin Optical Instrument Factory, Harbin, China) and a professional tapeline for every tree (Table 1). The measured trees were marked with paint to guarantee that no repetition was made. All species were identified by 2 observers working together by randomly selecting five 1 m × 1 m quadrants in each plot. Litter, herb and shrubs biomass, including roots, were also harvested from these 5 subplots. All samples including tree tissues (leaves, branches and stems), herbs, shrubs and litter were weighed and oven-dried at 75 °C to a constant weight in the lab and reweighed through wet-to-dry mass conversion factors.

In each plot, 5 soil cores (5 cm in diameter) were randomly collected at 0–20, 20–40, and 40–60 cm depth. In view of the homogeneous soil within the stand, soil samples within the same layer in one plot were thoroughly mixed into a homogenized sample. Another set of soil samples from 0–20, 20–40, and 40–60 cm depths were separately and intactly collected by inserting a steel cylinder of known volume (5 cm high and with a 5 cm inner diameter) for bulk density determination (ratio of dry mass to sample volume). All soil samples were put into labelled airtight plastic bags and taken back to the lab. A total of 60 soil samples and 60 bulk density samples were collected (5 stand ages × 4 plots × 3 depths). Moreover, a part of field samples was oven-dried at 105 °C for 24 h to determine soil moisture content gravimetrically. After removing the plant roots, fauna, and debris by hand, the soils were air-dried at room temperature, and then ground and passed through a 2-mm sieve for measurement of carbon contents. The carbon concentrations of the components from the tree, ground vegetation, forest floor and soil organic carbon (SOC) were determined by the potassium dichromate oxidation method.

Table 1. Characteristics of the 9-, 11-, 13-,15- and 22-year-old *T.* Zhongshanshan stands.

Age (a)	Site	Location	Altitude (m)	*H* (m)	*DBH* (cm)	Density (Stems ha^{-1})
9	Nanjing	118°49′35″ E 32°10′59″ N	15	8.4	13.6	1111
11	Jingzhou	112°13′37″ E 30°19′34″ N	34	9.6	14.1	1111
13	Chonqing	108°27′3″ E 30°45′58″ N	245	11.6	17.6	920
15	Yunnan	102°46′41″ E 24°49′43″ N	1891	12.9	23.2	830
22	Nanjing	118°47′40″ E 32°20′13″ N	17	14.6	33.1	410

H: height; *DBH*: diameter at breast height.

2.4. Height-Diameter at Breast Height Function Development

A nonlinear function below was used to model *H* for the sample tree measured for both *H* and *DBH* [13,14]. This function had the flexibility to produce satisfactory curves under most circumstances.

$$H = 1.3 + a \times [\exp(-b/DBH)]$$

where a and b are the parameters to be estimated.

2.5. Tree Biomass Model Development

In this study, the direct prediction of the tree biomass from measurement variables (*DBH* and *H*) was used to estimate the tree biomass of *T.* Zhongshanshan. Crown diameter is often quite irregular, even for a given species, depending on ecological conditions, and was not exploited further at this stage. We modeled tree biomass (*M*, in kg) as a function of *H* (in meters) and *DBH* (in cm) with the following 3 age-independent equations:

$$M = a \times (\mathrm{DBH})^b \quad (1)$$

$$M = a \times (\mathrm{DBH})^b \times H^c \quad (2)$$

$$M = a \times [(\mathrm{DBH})^2 \times H]^b \quad (3)$$

It is expedient to take logarithms for fitting the models and dealing with heterocedasticity. Therefore, Equations (1)–(3) can be linearized using logarithms in the following equations:

$$\mathrm{In}(M) = a + b \times \mathrm{In}(DBH) \quad (4)$$

$$\mathrm{In}(M) = a + b \times \mathrm{In}(DBH) + c \times \mathrm{In}(H) \quad (5)$$

$$\mathrm{In}(M) = a + b \times \mathrm{In}[(DBH)^2 \times H)] \quad (6)$$

where M is tree biomass, and a, b and c are the parameters to be estimated.

The coefficient of adjusted determination (R^2) is the most widely used criterion in the biomass model literature. The mean absolute prediction error (*MAPE*) was applied as the primary metric to evaluate the performance of models, whose statistical characteristics are proverbial and frequently used in ecology and environment assessment. The selection of our final model was based on high adjusted R^2 and low *MAPE*. The R^2 and *MAPE* were computed as follows:

$$R^2 = 1 - \frac{\sum_{i=1}^{n}\left(M_i - \hat{M}_i\right)^2}{\sum_{i=1}^{n}\left(M_i - M_i\right)^2}$$

$$MAPE = \frac{1}{n}\sum_{i=1}^{n}\frac{\left|M_i - \hat{M}_i\right|}{M_i}$$

where M_i is observed biomass, $\hat{M}_i$ is predicted biomass, M is the mean of observed biomass, and n is the number of trees.

In this study, we first used the above Equations (4)–(6) to estimate the biomass of tree components, including leaf, branch and stem. Then, we selected the best equation according to the evaluation statistics (R^2 and *MAPE*). Next, a linear seemingly unrelated regression approach (SUR) was used to fit the system of selected model that ensured the additivity property. The additivity of the linear equations is enforced by setting a constraint on the regression coefficients. The primary result showed that Equation (5) can better improve the fitting effect and performance of the model. Therefore, the following additive model system was constructed through a linear SUR based on the log-transformed data:

$$\begin{cases} \text{In}(M_l) = a_l + b_l \times \text{In}(DBH) + c_l \times \text{In}(H) + \varepsilon_l & (7) \\ \text{In}(M_s) = a_s + b_s \times \text{In}(DBH) + c_s \times \text{In}(H) + \varepsilon_s & (8) \\ \text{In}(M_b) = a_b + b_b \times \text{In}(DBH) + c_b \times \text{In}(H) + \varepsilon_b & (9) \\ \text{In}(M_{AGB}) = \text{In}(M_l) + \text{In}(M_s) + \text{In}(M_b) + \varepsilon_{AGB} & (10) \end{cases}$$

where the subscripts l, s, b and AGB stand for leaf, stem, branch and aboveground, respectively. ε is the model correction factor.

Root biomass in different *T*. Zhongshanshan stands was estimated using a fixed root to shoot ratio (0.26) as described by Ravindranath and Ostwald [15]. Biomass in all of the ecosystem components was extrapolated and scaled to a per hectare basis.

2.6. Estimation of Ecosystem Carbon Storage

We determined the carbon storages in plants by multiplying carbon concentration with dry mass amount. SOC storage up to 60 cm depth was calculated using SOC concentration, bulk density, and soil depth as follows:

$$\text{SOC storage (t ha}^{-1}\text{)} = \text{SOC} \times \text{BD} \times \text{T} \times 100$$

where SOC is soil organic carbon (%), BD is bulk density (Mg m^{-3}) and T is the soil thickness (m). The ecosystem carbon storage was calculated by adding biomass carbon storage (*T*. Zhongshanshan plants, herbs, shrubs, and litter) and SOC storage.

2.7. Statistical Analyses

The log-transformed linear regression procedure in SPSS 19.0 (IBM Corporation, Somers, NY, USA) software was used to fit the *H-DBH* function and allometric models' parameters. The procedure fits model parameters and variance parameters simultaneously by applying the maximum likelihood regression approach. This category of procedure was used due to its flexibility to work with equations forms and its recognized robustness over nonlinear models with additive error and log-transformed models. The SUR in the SAS/ETS Model Procedure (SAS Institute, Inc., Cary, NC, USA, 2011) [16] was used to fit the system of biomass equations for *T.* Zhongshanshan, in which the coefficients of the tree component biomass models were simultaneously estimated. The plotting software was Origin 2021 (Origin Lab, Northampton, MA, USA).

3. Results

3.1. Height-Diameter at Breast Height Function

The *H-DBH* function derived from the harvest trees and 5 stands along the Yangtze River were presented in Figure 2, with a correlation coefficient (r) of 0.934. Maximum tree heights occurred at *DBH*s of 35–45 cm, and all were of 22-year-old *T.* Zhongshanshan plantation. The predicted *H-DBH* function curve revealed that tree *H* generally showed a positive correlation with the increase in *DBH*, regardless of the site. This *H-DBH* function overestimated trees larger than *DBH* of about 35 cm for approximately 0.2 m in *H*.

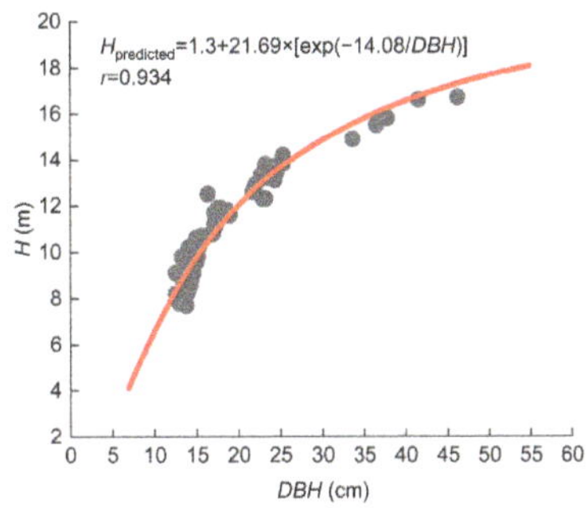

Figure 2. Observed scatter and predicted *H-DBH* function curve of *T.* Zhongshanshan plants.

3.2. Each Component Biomass Model Selection

The allometric equation in the form of a polynomial function is applicable to calculate the biomass of individual tree components. For *T.* Zhongshanshan stands, Equation (5) performed the best in estimating the leaf, stem and branch biomass compared to the other two models in terms of the R^2 and *MAPE* (Table 2). Thus, we used the Equation (5) to model the leaf, stem and branch biomass. The second performance model is Equation (6), with R^2 and *MAPE* slightly lower and higher than Equation (5), respectively.

Table 2. Equation parameters (a, b and c) and evaluation statistics (R^2 and *MAPE*) of allometric models for predicting individual components of *T.* Zhongshanshan plants.

Equation	Model	Component	Estimated Coefficients			R^2	MAPE
			a	b	c		
(4)	$\text{In}(M) = a + b \times \text{In}(DBH)$	leaf	−2.08	1.52	-	0.847	0.0999
		stem	−1.88	2.15	-	0.972	0.0255
		branch	−2.95	2.22	-	0.971	0.0278
(5)	$\text{In}(M) = a + b \times \text{In}(DBH) + c \times \text{In}(H)$	leaf	−3.57	0.16	2.47	0.853	0.0970
		stem	−4.02	0.18	3.57	0.979	0.0225
		branch	−5.63	0.01	4.01	0.980	0.0276
(6)	$\text{In}M = a + b \times \text{In}[(DBH)^2 \times H)]$	leaf	−2.45	0.59	-	0.850	0.0992
		stem	−2.40	0.84	-	0.975	0.0247
		branch	−3.49	0.87	-	0.975	0.0275

3.3. Additive Biomass Equations

For the SUR approach, the model for biomass components was fitted to enforce the additivity of the total tree biomass, including the leaf, stem and branch (Table 3). The SUR approach consisted first of fitting and selecting the best model (Equation (5) in this study) for each tree component. In *T.* Zhongshanshan plantations, the fit accuracy of the leaf and stem biomass was higher than Equation (5) by higher adjusted R^2 and lower *MAPE*. However, the branch biomass prediction using SUR method performed worse than Equation (5), which was accomplished by the maximum likelihood regression approach. Based on the SUR model for each component analyzed, we used M_t (total biomass) = M_l + M_s + M_b + M_r (root biomass) to calculate each component tree biomass. For leaf, stem and branch, the biomass values predicted by the maximum likelihood regression were close to the observed values, and the generalized SUR approach had similar estimations of leaf and stem but superior evaluation statistics (Figure 3).

Table 3. Parameter estimates (*a*, *b* and *c*) and evaluation statistics (R^2 and *MAPE*) of Equation (5) for different components using SUR approach.

Component	Estimated Coefficients			R^2	*MAPE*
	a	*b*	*c*		
leaf	−3.66	0.69	2.15	0.892	0.0927
stem	−3.92	0.18	3.57	0.981	0.0197
branch	−4.62	0.05	3.59	0.953	0.0495

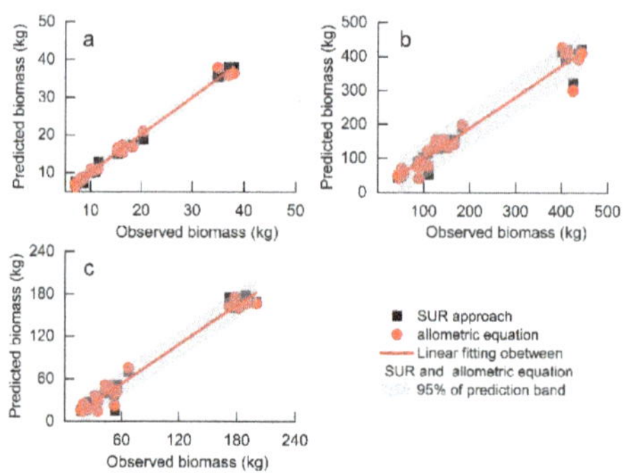

Figure 3. Comparison of the predicted leaf (**a**), stem (**b**) and branch (**c**) biomass against the corresponding observed biomass in *T.* Zhongshanshan plantations by SUR approach and allometric equation, respectively.

3.4. Individual Tree Biomass and Allocation

Adopting the equations established by SUR, the biomass of individual tree components was determined in the 9-, 11-, 13-, 15-, and 22-year-old *T.* Zhongshanshan plants (Figure 4). With increasing stand age, the biomass of all tree components increased. The total individual tree, leaf, stem, branch and root biomass was 156.80, 15.27, 64.20, 23.64 and 26.82 kg of the 9-year-old *T.* Zhongshanshan plants and increased to1245.72, 93.5, 545.50, 181.01 and 213.08 kg of the 22-year-old plants. With increasing stand age, the proportion of leaf biomass gradually decreased with values of 9.74%, 8.40% and 7.36% in the 9-, 11- and 13-year-old *T.* Zhongshanshan plants. For all *T.* Zhongshanshan plants, the stem had the largest proportion of biomass among all components, whose contribution to total tree biomass was 40.94%, 41.95%, 43.00%, 43.38% and 43.79% along the chronosequence, showing a slightly increasing trend. Similar to the stem, individual branch biomass also exhibited a slightly increasing tendency with increasing stand age.

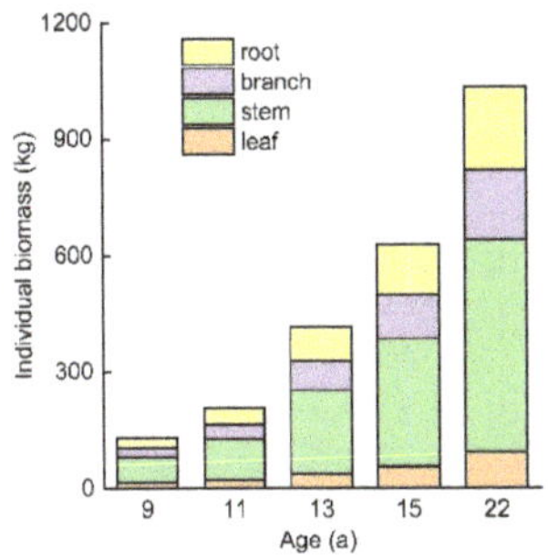

Figure 4. Individual biomass of each component of the 9-, 11-, 13-, 15-, and 22-year-old *T.* Zhongshanshan plants.

According to the plant density, the biomass of different ecosystem components was established in Table 4. As the stand aged, the biomass of most tree components increased, including total tree, leaf, stem, branch and root biomass, except for the 22-year-old *T.* Zhongshanshan stand. Since the density of the above-mentioned stand was 410 stems·ha^{-1}, which was less than half of the other stands. The total tree biomass was 53.43, 84.87, 140.67, 192.71 and 156.65 t ha^{-1} in the 9-, 11-, 13-, 15-, and 22-year-old *T.* Zhongshanshan stands. Total understory biomass exhibited a slight increment from 9 to 15-year-old stand and was 3.68 times compared to 15-year-old stand. Litter biomass decreased from 9-year-old stand (3.26 t ha^{-1}) to 13-year-old stand (3.12 t ha^{-1}) and increased in later stages (6.21 and 6.12 t ha^{-1}, respectively, in 15- and 22- year-old stands).

Table 4. Biomass of different ecosystem components in the 9-, 11-, 13-, 15-, and 22-year-old *T.* Zhongshanshan stands.

Components	Biomass (t ha^{-1})				
	9-Year-Old	**11-Year-Old**	**13-Year-Old**	**15-Year-Old**	**22-Year-Old**
Leaf	6.28	8.60	12.50	17.12	14.12
Stem	26.39	42.95	72.99	100.86	82.75
Branch	9.74	15.81	26.19	34.97	27.46
Root	11.03	17.51	29.00	39.77	32.32
Total tree	53.43	84.87	140.67	192.71	156.65
Total understories	1.82	1.99	2.21	2.22	8.15
Litter	3.27	2.14	3.12	6.22	6.12
Total	58.52	88.99	146.01	201.15	170.92

The total ecosystem biomass of *T.* Zhongshanshan stands over the 5 age groups was highest in the stem, followed by (in decreasing order) the root, the branch, the leaf, understories or litter. The proportion of the biomass from stem increased with forest age (except for the 22-year-old *T.* Zhongshanshan stand), while that in the leaf and understories declined. The stem accounted for 48.87%, 50.80%, 51.95% and 58.93% in the 9-, 11-, 13-, 15-year-old stand, respectively. The leaf accounted for 10.73%, 9.67%, 8.56% and 8.51%, the understory for 3.12%, 2.23%, 1.51% and 1.10% from 9 to 15-year-old stand, respectively. Although the root biomass of *T.* Zhongshanshan was derived from fixed root to shoot ratio (0.26), the highest proportion occurred in 15-year-old with 19.99%, which was seemingly independent of tree age in this study. Since root biomass proportion of total ecosystem was 18.84%, 19.68%, 19.86%, 19.77% and 18.91% from 9 to 22-year-old stand, respectively (Figure 5).

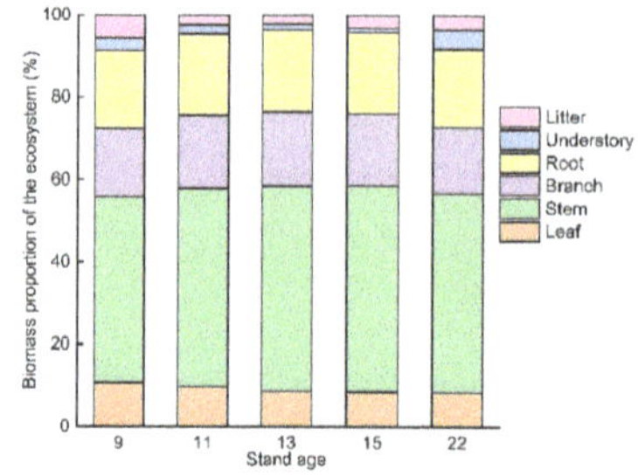

Figure 5. Biomass proportion of the ecosystem from leaf, stem, branch, root, understory and litter in the 9-, 11-, 13-, 15-, and 22-year-old *T.* Zhongshanshan stands.

3.5. Ecosystem Carbon Storage and Allocation

The aboveground (leaf, stem, branch, understory and litter), belowground (root), and total ecosystem carbon storage for each of the 5 stands were provided in Table 5. The carbon storage of the total tree, soil, and total ecosystem increased with increasing stand age (except for the 22-year-old *T.* Zhongshanshan stand), from 25.32, 96.89 and 124.76 t ha^{-1} in the 9-year-old stand, respectively, to 90.89, 117.12, and 212.22 t ha^{-1}, respectively, in the 15-year-old stand. In the 22-year-old *T.* Zhongshanshan stand, carbon storage was 73.85, 136.66 and 217.64 t ha^{-1}, respectively, from total tree, soil and total ecosystem. For total tree carbon storage manifested a sigmoidal schema, first promptly rising and then gradually decreasing. The carbon sequestration rates of the ecosystem were 10.26 and 12.26 t ha^{-1} $year^{-1}$ during the 9th–13th and 13th–15th years, respectively. Compared with the 9-year-old stand, the carbon storage of the 0–60 cm soil layer increased from 11 to 22-year-old stand, indicating an evident accumulation course of organic carbon after afforestation, with an annual accumulation rate of 3.05 t ha^{-1}. The understory carbon storage (4.07 t ha^{-1}) of the 22-year-old stand was much higher than that of the other stands (0.91–1.10 t ha^{-1}), indicating an obvious increase in floor vegetation. The carbon storage of the litter exhibited a distinct relationship with stand age, with lower values (1.06–1.63 t ha^{-1}) and higher values (3.06–3.10 t ha^{-1}) occurring in the 9 to 13-year-old stands and 15 to 22-year-old stands, respectively.

Table 5. Carbon storage of different ecosystem components in the 9-, 11-, 13-, 15-, and 22-year-old *T.* Zhongshanshan stands.

Components	Carbon Storage (t ha^{-1})				
	9-Year-Old	**11-Year-Old**	**13-Year-Old**	**15-Year-Old**	**22-Year-Old**
Leaf	3.26	4.47	6.50	8.90	7.34
Stem	11.59	18.85	32.04	44.28	36.33
Branch	4.74	7.70	12.75	17.03	13.37
Root	5.73	9.11	15.08	20.68	16.81
Total tree	25.32	40.13	66.37	90.89	73.85
Understory	0.91	0.99	1.11	1.11	4.07
Litter	1.63	1.07	1.56	3.11	3.06
Soil	96.89	104.36	116.32	117.12	136.66
Total ecosystem	120.08	156.68	211.74	263.11	231.49

In all 5 *T.* Zhongshanshan stands, the proportion of carbon storage in the soil was more than half, with specific values of 77.66%, 71.21%, 62.75%, 55.19%, and 62.79%, respectively (Figure 6). Except for in the 22-year-old stand, the proportion of carbon stored in the soil decreased with increasing age and the proportion stored in vegetation increased correspondingly. Of the vegetation parts, stem and root were the two largest contributors to the total ecosystem carbon pool in all 5 stands. Understory and litter contributed little to the total site carbon storage, accounting from 0.52% to 1.87%, and 0.73% to 1.46% within these stands. The contribution of the tree root increased with stand aging from 4.60% in the 9-year-old stand to 8.14% in the 13-year-old stand, and then decreased to 9.74% and 7.72% in the 15- and 22-year-old stands.

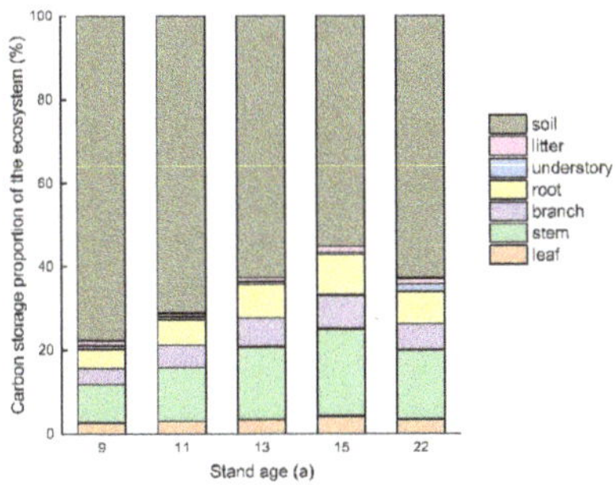

Figure 6. Carbon storage proportion of the ecosystem from leaf, stem, branch, root, understory and litter in the 9-, 11-, 13-, 15-, and 22-year-old *T.* Zhongshanshan stands.

4. Discussion

4.1. Tree Biomass Development

China, especially the YRB area, has a stringent policy of forest protection [17,18] and it was, hence, impossible to log trees for our study to establish site-specific allometric models. Moreover, the number of logged *T.* Zhongshanshan trees was low, owing to the conservation of germplasm resources limitation and scarcity of *T.* Zhongshanshan older than 20 years. In this study, only a limited number of samples of *T.* Zhongshanshan material was available for the construction of the biomass equations across the different stages of stand development. The quantity of tree samples used to develop allometric equations is quite multitudinous in the literature. Xue et al. [19] harvested 72 *Casuarina equisetifolia* trees on Hainan Island; Mugasha et al. [20] reported 30 samples for the wet lowland forests equations; Zhang et al. [21] established allometric biomass equations based on 8 Mongolian Pine trees. The precision of allometric models is usually determined by pivotal factors, such as tree species, *DBH* homogeneity, or the amount of sampled trees. For *T.* Zhongshanshan stands, artificially planted forests have consistent age and growth performance, which can make up for the shortage of quantity to a certain extent.

Previous studies have demonstrated that a single *DBH*-based allometric model can reasonably predict biomass in several plants including *Pentaclethram acroloba* [22], *Pinus tabuliformis* [23] and *Cunninghamia lanceolata* [24]. As before, ordinary surveys of diameter were not only simpler to implement in the field but were also more often to exist in historical data. For *T.* Zhongshanshan plants, although *DBH* evidenced to be a fairly good predictor of biomass, the selected 3-parameter polynomial model performed better. On this basis, the implementation of SUR approach provided further accurate estimates of biomass component fractions. Such results are consistent with many previous studies, which showed aboveground biomass models with combined *DBH* and *H* as the most suitable predictors [25–27]. Chave et al. [25] and Mensah et al. [26] both verified that the inclusion of *H* reduced the error of aboveground biomass estimates by 6.70% and 35.30% in predicting the biomass of forests, since *H-DBH* relevance varied across a range of ecological conditions. In West African savanna ecosystems, Ganamé et al. [28] narrowed the confidence intervals of the biomass estimation and subsequently increases the accuracy of the predictions by applying SUR approach. Diverse from Mbow et al. [29], who refrained from relying heavily on tree *H* and hold the opinion that the accuracy of tree measurement was generally much lower than that of *DBH*. This is often due to the approximate methods adopted to estimate *H* in the early field investigation, simultaneously the rather arbitrary condition to consider, such as diminutive and isolated branches stretching out of the canopy.

With the improvement of *H* measurement accuracy, the estimation of tree biomass, as well as the description of stands and their performance over time, relies largely on an accurate *H-DBH* relationship [30]. We also developed *H-DBH* function derived from the harvest trees and 5 stands along the Yangtze River. This is particularly beneficial to *T.* Zhongshanshan plantations in the cases where the actual measurements of *H* growth are not available, and follow-up estimating biomass across large spatial scales using forest inventory data.

4.2. Biomass Distribution

Age is the most vital factor influencing the magnitude and distribution of biomass in plants. In this study, the total individual tree, leaf, stem, branch and root biomass increased promptly from the 9th to the 22nd year, where a linear positive correlation could be extracted for this species between forest age and total tree biomass. The proportion of leaf biomass gradually decreased with increased stand age. This result was consistent with the reports of many previous studies [31,32]. Acting as a valuable component, the leaf is highly correlated with forest performance in young *T.* Zhongshanshan stands (<10-year-old). As stands age, tree magnitude increases during ontogeny and more carbohydrate resources are distributed for stem growth [19]. Hence, leaf biomass increases proportionally less than stem mass. The proportion of stem accounted for most, and remained stable at later stages, indicating that stem is a vital composition when accounting biomass partitioning for *T.* Zhongshanshan plants [33].

Extrapolated by the polynomial growth equation, the total tree biomass of *T.* Zhongshanshan increased rapidly in the first 2 or 3 decades and increased slowly later, showing a sigmoidal pattern. Such a pattern after afforestation could also be found in many prior studies [23,24,29]. From the 9- to the 15-year-old stand in this study, the *T.* Zhongshanshan ecosystem biomass increased steadily, and there was a slight decline in the 22-year-old stand, which could be owing to a smaller forestation density (approximately 5.4 m and 4.5 m spacing between and within rows). In *P. strobus* plantations, the biomass in the 42-year-old stand corresponds to that of the 65-year-old, implying that tree density played an important role in ecosystem biomass accumulation [34]. Thereupon, understory biomass in the 22-year-old stand was nearly 4 times that of the 9- to 15-year-old stands averagely. As demonstrated by Xu et al. [35], the coverage and biomass of the understory vegetation increased significantly with the growth of the stand. These consistent relationships between understory biomass and age might be due to the significant increase in canopy closure, which may result in species diversity and abundance. As forest aged, a smaller initial tree density might increase the light transmittance for understory plants and lead to lesser competition. A more spacious interval coupled with light penetration help to increase temperature and ventilation, which is instrumental to understory growth [36]. The total tree biomass was 192.71 and 156.65 t ha^{-1} in the 15- and 22-year-old *T.* Zhongshanshan stands. In comparison with 15-year-old *P. massoniana* [37] and 22-year-old *P. strobus* stands [34], whose tree biomass was 78.5 and 128.0 t ha^{-1}, *T.* Zhongshanshan stands of the same age exhibited higher biomass accumulation. Although Moriondo et al. [38] reported that tree biomass accumulation depends on growth habitat, the soil on which plants are growing, disturbance regime and interaction with belowground vegetation, our study proved that *T.* Zhongshanshan is a fast-growing variety to some extent.

4.3. Ecosystem Carbon Storage

Ecosystem carbon storage and portioning accorded well with biomass accumulation in the vegetation component and were age-dependent in the soil component. The total ecosystem carbon storage of 9- to 22-year-old *T.* Zhongshanshan stands ranged from 124.76 to 217.64 t ha^{-1}, of which total tree ranges from 25.32 to 90.89 t ha^{-1}. These values were well within the range of above-ground biomass carbon storage (4.5–462.5 t ha^{-1}, averagely 61.9 t ha^{-1}) reported by a case study of carbon sequestration following reforestation in the YRB area [39]. *T.* Zhongshanshan plantations had a substantial potential for carbon sequestration as the total land area under these plantations was consecutively expanding, most of which was still immature. The carbon storage of the 0–60 cm soil layer increased from 11- to 22-year-old stand, with an annual accumulation rate of 3.05 t ha^{-1}, implying substantial amounts of carbon accumulation during the transition period from young to near-mature (20- to 30-year-old) *T.* Zhongshanshan stands. The older age of tree stands produces more litter and root biomass which ultimately supplies more organic matter to the soil. Our results of SOC storage were comparable to the *T. distichum* forests (123.3 t ha^{-1}) of South Caspian Sea in similar soil depth [40] and were higher than the national average in China (96.0 t ha^{-1}) [41] but lower than upper reaches of the YRB (164 t ha^{-1}) [42]. Probably owing to the regional climatic patterns, a decreasing temperature from west to east in the YRB led to a slower decomposition of SOM and a lower soil respiration rate in the upper reaches. Corresponding to other investigations conducted in forest ecosystems [43], the highest carbon storage occurred in the soil component of *T.* Zhongshanshan ecosystem but contradicted the findings of Vesterdal et al. [44] and Lü et al. [45], who found that soils only contributed about one third in an afforestation ecosystem. Gogoi et al. [46] proclaimed that the continuous disturbance by human interference declines SOC storage. The starting point in soil carbon storage also possibly determined the sequestration potential.

5. Conclusions

The results described by the selected 3-parameter polynomial model and SUR approach pave the way for a more systematic estimation of *T.* Zhongshanshan biomass. The fitted *H-DBH* function is quite capable of accounting for the relationship between *H* and *DBH* of *T.* Zhongshanshan plants. *T.* Zhongshanshan is a fast-growing variety, as compared tree biomass with other trees of the same stand age in similar subtropical areas. In conclusion, the present study revealed that the *T.* Zhongshanshan stands in the YRB area can store 124.76 to 217.64 t ha^{-1} carbon, of which total tree ranges from 25.32 to 90.89 t ha^{-1}, with 55.19% to 77.66% storing in the soil. Large *T.* Zhongshanshan trees contributed greatly to carbon storage in living biomass and may be the main reason accounting for having a persistent potency. A potential limitation involved with this study was the absence of tree age and taper as explanatory variables. The performance of the generalized approaches in subsequent studies could be marginally improved on a large enough and sufficiently representative set of individual plants by combining more explanatory variables. Multisite studies are further required to fully elaborate the patterns of biomass and carbon storage of individual trees and the ecosystem of *T.* Zhongshanshan plantations by stand age.

Author Contributions: Q.S.: conceptualization, data collection, formal analysis, writing—original draft, editing; J.H.: conceptualization, writing—review and editing; D.C.: writing—review and editing; Y.Y.: writing—review and editing. All authors have read and agreed to the published version of the manuscript.

Funding: This research was funded by the Innovation Project of Plant Germplasm Resources of Chinese Academy of Sciences (kfj-brsn-2018-6-003), Jiangsu Special Fund on Technology Innovation of Carbon Dioxide Peaking and Carbon Neutrality (BE2022420), and Jiangsu Long-term Scientific Research Base for Taxodium Rich. Breeding and Cultivation (LYKJ [2021]05).

Data Availability Statement: Publicly available datasets were analyzed in this study.

Conflicts of Interest: The authors declare no conflict of interest.

References

1. Pugh, T.A.; Lindeskog, M.; Smith, B.; Poulter, B.; Arneth, A.; Haverd, V.; Calle, L. Role of forest regrowth in global carbon sink dynamics. *Proc. Natl. Acad. Sci. USA* **2019**, *116*, 4382–4387. [CrossRef] [PubMed]
2. Muttaqin, M.Z.; Alviya, I.; Lugina, M.; Hamdani, F.A.U. Developing community-based forest ecosystem service management to reduce emissions from deforestation and forest degradation. *For. Policy Econ.* **2019**, *108*, 101938. [CrossRef]
3. Daba, D.E.; Soromessa, T. The accuracy of species-specific allometric equations for estimating aboveground biomass in tropical moist montane forests: Case study of *Albizia grandibracteata* and *Trichilia dregeana*. *Carbon Bal. Manag.* **2019**, *14*, 18. [CrossRef] [PubMed]

4. Hirigoyen, A.; Resquin, F.; Cerrillo, R.N.; Franco, J.; Casnati, C.R. Stand biomass estimation methods for *Eucalyptus grandis* and *Eucalyptus dunnii* in Uruguay. *BOSQUE* **2021**, *42*, 53–66. [CrossRef]
5. Brandeis, T.J.; Delaney, M.; Parresol, B.R.; Royer, L. Development of equations for predicting Puerto Rican subtropical dry forest biomass and volume. *For. Ecol. Manag.* **2006**, *233*, 133–142. [CrossRef]
6. Zhao, D.; Kane, M.; Markewitz, D.; Teskey, R.; Clutter, M. Additive tree biomass equations for midrotation loblolly pine plantations. *For. Sci.* **2015**, *61*, 613–623. [CrossRef]
7. Xie, L.; Fu, L.; Widagdo, F.R.A.; Dong, L.H.; Li, F.R. Improving the accuracy of tree biomass estimations for three coniferous tree species in Northeast China. *Trees* **2022**, *36*, 451–469. [CrossRef]
8. Huang, H.; Li, Z.; Gong, P.; Cheng, X.; Clinton, N.; Cao, C.; Ni, W.; Wang, L. Automated methods for measuring DBH and tree heights with a commercial scanning lidar. *Photogramm. Eng. Remote Sens.* **2011**, *77*, 219–227. [CrossRef]
9. Hua, J.F.; Han, L.W.; Wang, Z.Q.; Gu, C.S.; Yin, Y.L. Morpho-anatomical and photosynthetic responses of *Taxodium* hybrid 'Zhongshanshan' 406 to prolonged flooding. *Flora* **2017**, *231*, 29–37. [CrossRef]
10. Wang, Z.Q.; Yin, Y.L.; Hua, J.F.; Fan, W.C.; Yu, C.G.; Xuan, L.; Yu, F.Y. Cloning and characterization of *ThSHRs* and *ThSCR* transcription factors in *Taxodium* hybrid 'Zhongshanshan 406'. *Genes* **2017**, *8*, 185. [CrossRef]
11. Xuan, L.; Hua, J.F.; Zhang, F.; Wang, Z.Q.; Pei, X.X.; Yang, Y.; Yin, Y.L.; Creech, D.L. Identification and functional analysis of *ThADH1* and *ThADH4* genes involved in tolerance to waterlogging stress in *Taxodium* hybrid 'Zhongshanshan 406'. *Genes* **2021**, *12*, 225. [CrossRef] [PubMed]
12. Shi, Q.; Yin, Y.L.; Wang, Z.Q.; Creech, D.; Hua, J.F. Influence of soil properties on the performance of the *Taxodium* hybrid 'Zhongshanshan 407' in a short-term pot experiment. *Soil Sci. Plant Nutr.* **2017**, *63*, 145–152. [CrossRef]
13. Buford, M. Height-diameter relationships at age 15 in loblolly pine seed sources. *For. Sci.* **1986**, *32*, 812–818.
14. Huang, S.; Titus, S.J.; Wiens, D.P. Comparison of nonlinear height–diameter functions for major Alberta tree species. *Can. J. For. Res.* **1992**, *22*, 1297–1304. [CrossRef]
15. Ravindranath, N.H.; Ostwald, M. *Carbon Inventory Methods: Handbook for Greenhouse Gas Inventory, Carbon Mitigation and Roundwood Production Projects*; Springer Science and Business Media: Dordrecht, The Netherlands, 2007.
16. Dong, L.; Zhang, L.; Li, F.R. Developing additive systems of biomass equations for nine hardwood species in Northeast China. *Trees* **2015**, *29*, 1149–1163. [CrossRef]
17. Brandt, J.S.; Butsic, V.; Schwab, B.; Kuemmerle, T.; Radeloff, V.C. The relative effectiveness of protected areas, a logging ban, and sacred areas for old-growth forest protection in southwest China. *Biol. Conserv.* **2015**, *181*, 1–8. [CrossRef]
18. Liu, N.; Sun, P.; Caldwell, P.V.; Harper, R.; Liu, S.; Sun, G. Trade-off between watershed water yield and ecosystem productivity along elevation gradients on a complex terrain in southwestern China. *J. Hydrol.* **2020**, *590*, 125449. [CrossRef]
19. Xue, Y.; Yang, Z.; Wang, X.; Lin, Z.; Li, D.; Su, S. Tree biomass allocation and its model additivity for *Casuarina equisetifolia* in a tropical forest of Hainan Island, China. *PLoS ONE* **2016**, *11*, e0151858. [CrossRef]
20. Mugasha, W.A.; Mwakalukwa, E.E.; Luoga, E.; Malimbwi, R.E.; Zahabu, E.; Silayo, D.S.; Sola, G.; Crete, P.; Henry, M.; Kashindye, A. Allometric models for estimating tree volume and aboveground biomass in lowland forests of Tanzania. *Inter. J. For. Res.* **2016**, *2016*, 8076271. [CrossRef]
21. Zhang, X.; Zhang, X.L.; Han, H.; Shi, Z.J.; Yang, X.H. Biomass accumulation and carbon sequestration in an age-sequence of Mongolian pine plantations in Horqin sandy land, China. *Forests* **2019**, *10*, 197. [CrossRef]
22. Segura, M.; Kanninen, M. Allometric models for tree volume and total aboveground biomass in a tropical humid forest in Costa Rica. *Biotropica* **2005**, *37*, 2–8. [CrossRef]
23. Cao, J.; Wang, X.; Tian, Y.; Wen, Z.; Zha, T. Pattern of carbon allocation across three different stages of stand development of a Chinese pine (*Pinus tabulaeformis*) forest. *Ecol. Res.* **2012**, *27*, 883–892. [CrossRef]
24. Xiang, W.H.; Li, L.H.; Ouyang, S.; Xiao, W.F.; Zeng, L.X.; Chen, L.; Lei, P.F.; Deng, X.W.; Zeng, Y.L.; Fang, J.P. Effects of stand age on tree biomass partitioning and allometric equations in Chinese fir (*Cunninghamia lanceolata*) plantations. *Eur. J. For. Res.* **2021**, *140*, 317–332. [CrossRef]
25. Chave, J.; Andalo, C.; Brown, S.; Cairns, M.A.; Chambers, J.Q.; Eamus, D.; Fölster, H.; Fromard, F.; Higuchi, N.; Kira, T.; et al. Tree allometry and improved estimation of carbon stocks and balance in tropical forests. *Oecologia* **2005**, *145*, 87–99. [CrossRef] [PubMed]
26. Mensah, S.; Veldtman, R.; Seifert, T. Allometric models for height and aboveground biomass of dominant tree species in South African Mistbelt forests. *South. For.* **2017**, *79*, 19–30. [CrossRef]
27. Picard, N.; Rutishauser, E.; Ploton, P.; Ngomanda, A.; Henry, M. Should tree biomass allometry be restricted to power models? *For. Ecol. Manag.* **2015**, *353*, 156–163. [CrossRef]
28. Ganamé, M.; Bayen, P.; Ouédraogo, I.; Larba, H.B.; Adjima, T. Allometric models for improving aboveground biomass estimates in West African savanna ecosystems. *Trees For. People* **2021**, *4*, 100077. [CrossRef]
29. Mbow, C.; Verstraete, M.M.; Sambou, B.; Diaw, A.T.; Neufeldt, H. Allometric models for aboveground biomass in dry savanna trees of the Sudan and Sudan-Guinean ecosystems of Southern Senegal. *J. For. Res.* **2014**, *19*, 340–347. [CrossRef]
30. Baral, S.; Neumann, M.; Basnyat, B.; Sharma, R.P.; Silwal, R.; Shrestha, H.L.; Subedi, T.; Vacik, H. Volume functions for *Shorea robusta* Gaertn. in Nepal. *Forestry* **2022**, *95*, 405–415. [CrossRef]
31. Claus, A.; George, E. Effect of stand age on fine-root biomass and biomass distribution in three European forest chronosequences. *Can. J. For. Res.* **2005**, *35*, 1617–1625. [CrossRef]

32. Du, H.; Zeng, F.; Peng, W.; Wang, K.; Zhang, H.; Liu, L.; Song, T. Carbon storage in a *Eucalyptus* plantation chronosequence in Southern China. *Forests* **2015**, *6*, 1763–1778. [CrossRef]
33. Panzoua, G.J.L.; Fayolle, A.; Feldpausch, T.R.; Ligot, G.; Doucet, J.L.; Forni, E.; Zombo, I.; Mazengue, M.; Loumeto, J.J.; Gourlet-Fleury, S. What controls local-scale aboveground biomass variation in central Africa? Testing structural, composition and architectural attributes. *For. Ecol. Manag.* **2018**, *429*, 570–578. [CrossRef]
34. Peichl, M.; Arain, M.A. Above- and belowground ecosystem biomass and carbon pools in an age-sequence of temperate pine plantation forests. *Agric. For. Meteorol.* **2006**, *140*, 51–63. [CrossRef]
35. Xu, M.P.; Lu, X.Q.; Xu, Y.D.; Zhong, Z.K.; Zhang, W.; Ren, C.J.; Han, X.H.; Yang, G.H.; Feng, Y.Z. Dynamics of bacterial community in litter and soil along a chronosequence of *Robinia pseudoacacia* plantations. *Sci. Total Environ.* **2020**, *703*, 135613. [CrossRef] [PubMed]
36. Nunes, M.H.; Camargo, J.L.C.; Vincent, G.; Calders, K.; Oliveira, R.S.; Huete, A.; Mendes de Moura, Y.; Nelson, B.; Smith, M.N.; Stark, S.C. Forest fragmentation impacts the seasonality of Amazonian evergreen canopies. *Nat. Commun.* **2022**, *13*, 917. [CrossRef]
37. Chen, Z. Effects of stand density on the biomass and productivity of *Pinus massoniana* air-sowing stands. *J. Cent. South For. Univ.* **2001**, *21*, 44–47, (In Chinese with English Summary).
38. Moriondo, M.; Leolini, L.; Brilli, L.; Dibari, C.; Tognetti, R.; Giovannelli, A.; Rapi, B.; Battista, P.; Caruso, G.; Gucci, R. A simple model simulating development and growth of an olive grove. *Eur. J. Agron.* **2019**, *105*, 129–145. [CrossRef]
39. Wang, J.; Delang, C.O.; Hou, G.; Gao, L.; Yang, X.; Lu, X. Carbon sequestration in biomass and soil following reforestation: A case study of the Yangtze River Basin. *J. For. Res.* **2022**, *33*, 1663–1690. [CrossRef]
40. Eslamdoust, J.; Sohrabi, H. Carbon storage in biomass, litter, and soil of different native and introduced fast-growing tree plantations in the South Caspian Sea. *J. For. Res.* **2018**, *29*, 449–457. [CrossRef]
41. Yu, D.S.; Shi, X.Z.; Wang, H.J.; Sun, W.X.; Warner, E.D.; Liu, Q.H. National scale analysis of soil organic carbon storage in China based on Chinese soil taxonomy. *Pedosphere* **2007**, *17*, 11–18. [CrossRef]
42. Wang, X.G.; Zhu, B.; Hua, K.K.; Luo, Y.; Zhang, J.; Zhang, A.B. Assessment of soil organic carbon stock in the upper Yangtze River basin. *J. Mt. Sci.* **2013**, *10*, 866–872. [CrossRef]
43. Dignac, M.F.; Derrien, D.; Barre, P.; Barot, S.; Cécillon, L.; Chenu, C.; Chevallier, T.; Freschet, G.T.; Garnier, P.; Guenet, B. Increasing soil carbon storage: Mechanisms, effects of agricultural practices and proxies. A review. *Agron. Sustain. Dev.* **2017**, *37*, 14. [CrossRef]
44. Vesterdal, L.; Rosenqvist, L.; Salm, C.; Hansen, K.; Groenenberg, B.J.; Johansson, M.B. Carbon sequestration in soil and biomass following afforestation: Experiences from oak and Norway spruce chronosequences in Denmark, Sweden and the Netherlands. In *Environmental Effects of Afforestation in North-Western Europe*; Springer: Berlin/Heidelberg, Germany, 2007; pp. 19–51.
45. Lü, X.; Yin, J.; Jepsen, M.R.; Tang, J. Ecosystem carbon storage and partitioning in a tropical seasonal forest in Southwestern China. *For. Ecol. Manag.* **2010**, *260*, 1798–1803. [CrossRef]
46. Gogoi, A.; Ahirwal, J.; Sahoo, U.K. Evaluation of ecosystem carbon storage in major forest types of Eastern Himalaya: Implications for carbon sink management. *J. Environ. Manag.* **2022**, *302*, 113972. [CrossRef]

forests

MDPI

Article

Research on the Impact of Cooperative Membership on Forest Farmer Household Income and Assets—Case Study from Liaoning Herbal Medicine Planting Cooperatives, China

Jingyu Wang [1,*], Zhe Zhao [2] and Lei Gao [3]

[1] School of Economic and Management, Zhejiang A&F University, Hangzhou 311300, China
[2] School of Economics, Faculty of Economics, Liaoning University, Shenyang 110136, China; zhaozhe@lnu.edu.cn
[3] School of Economics and Management, Yanshan University, Qinhuangdao 066004, China; gaolei@ysu.edu.cn
* Correspondence: wangjy@zafu.edu.cn; Tel.: +86-182-1057-9605

Abstract: Improving the income and assets of forest farmers is the basis for realizing the sustainable development of forestry. In this paper, we tested the impact of membership in herbal medicine planting cooperatives on forest farmer household income and assets using the propensity score matching (PSM) method and household surveys of the study area. The results showed that cooperative membership can greatly improve forest farmer household income and assets; the higher the accumulation of forest farmer household social capital and human capital, the more farmers were inclined to participate in cooperatives. Householders who were migrant workers were more likely to make the decision to participate in cooperatives compared with those without migrant work experiences. The results of ATT further verified the conclusion that cooperative membership can significantly improve income and assets, which increased by 7.04% and 4.19%, respectively. In addition, according to the survey, the current development of cooperatives in the forestry area experienced problems such as inconsistent quantitative and qualitative development, insufficient driving force, irregular operating mode, inaccurate policy support, and inadequate guidance mechanisms. This paper focused on innovating cooperation mechanisms, enriching joint forms, enhancing driving capacity, stimulating internal driving forces, strengthening system construction, improving standards, enhancing guidance services, and strengthening institutional guarantees. These recommendations have been put forward to guide policy for sustainable forest development.

Keywords: forest farmer; specialized cooperatives; farmer household income; farmer household assets; PSM model

check for updates

Citation: Wang, J.; Zhao, Z.; Gao, L. Research on the Impact of Cooperative Membership on Forest Farmer Household Income and Assets—Case Study from Liaoning Herbal Medicine Planting Cooperatives, China. *Forests* **2023**, *14*, 1725. https://doi.org/10.3390/f14091725

Academic Editor: Diana Feliciano

Received: 19 July 2023
Revised: 31 July 2023
Accepted: 10 August 2023
Published: 27 August 2023

1. Introduction

Specialized farmer cooperatives (hereafter, "cooperatives") are cooperative economic organizations of mutual support and democratic management on the basis of household contract management. These cooperatives not only overcome the contradiction between "small production" in the small-scale peasant economy and "large demand" in the market economy, but also meet consumer requirements for the quality and safety of agricultural products. They represent an important organizational means of realizing the organic connection between small farmers and modern agricultural development [1–3]. Considering the basic conditions of "big country and smallholders," smallholder management will remain the basis of China's agriculture [4]. Cooperatives not only allow farmers to participate in agricultural production and professional cooperation, but also allow them to benefit by effectively connecting scattered individual farmers with agricultural modernization. In this way, cooperatives make a significant contribution to farmers by promoting agricultural modernization [5–7]. Consequently, the government has actively taken various measures to support the development of cooperatives and promote economic cooperation and market integration [8].

From the perspective of national legislation, the Agricultural Law of the People's Republic of China, which came into effect in March 2003, stipulated that "farmers are encouraged to voluntarily form all kinds of professional cooperative economic organizations on the basis of household contract operations." According to the Law of the People's Republic of China on Specialized Farmer Cooperatives, "the State shall promote the development of farmer specialized cooperatives through financial support, preferential taxation, support for finance, science and technology, personnel, and guidance of industrial policies and other measures." According to the Rural Revitalization Promotion Law of the People's Republic of China, "the State supports farmer specialized cooperatives, family farms, agriculture-related enterprises, e-commerce enterprises, and specialized agricultural social service organizations to establish a close interest linkage mechanism with farmers in a variety of ways, so that farmers can share the value-added benefits of the whole industrial chain." On this basis, the government has also formulated specific policies to support the development of cooperatives, especially providing important financial support [9,10]. In terms of allocating public budget funds, the central government generally allocates funds to provincial governments to meet local needs to support the development of cooperatives [11]. This practice has laid a financial foundation for the development of cooperatives in various regions. In addition, local governments below the provincial level also often increase financial support for cooperatives from their budgets.

In recent years, with strong support from government departments at all levels, cooperatives have grown rapidly, totaling 2.2 million nationwide by 2020. However, whether the development of cooperatives is conducive to the realization of relevant government goals (such as increasing farmers' income, promoting the effective connection between small farmers and modern agriculture, narrowing the income gap, etc.) has become the focus of attention of scholars and government departments. A large number of existing studies have shown that the impact of cooperative membership on farmers' household income is significantly positive [12–14]. Cooperatives and enterprises are important market suppliers of agricultural socialized services, and it is likely that the agricultural socialized services market will gradually form a multi-subject competitive supply pattern. As the strength of cooperatives increases, they can compete for the market share of enterprises and improve the welfare of farmers through lower prices and higher utility effects [15]. At the same time, they force enterprises to reduce the price of production means, improve the welfare of participating farmers, and contribute to the improvement of the overall welfare of farmers. Furthermore, many studies consider that the increased income derived from cooperatives differs for different types of households. Some researchers believe that this effect is more obvious for large-scale and high-income households [16], while others believe that the effect is more obvious for low-income and poor households [17]. Yet others consider that the efficiency of cooperatives is relatively low and the service function needs to be strengthened [18].

Overall, further research is still necessary to examine the following aspects in greater depth. Firstly, most of the existing research has focused on the impact of cooperative membership on household income; however, it has ignored the impact on household assets, which also form an important part of family welfare. Secondly, most of the existing research has focused on the macro level and the theoretical level. Therefore, there is a lack of a practical approach to examine specific regions, especially quantitative analysis based on survey data. Due to differences in cooperative types, industrial characteristics, regional resource endowments, socioeconomic development levels, research perspectives, and other factors, the conclusions drawn will inevitably be different. Therefore, it is difficult to determine a development path that is suitable for the actual situation of specific regions. Thirdly, most of the existing research has focused on the whole industry, and studies are lacking research on the characteristic industrial cooperatives that rely on regional resource endowment, which has played a huge role in industrial poverty alleviation.

This paper selected the farmer households that plant herbal medicine in the eastern mountainous forest area of Liaoning province as research object. We empirically tested the impact of membership in herbal medicine planting cooperatives on forest farmer household income and assets through the combination of theory and demonstration. We also explored the potential impact of cooperatives on continuously promoting increasing household income and guaranteeing the stable and sustainable development of forestry area, in order to obtain new supporting evidence.

2. Theoretical Analysis

2.1. The Impact Mechanism of Cooperative Membership on Farmer Household Income

Firstly, economies of scale come from collective action. Compared with investor-owned enterprises, cooperatives are more conducive to saving transaction costs by trading with farmers [19,20]. The central position of cooperatives in the agricultural organization system is determined by the collective action of farmers through cooperatives in an ideal way. Moreover, rural elites with Party member farmers as the core play a leading and exemplary role; they organize ordinary farmers through brand building, deep processing, market negotiation, signing orders, unified purchase of agricultural products, cooperative inspection, coordinated sales, unified management, and other ways, to achieve the high added value of agricultural products that are difficult for individual farmers to achieve and can improve the level of farmer household income [21]. The core members of cooperatives rely on the contribution of ordinary farmers to the quantity and quality of products so that they can achieve a scale effect in quantity and a value-added impact on quality, to save transaction costs and obtain corresponding bargaining power in the market [22]. Furthermore, cooperatives with strong profitability have great development potential. They will have obvious driving effects and will continue to follow the path of specialized joint development and organize family farms and cooperatives in villages and towns to form joint cooperatives, which will further benefit from economies of scale [23]. Cooperatives actively promote the organic integration of agricultural production, the processing and marketing of agricultural products, catering, leisure, rural culture, tourism, education, fitness, and other emerging industries, to extend the agricultural industrial chain, realize the integrated development of three sectors, drive increases in the village's collective income, and help farmers get rich [24].

Secondly, contract scales, standardization, and brand development construction also have an obvious effect on increasing income. Cooperatives sell their products in bulk through sales contracts signed before harvest, reducing the market risk of price fluctuations of agricultural products [25]. This benefit has been fully demonstrated by cooperatives in China and many other countries. Cooperatives that seek to maximize the welfare of their members have more incentive than investor-owned cooperatives to invest in innovations aimed at improving quality, thereby improving the nature of product differentiation and market structure. Branding the agricultural products of cooperatives can make them more distinctive, and the prices will be higher. Cooperative members can earn 20% more than non-member farmers by promoting the standardization and branding of agricultural products in China [26].

Thirdly, according to the provisions of the Law of Cooperatives, the surplus distribution of the cooperative is based on the agreement between the members and the cooperatives, and the distributable surplus is mainly returned in line with the proportion of the trading volume (amount) between the members and the cooperatives, and not less than 60% of the total amount. After the return, the remaining part is evenly quantified to the members based on the amount of capital contribution recorded in the members' accounts and the share of the provident fund, as well as the property formed by the direct subsidies from the state finance and donations from others, and is distributed to the members in proportion.

Finally, cooperatives have access to many kinds of government support. According to the relevant provisions of the Law of Cooperatives, the central and local financial departments shall separately allocate funds to provide support for cooperatives to carry out such services as information, training, standards and certification of agricultural products (quality), construction of agricultural production infrastructure, marketing, and technology promotion [27]. And cooperatives enjoy tax concessions in agricultural production, processing, distribution, services, and many other agriculture-related economic activities [28].

2.2. The Impact Mechanism of Cooperative Membership on Farmer Household Assets

On the one hand, the advantages of mechanization and scale and the supporting role of the government's agricultural machinery purchase subsidy will encourage farmers to buy advanced and applicable agricultural machinery. Farmers can buy agricultural machinery supported and popularized by governments at subsidized prices and participate in cooperatives to carry out mechanized production with agricultural machinery. This can not only improve agricultural production efficiency but also reduce labor input, thereby liberating a large number of labor forces [29]. Additionally, by participating in cooperatives that provide large-scale, professional, and social services, farmers can not only earn service fee income but also obtain additional government subsidies. Increasing income can further encourage farmers to purchase agricultural machinery that is more advanced and applicable and ensure that they can afford to buy it [30].

On the other hand, the purchase price of agricultural machinery is generally high, and the government subsidy for farmers to buy agricultural machinery is usually less than 30%, which stimulates a large amount of capital investment by cooperative members (the funds come from the farmers' own funds and bank loans, etc.) [31]. Moreover, the local agricultural machinery purchase subsidy limits the quantity of agricultural machinery purchased by cooperatives more loosely, resulting in a higher asset effect.

3. Materials and Methodology

3.1. Variables

3.1.1. Treatment Variable

This paper selected membership in herbal medicine planting cooperatives as a treatment variable, and a counterfactual hypothesis was adopted to analyze the treatment effect of cooperative membership on farmer household income and assets. Based on this, cooperative membership was set as a dummy variable. If the sample household participated in a cooperative, it was assigned a value of 1 (as the treatment group). If not, it was assigned a value of 0 (as the control group).

3.1.2. Consequence Variable

According to the research objective, this paper selected farmer household income and farmer household assets as consequence variables from survey data. And the logarithm of the total income of the farmer household and the valuation of agricultural operating fixed assets owned by the farmer household were taken as specific indexes.

3.1.3. Matching Variable

Based on the existing research, this paper selected household social capital, human capital, land area, labor scale, householder age, the education level of the householder, and the migrant work experience of the householder as matching variables [17,24,32]. The definitions of variables are shown in Table 1. The social capital (Table 2) and human capital of farmers were obtained by referring to the research of Liu et al. (2020) [33] and calculated by principal component factor analysis score.

Table 1. The definitions of variables.

Variables	Definitions
Participating cooperatives (PC)	Participating (as treatment group) = 1; Non- participating (as control group) = 0.
Farmer household income (income)	The logarithm of the total income related to herbal medicine planting and processing of the farmer's household in 2021 (in CNY).
Farmer household assets (assets)	The logarithm of the valuation of agricultural operating fixed assets owned by the farmer's household in 2021 (in CNY).
Farmer household social capital (social)	Principal component factor analysis score.
Farmer household human capital (human)	Principal component factor analysis score.
Farmer household land area under operation (land)	The total area of land planted with herbal medicine by the farmer's household.
Farmer household labor scale (labor)	The number of labor forces participating in herbal medicine planting and processing.
Householder age (age)	-
The education level of the householder (education)	Primary school = 1; junior high school = 2; high school = 3; college = 4.
The migrant work experience of the householder (migrant)	Yes = 1; No = 2.

Table 2. Composition of farmer household social capital and human capital.

Variables	Dimension	Definitions
Farmer household social capital (social)	Social network	Relationship with relatives and friends
		Relationship with village leaders
	Social trust	The number of people you can trust in your village
		The degree of trust in the village committee
		Possibility of finding something lost in the village
	Social prestige	The importance of individual opinions in village collective decision making
		Probability of others asking you for help when they are in trouble
		Frequency of mentoring others
		Frequency of resolving conflicts in the village
	Social participation	Willingness to participate in village collective activities
Farmer household human capital (human)	Knowledge	Average education level of the household workforce
	Capacity	Average age of the household workforce
		Average health level of the household

3.2. Data

The data used in this paper were all taken from a micro survey in Liaoning province. The research group designed the questionnaire according to the characteristics of agricultural development and agricultural resource endowment in Liaoning province and adjusted the questionnaire on the basis of pre-investigation. The micro survey was carried out in the eastern mountains in June 2022. During the survey process, the interviewers asked questions, and the farmers answered them. And the information was not filled in the questionnaire until the interviewers confirmed that it was correct and reviewed the questionnaire, thus ensuring the accuracy of the survey information. The survey covered farmers' operational expenses and income, operation and management, natural risks, market risks, family assets, financial services, and other aspects. After eliminating the questionnaires that omitted key information or provided contradictory information, 511 valid questionnaires were obtained, with an effective rate of 95.93%. In this paper, 248 farmers who grow herbal medicine in forest area were selected as the specific analysis objects. The basic information of the sample is shown in Table 3.

Table 3. Data description.

	Sample N = 248		Treatment Group N = 71		Control Group N = 177	
	Mean	Standard	Mean	Standard	Mean	Standard
income	63,884	647,331	203,673	1204,276	7810	14,802
assets	330,227	3,224,564	1,013,752	6,000,990	56,044	73,455
social	0.573	0.173	0.651	0.166	0.541	0.166
human	0.501	0.163	0.562	0.147	0.477	0.163
land	57.218	503.425	157.359	936.444	17.048	35.053
labor	2.315	0.925	2.366	1.072	2.294	0.862
age	57.649	9.632	54.690	9.503	58.836	9.452
education	1.831	0.739	1.972	0.755	1.774	0.727

It can be seen from Table 3 that 71 households participated in cooperatives and 177 households did not. Without controlling for other socioeconomic characteristics, the between-group differences in farmer household income and farmer household assets for participating in cooperatives or not were significantly positive at a 1% level. Additionally, there were also significant differences in farmer household social capital, farmer household human capital, farmer household land area under operation, farmer household labor scale, householder age, the education level of the householder, and migrant work experience of the householder among the farmer households that participated in cooperatives or not. However, a simple comparison of means can only reflect the differences between participating and not participating in terms of correlation. To investigate the impact of cooperative membership on farmer household income and assets from the perspective of causality, a more rigorous econometric analysis method was needed.

3.3. Methodology

This paper selected the estimation method of propensity score matching (PSM) to quantitatively analyze the impact of cooperative membership on farmer household income and assets. The specific steps were as follows:

Firstly, we regarded whether the ith farmer household did or did not have a cooperative membership as a binary random variable. Where, $PC_i = 1$ represents the farmer with a cooperative membership; $PC_i = 0$ represents the farmer without a cooperative membership; $income_i$ and $assets_i$ represent the ith farmer household income and assets, respectively; $income_{1i}$ and $assets_{1i}$ represent the farmer household income and assets of $PC_i = 1$, $income_{0i}$ and $assets_{0i}$ represent the farmer household income and assets of $PC_i = 0$. Therefore, the change in farmer household income and assets resulting from cooperative membership can be represented by $income_{1i} - income_{0i}$ and $assets_{1i} - assets_{0i}$. As the farmer household income and assets of participating or non-participating cooperatives cannot be observed simultaneously, they can be defined as follows:

$$income_i = (1 - PC_i) \times income_{0i} + PC_i \times income_{1i} = income_{0i} + PC_i \times (income_{1i} - income_{0i})$$

$$assets_i = (1 - PC_i) \times assets_{0i} + PC_i \times assets_{1i} = assets_{0i} + PC_i \times (assets_{1i} - assets_{0i})$$

Secondly, as the farmer households were non-random and self-selecting when choosing whether to participate in cooperatives or not. Moreover, since the data of participating and non-participating cooperatives cannot be obtained at the same time, direct comparison easily produced endogeneity. The propensity score matching (PSM) model can effectively solve the problem of endogeneity through constructing a counterfactual hypothesis to reduce the multidimensional information of farmer households to a factor and match the farmers who participated in cooperatives or not in multiple dimensions. Specifically, in the set of farmer households that did not participate in cooperatives, one or some will be selected to match each farmer household that did participate in a cooperative. When the analysis sample was restricted to individuals that had already received "treatment", the Average Treatment Effects on the Treated (ATT) was obtained, that is, the impact of cooperative membership on farmer household income and assets.

$$E(income1i - income_{0i}|PC_i = 1)$$

$$E(assets_{1i} - assets_{0i}|PC_i = 1)$$

Under the assumption of conditional expectation independence, the ATT can be estimated by the PSM model through matching the different farmer households with similar characteristics that participated or did not participate in cooperatives through estimating each individual's propensity score, namely, $P(X_i)$.

$$ATTincome = \frac{1}{N1}\sum_{i=1}^{N1}(income_{1i} - \sum_{j \in PC(pi)} wijincome_{0i})$$

$$ATTassets = \frac{1}{N1}\sum_{i=1}^{N1}(assets_{1i} - \sum_{j \in PC(pi)} wijassets_{0i})$$

where N_1 represents the number of individuals in the treatment group; $PC(pi)PC(pi)$ represents the paired group of the ith individual in the treatment group; $wijwij$ represents the weight of each individual in the paired group of the ith individual, and $\sum_{j \in PC(pi)} wij = 1 \sum_{j \in PC(pi)} wij = 1$. Moreover, $PC(pi)PC(pi)$ and $wijwij$ will definitely be different under different matching methods. On the basis of reference to existing research, the four PSM methods of Nearest Neighbor Matching (NNM), Local Linear Regression Matching (LLR), Kernel Matching (KM), and Radius Matching (RM) were selected to estimate the ATT in this paper.

4. Results

4.1. Baseline Regression Analysis

Tables 4 and 5 show the baseline regression results of farmer household income and assets. It can be seen that the coefficients of determination R^2 were both higher than 15%, indicating that the model had a high level of fitting, and the independent variables greatly explained the dependent variable, which can estimate the impact of cooperative membership on farmer household income and assets accurately. The coefficients of the variable PC were also both significantly positive at a 1% level, indicating that cooperative membership can greatly improve farmer household income and assets.

The variables of farmer household human capital and farmer household land area under operation passed the significance test, as seen in Tables 4 and 5, indicating that the accumulation of social capital and scale operation both helped to increase farmer household income and assets, and the research of Khan et al. (2022) supported this result [34]. Additionally, the migrant work experience of householders had a positive effect on farmer household income, while the farmer household labor scale had a positive effect on farmer household assets. However, the householder age had a negative effect on farmer household assets, which may be due to a decrease in the willingness to accept new technology and machinery as people get older. Zhang et al. (2020) pointed out that age was an important factor affecting the popularization and application of new machinery [35].

Table 4. The baseline regression results of farmer household income.

Explanatory Variables	Coefficient	Standard Error	T Value	*p* Value
PC	0.850 ***	0.207	4.110	0.000
social	0.896 *	0.464	1.930	0.054
human	1.204	0.864	1.390	0.165
land	0.001 ***	0.000	6.600	0.000
labor	0.025	0.089	0.280	0.780
age	−0.006	0.012	−0.510	0.612
education	−0.072	0.146	−0.500	0.621
job	−0.430 **	0.164	−2.620	0.009
Constant	7.842 ***	0.980	8.000	0.000
$PseudoR^2$	0.326			
F value	12.441			

Note: *, **, and *** represent 10%, 5%, and 1% significance, respectively.

Table 5. The baseline regression results of farmer household assets.

Explanatory Variables	Coefficient	Standard Error	T Value	*p* Value
PC	0.603 ***	0.169	3.570	0.000
social	0.731 *	0.404	1.810	0.072
human	0.414	0.717	0.580	0.564
land	0.001 ***	0.000	7.490	0.000
labor	0.388 ***	0.082	4.730	0.000
age	−0.032 ***	0.010	−3.040	0.003
education	−0.009	0.117	−0.080	0.937
job	−0.009	0.129	−0.070	0.946
Constant	10.842 ***	0.831	13.040	0.000
$PseudoR^2$	0.326			
F value	22.180			

Note: * and *** represent 10% and 1% significance, respectively.

4.2. Propensity Score Matching Analysis

In order to achieve sample matching, this paper selected the Logistic model to estimate the propensity score, referring to the research of Ma and Abdulai (2017), Meng et al. (2020), and Hoang (2021) [17,24,32], and sought the farmer households without cooperative membership, but that had similar economic characteristics with farmer households with cooperative membership for analysis. The model included the variables of farmer household social capital, farmer household human capital, farmer household land area under operation, farmer household labor scale, householder age, the education level of the householder, and migrant work experience of the householder.

Tables 6 and 7 show the results of propensity score matching analysis on farmer household income and assets, where the chi^2 values were 40.18 and 40.64, respectively, and the probability of being less than the *p* value for both was 0. Therefore, the null hypothesis was rejected, indicating that the model had a high level of fitting, while the independent variables greatly explained the dependent variable. Additionally, the consistent results presented in Tables 5 and 7 also showed that the higher the accumulation of farmer household social capital and human capital, the more farmers were inclined to participate in cooperatives. He et al. (2022) obtained a similar conclusion [36]. Additionally, compared with householders without migrant work experiences, householders who had been migrant workers were more likely to make the decision to participate in cooperatives.

Table 6. The results of propensity score matching analysis of farmer household income.

Explanatory Variables	Coefficient	Standard Error
social	3.688 ***	0.990
human	4.345 *	2.524
land	0.002	0.002
labor	−0.075	0.178
age	−0.008	0.032
education	−0.376	0.333
job	−0.711 **	0.348
N	248	
pseudoR^2	0.1364	
chi2	10.18	
Prob > chi2	0.000	

Note: *, **, and *** represent 10%, 5%, and 1% significance, respectively.

Table 7. The results of propensity score matching analysis of farmer household assets.

Explanatory Variables	Coefficient	Standard Error
social	3.681 ***	0.988
human	4.367 *	2.525
land	0.002	0.002
labor	−0.077	0.178
age	−0.009	0.032
education	−0.379	0.333
job	−0.674 **	0.345
N	248	
pseudoR^2	0.1368	
chi2	40.63	
Prob > chi2	0.000	

Note: *, **, and *** represent 10%, 5%, and 1% significance, respectively.

An important assumption in propensity score matching analysis was the assumption of balance, which required that there were no systematic differences in the matching variables between the treatment group and the control group after matching. In order to ensure the robustness of results, the balance test of propensity score matching was further tested by the methods of Psmatch2 and Pstest based on the existing research of Ma and Abdulai (2017) [17].

Tables 8 and 9 show consistent results; the deviation values of variables were all lower than 10%, indicating that the deviation was within the acceptable range. At the same time, the *p* values of all variables were greater than 10%, accepting the null hypothesis, indicating that after matching, there was no significant difference in the variables of economic characteristics between farmer households with cooperative membership and those without. The propensity score matching passed the balance test, and the matching effect was good.

Table 8. The results of the balance test of propensity score matching for farmer household income.

Variables	Mean		Deviation/%	T Value	*p* Value
	Participating	Non-Participating			
social	0.631	0.621	6.30	0.390	0.699
human	0.552	0.550	1.80	0.120	0.908
land	16.414	31.010	−2.20	−1.770	0.179
labor	2.281	2.418	−11.00	−0.840	0.403
age	55.031	54.910	1.30	0.080	0.938
education	1.938	1.910	3.70	0.210	0.832
job	0.313	0.297	3.30	0.190	0.849

Table 9. The results of the balance test of propensity score matching for farmer household assets.

Variables	Mean		Deviation/%	T Value	p Value
	Participating	Non-Participating			
social	0.635	0.625	5.9	0.37	0.712
human	0.557	0.549	5.2	0.35	0.73
land	16.394	32.091	−2.4	−1.87	0.064
labor	2.273	2.401	−10.0	−0.79	0.433
age	54.667	54.765	−1.0	−0.06	0.951
education	1.939	1.860	9.7	0.64	0.522
job	0.318	0.311	1.6	0.09	0.926

4.3. Average Treatment Effect Analysis

After propensity score matching, the ATT results of the impact of cooperative membership on farmer household income and assets are shown in Tables 10 and 11. It can be seen that the T value estimated by the four PSM methods of NNM, LLR, KM, and RM were all less than the results of baseline regression, and the overall effect was significant at a 1% level, which further verified the conclusion that cooperative membership can improve farmer household income and assets greatly, improvements that reached 7.04% and 4.19%, respectively.

Table 10. The ATT results of the impact of cooperative membership on farmer household income.

Methods	Treatment Group	Control Group	ATT	Standard Error	T Value	△ (%)
NNM	9.239	8.658	0.581	0.229	2.54 ***	6.71%
LLR	9.239	8.583	0.656	0.261	2.51 ***	7.64%
KM	9.239	8.618	0.621	0.216	2.88 ***	7.21%
RM	9.217	8.645	0.571	0.232	2.47 ***	6.62%
Average	9.234	8.626	0.607			7.04%

Note: *** represent 1% significance.

Table 11. The ATT results of the impact of cooperative membership on farmer household assets.

Methods	Treatment Group	Control Group	ATT	Standard Error	T Value	△ (%)
NNM	11.187	10.742	0.444	0.199	2.23 ***	4.14%
LLR	11.187	10.696	0.491	0.250	1.96 ***	4.49%
KM	11.187	10.737	0.449	0.185	2.42 ***	4.19%
RM	11.192	10.779	0.413	0.204	2.02 ***	3.83%
Average	11.188	10.739	0.449			4.19%

Note: *** represent 1% significance.

5. Conclusions and Discussion

In this paper, we selected the estimation method of propensity score matching (PSM) to quantitatively analyze the impact of membership in herbal medicine planting cooperatives on farmer household income and assets on the basis of an in-depth survey of the study area. We found that, firstly, cooperative membership can greatly improve farmer household income and assets, the migrant work experience of householders had a positive effect on farmer household income, the farmer household labor scale had a positive effect on farmer household assets, and the householder age had a negative effect on farmer household assets. Secondly, the higher the accumulation of farmer household social capital and human capital, the more farmers were inclined to participate in cooperatives. And compared with householders without migrant work experiences, householders who had been migrant workers were more likely to make the decision to participate in cooperatives. Finally, the

ATT results further verified the conclusion that cooperative membership can improve farmer household income and assets greatly, improvements that reached 7.04% and 4.19%, respectively.

It can be seen from the results that cooperative membership can greatly improve the income and assets of farmers. Here, the leading role of industrial development cannot be ignored. A large amount of research has verified the basic role of industries, especially the characteristic industries relying on regional resource endowment in economic development and farmers' income increase [7,37,38]. The herbal medicine industry selected in this paper was the main characteristic industry in Liaoning province. The existing varieties of herbal medicine mainly include understory ginseng, codonopsis, schisandra chinensis, asarum, gentian, acanthopanax, astragalus, etc. Huairen county in this region was named "the hometown of China's acanthopanax and understory ginseng", and "Fushun's schisandra chinensis" and "Fushun's understory ginseng" were certified as national geographic indications. Gentian and asarum occupied 80% and 70% of the national market sales, respectively.

In this context, cooperatives had become an important carrier of rural industrial prosperity, and only the coordinated development of cooperatives and industries can better serve rural revitalization. However, through survey and investigation, it was found that with the strong support of government departments at all levels, cooperatives in this region have developed rapidly, but there were still many problems, which made it difficult to meet the requirements of rural revitalization and agricultural and rural modernization in the new development stage. Overall, it mainly included five aspects:

Firstly, the development of quantity and quality was inconsistent. Despite the rapid development of cooperatives in recent years, their scale development and quality improvement were not synchronized, and the problem of "small, scattered, weak and empty" still existed. Survey data showed that the number of cooperatives in real operation was less than one-third of the total. Additionally, cooperatives with registered trademarks and quality certification of agricultural products only accounted for 4.33% and 1.32% of the total number in this region, and there were problems such as lagging brand building, standardization of agricultural products, and insufficient deep processing.

Secondly, the driving ability was insufficient. By the end of 2021, enterprise-led cooperatives accounted for less than 5% of the total in this region, and only 12.15% and 8.03% of cooperatives that uniformly organized the sale of agricultural products and purchased agricultural production inputs reached more than 80%. Overall, the radiation driving ability of cooperatives also needs to be improved.

Thirdly, the operating mechanism was irregular. On the one hand, the organizational structure was not perfect, and the internal interests of cooperatives were loosely connected. On the other hand, the financial management mechanism was not reasonable enough. Additionally, the cooperative surplus distribution was not standard enough. Most cooperatives mainly used price rebates instead of surplus distribution. In 2021, the number of cooperatives with surplus distribution returned by transaction volume was 4092, among which 2934 cooperatives returned more than 60%, accounting for only 5.23% of the total.

Fourthly, the support policy was inaccurate. In 2021, the total amount of government support funds in this region increased by 4.26% year-on-year, but only 3.68% of cooperatives received financial support funds, and less than 8% of the total undertook national agriculture-related projects. Additionally, although agriculture-related financial institutions had increased their credit support for "agriculture, rural areas and farmers", most financing products had certain limitations in terms of quota, use, and time, and problems such as difficult and expensive financing still existed.

Finally, the guidance service was inadequate. After the institutional reform in 2018, the guidance service work of cooperatives originally undertaken by agricultural economic stations (offices) was assigned to agricultural and rural administrative departments. However, the full-time staff of agricultural and rural administrative departments at the county level in this region was generally no more than 3, resulting in insufficient guidance service work for cooperatives.

The political implications of this research include the following aspects: Firstly, innovating cooperation mechanisms and enriching joint forms by exploring different modes of cooperation among farmers using resource elements such as the contracted forestland and encouraging cooperatives closely tied to the industry to reorganize resources through mergers and amalgamations on a voluntary basis. Secondly, strengthening driving ability and stimulating endogenous driving forces through developing rural industries, and encouraging cooperatives to build specialized villages and towns with high quality, good benefits, and obvious advantages in leading industries, relying on local resource endowments, so as to form a development pattern with strong competitiveness, distinctive characteristics, and appropriate scale. Thirdly, enhancing driving capacity and stimulating internal driving forces by improving the organizational structure, standardizing financial management, and managing income distribution. Fourthly, strengthening system construction and improving standard level through increasing support for fiscal projects, innovating financial services, and focusing on digital empowerment. Finally, enhancing guidance services and strengthening institutional guarantees by establishing and improving a comprehensive and coordinated working mechanism, strengthening talent support to accelerate the development of basic institutions.

Author Contributions: Conceptualization, J.W. and. Z.Z.; methodology, Z.Z.; writing—original draft preparation, J.W.; writing—review and editing, L.G.; supervision, J.W. All authors have read and agreed to the published version of the manuscript.

Funding: This study was financially supported by the Youth Project of Liaoning Provincial Philosophy and Social Science Planning Fund (L21CJY012); the Commissioned Project of Young Talents Training Object of Philosophy and Social Sciences in Liaoning Province in 2022 (2022lslqnrcwtkt-34); and Fundamental Research Funds for the Provincial Universities of Zhejiang (22FR013).

Institutional Review Board Statement: Not applicable.

Informed Consent Statement: Not applicable.

Data Availability Statement: Not applicable.

Conflicts of Interest: The authors declare no conflict of interest.

References

1. Hu, Z.; Zhang, Q.F.; Donaldson, J.A. Farmers' cooperatives in China: A typology of fraud and failure. *China J.* **2017**, *78*, 1–24. [CrossRef]
2. Zhang, J.; Wu, J.; Simpson, J.; Arthur, C.L. Membership of Chinese Farmer Specialized Cooperatives and Direct Subsidies for Farmer Households: A Multi-Province Data Study. *Chin. Econ.* **2019**, *52*, 400–421. [CrossRef]
3. Yu, L.; Huang, W. Non-economic societal impact or economic revenue? A performance and efficiency analysis of farmer cooperatives in China. *J. Rural. Stud.* **2020**, *80*, 123–134.
4. Cui, Z.; Zhang, H.; Chen, X.; Zhang, C.; Ma, W.; Huang, C.; Zhang, W.; Mi, G.; Miao, Y.; Li, X.; et al. Pursuing sustainable productivity with millions of smallholder farmers. *Nature* **2018**, *555*, 363–366. [CrossRef]
5. Shen, M.; Shen, J. Evaluating the cooperative and family farm programs in China: A rural governance perspective. *Land Use Policy* **2018**, *79*, 240–250. [CrossRef]
6. Ji, C.; Jin, S.; Wang, H.; Ye, C. Estimating effects of cooperative membership on farmers' safe production behaviors: Evidence from pig sector in China. *Food Policy* **2019**, *83*, 231–245. [CrossRef]
7. Wang, C.; Wang, X.; Wang, Y.; Zhan, J.; Chu, X.; Teng, Y.; Liu, W.; Wang, H. Spatio-temporal analysis of human wellbeing and its coupling relationship with ecosystem services in Shandong province, China. *J. Geogr. Sci.* **2023**, *33*, 392–412.
8. Yang, Y.; Su, Y. Public voice via social media: Role in cooperative governance during public health emergency. *Int. J. Env. Res. Public Health* **2020**, *17*, 6840. [CrossRef]
9. Ajates, R. An integrated conceptual framework for the study of agricultural cooperatives: From repolitisation to cooperative sustainability. *J. Rural. Stud.* **2020**, *78*, 467–479. [CrossRef]
10. Ferreri, M.; Vidal, L. Public-cooperative policy mechanisms for housing commons. *Int. J. Hous. Policy* **2022**, *22*, 149–173. [CrossRef]
11. Liu, C.; Dou, X.; Li, J.; Cai, L.A. Analyzing government role in rural tourism development: An empirical investigation from China. *J. Rural. Stud.* **2020**, *79*, 177–188. [CrossRef]

12. Liu, M.Y.; Feng, X.L.; Wang, S.G.; Zhong, Y. Does poverty-alleviation-based industry development improve farmers' livelihood capital? *J. Integr. Agric.* **2021**, *20*, 915–926. [CrossRef]
13. Candemir, A.; Duvaleix, S.; Latruffe, L. Agricultural cooperatives and farm sustainability—A literature review. *J. Econ. Surv.* **2021**, *35*, 1118–1144. [CrossRef]
14. Li, X.; Ito, J. An empirical study of land rental development in rural Gansu, China: The role of agricultural cooperatives and transaction costs. *Land Use Policy* **2021**, *109*, 105621. [CrossRef]
15. Lin, J.; Zhang, Z.; Liu, Z.; Rommel, J. The impact of cooperatives' transportation services on farm income: Evidence from tobacco farmers in Guizhou, China. *Agribusiness* **2020**, *36*, 146–158. [CrossRef]
16. Deng, L.; Chen, L.; Zhao, J.; Wang, R. Comparative analysis on environmental and economic performance of agricultural cooperatives and smallholder farmers: The case of grape production in Hebei, China. *PLoS ONE* **2021**, *16*, e0245981. [CrossRef]
17. Ma, W.; Abdulai, A. The economic impacts of agricultural cooperatives on smallholder farmers in rural China. *Agribusiness* **2017**, *33*, 537–551. [CrossRef]
18. Dan, Y.; Zhang, H.W.; Liu, Z.M.; Qiao, Z. Do cooperatives participation and technology adoption improve farmers' welfare in China? A joint analysis accounting for selection bias. *J. Integr. Agric.* **2021**, *20*, 1716–1726.
19. Dong, Y.; Mu, Y.; Abler, D. Do farmer professional cooperatives improve technical efficiency and income? Evidence from small vegetable farms in China. *J. Agric. Appl. Econ.* **2019**, *51*, 591–605.
20. Ma, W.; Zheng, H.; Zhu, Y.; Qi, J. Effects of cooperative membership on financial performance of banana farmers in China: A heterogeneous analysis. *Ann. Public Coop. Econ.* **2022**, *93*, 5–27. [CrossRef]
21. Rois-Díaz, M.; Lovric, N.; Lovric, M.; Ferreiro-Domínguez, n.; Mosquera-Losada, M.R.; den Herder, M.; Graves, A.; Palma, J.H.N.; Paulo, J.A.; Pisanelli, A.; et al. Farmers' reasoning behind the uptake of agroforestry practices: Evidence from multiple case-studies across Europe. *Agrofor. Syst.* **2018**, *92*, 811–828.
22. Gong, T.C.; Battese, G.E.; Villano, R.A. Family farms plus cooperatives in China: Technical efficiency in crop production. *J. Asian Econ.* **2019**, *64*, 101129. [CrossRef]
23. Yun, S.; Qinghai, L.; Jing, Y. Farmers' Cooperatives' Poverty-Reducing Roles in Agricultural Supply Chain Finance. *China Econ.* **2020**, *15*, 76–91.
24. Meng, L.I.; Christopher, G.A.N.; Wanglin, M.A.; Jiang, W. Impact of cash crop cultivation on household income and migration decisions: Evidence from low-income regions in China. *J. Integr. Agric.* **2020**, *19*, 2571–2581.
25. Ficko, A.; Lidestav, G.; Dhubháin, Á.N.; Karppinen, H.; Zivojinovic, I.; Westin, K. European private forest owner typologies: A review of methods and use. *For. Policy Econ.* **2019**, *99*, 21–31. [CrossRef]
26. Zou, Y.; Wang, Q. Impacts of farmer cooperative membership on household income and inequality: Evidence from a household survey in China. *Agric. Food Econ.* **2022**, *10*, 17. [CrossRef]
27. Zhang, S.; Wolz, A.; Ding, Y. Is there a role for agricultural production cooperatives in agricultural development? *Evidence from China. Outlook Agric.* **2020**, *49*, 256–263. [CrossRef]
28. Mirzania, P.; Ford, A.; Andrews, D.; Ofori, G.; Maidment, G. The impact of policy changes: The opportunities of Community Renewable Energy projects in the UK and the barriers they face. *Energy Policy* **2019**, *129*, 1282–1296. [CrossRef]
29. Jiang, M.; Hu, X.; Chunga, J.; Lin, Z.; Fei, R. Does the popularization of agricultural mechanization improve energy-environment performance in China's agricultural sector? *J. Clean. Prod.* **2020**, *276*, 124210. [CrossRef]
30. Siméon, N.; Li, X.; Sangmeng, X. China's agricultural assistance efficiency to Africa: Two decades of Forum for China-Africa Cooperation creation. *J. Agric. Food Res.* **2022**, *9*, 100329. [CrossRef]
31. Tong, H.; Qian, X.D.; Chen, Y.; Wei, Q.; Xia, S.; Qian, N.; Li, J. Optimal decision of agricultural machinery product quality under the regulation of government subsidy policy. *Afr. Asian Stud.* **2020**, *19*, 218–244. [CrossRef]
32. Hoang, V. Impact of contract farming on farmers' income in the food value chain: A theoretical analysis and empirical study in Vietnam. *Agriculture* **2021**, *11*, 797. [CrossRef]
33. Liu, Y.; Zou, L.; Wang, Y. Spatial-temporal characteristics and influencing factors of agricultural eco-efficiency in China in recent 40 years. *Land Use Policy* **2020**, *97*, 104794. [CrossRef]
34. Khan, N.; Ray, R.L.; Kassem, H.S.; Ihtisham, M.; Siddiqui, B.N.; Zhang, S. Can Cooperative Supports and Adoption of Improved Technologies Help Increase Agricultural Income? Evidence from a Recent Study. *Land* **2022**, *11*, 361.
35. Zhang, E.; Zhou, Y.; Xia, X.; Wang, W.; Qiao, J. Study on Spatial Distribution Characteristics and Economic Effect of County Cooperatives in Underdeveloped Agricultural Area. *Front. Sustain. Dev.* **2021**, *1*, 7–16.
36. He, R.W.; Guo, S.L.; Deng, X.; Zhou, K. Influence of social capital on the livelihood strategies of farmers under China's rural revitalization strategy in poor mountain areas: A case study of the Liangshan Yi autonomous prefecture. *J. Moun. Sci.* **2022**, *19*, 958–973. [CrossRef]

37. Liu, J.; Zhang, G.; Zhang, J.; Li, C. Human capital, social capital, and farmers' credit availability in China: Based on the analysis of the ordered probit and PSM models. *Sustainability* **2020**, *12*, 1583. [CrossRef]
38. Wang, M.; He, B.; Zhang, J.; Jin, Y. Analysis of the Effect of Cooperatives on Increasing Farmers' Income from the Perspective of Industry Prosperity Based on the PSM Empirical Study in Shennongjia Region. *Sustainability* **2021**, *13*, 13172. [CrossRef]

forests

MDPI

Article

Ecological Restoration in the Loess Plateau, China Necessitates Targeted Management Strategy: Evidence from the Beiluo River Basin

Jiacheng Xing [1], Jianjun Zhang [1,2,*], Jing Wang [1], Mingjun Li [1], Shitan Nie [1] and Mingjie Qian [1,2]

1 School of Land Science and Technology, China University of Geosciences (Beijing), Beijing 100083, China; qianmingjie@cugb.edu.cn (M.Q.)
2 Key Laboratory of Land Consolidation and Rehabilitation, Ministry of Natural Resources, Beijing 100035, China
* Correspondence: j.zhang@cugb.edu.cn

Abstract: Vegetation on the Loess Plateau, China, has continuously improved thanks to certain ecological restoration (ER) strategies, including the integrated soil conservation project that began in the late 1970s and the "Grain for Green" project that began in the 1990s. The experience of these strategies in different geomorphological regions is of great value to ER worldwide. In this study, the evolution of the land-use transition (LUT) pathway and ecosystem service value (ESV) in four geomorphological regions of the Beiluo River Basin was analyzed using geo-informatic Tupu and the equivalent factor method with data from 1975 to 2015. The results indicated that, from 1975 to 2015, the proportion of forestland in the Beiluo River basin increased by 18.27%, while the areas of shrub, grassland, cultivated land, and water decreased by 1.03%, 0.16%, 18.23%, and 0.26%, respectively. In the past 40 years, the overall ESV of the basin increased by USD 3.209 billion (54.16%). The landform, vegetation cover, LUT, and ESV analysis indicated that the main ecological functions of the loess hilly and gully (LHG), loess plateau gully (LPG), rocky mountain (RM), and terrace and plain (TP) regions are soil and water conservation (SWC), SWC and food production, regulation and food production, respectively. ER projects enhanced the main ecological function of individual regions. In detail, the transition of "cultivated land → grassland" enhanced SWC function in the LHG region, and the transition of "grassland (shrub) → forestland" enhanced the regulating services of the RM and LPG regions. Moreover, the transition of "cultivated land to grassland" did not seriously lower the food production services of the TP and LPG regions, owing to the increase in grain yield per unit area. However, there were alternating transitions between cultivated land and ecological land types, implying a game between the peasant households' demands and the ER strategies. Conflicting demands between local households and the public necessitate precision ER strategies, including land planning, ecological compensation, training and employment for local residents, etc.

Keywords: precise ecological restoration; land-use/land-cover change (LUCC); ecosystem service value (ESV); Loess Plateau; geo-informatic Tupu (GT); China

Citation: Xing, J.; Zhang, J.; Wang, J.; Li, M.; Nie, S.; Qian, M. Ecological Restoration in the Loess Plateau, China Necessitates Targeted Management Strategy: Evidence from the Beiluo River Basin. *Forests* **2023**, *14*, 1753. https://doi.org/10.3390/f14091753

Academic Editor: Xibao Xu

Received: 16 July 2023
Revised: 26 August 2023
Accepted: 27 August 2023
Published: 30 August 2023

1. Introduction

In order to cope with climate change and control the deteriorating ecological environment, ecological restoration (ER) projects have been carried out on a global scale [1,2]. In 2010, the Strategic Plan for Biodiversity 2011–2020 was signed at the Conference of the Parties to the Convention on Biological Diversity. In 2012, under the auspices of the United Nations Environment Programme, the Intergovernmental Platform on Biodiversity and Ecosystem Services (ESs) was approved by the United Nations (http://www.ipbes.net (accessed on 10 July 2023)). The United Nations has put forward a series of actions, such as the Decade of Sustainable Development, the Decade of Ecological Restoration, and the Decade of Marine Science for Sustainable Development, to promote the restoration of global ecosystems [3,4]. The implementation of ER projects guarantees the restoration of ESs [5,6]. It is important to study the impact of ER, especially the existing problems, for environmental management [7]. In the context of global ER, countries around the world generally believe that the future will be a key period to improving human well-being and promoting sustainable development through ER, and the study of ER is a hot topic at present [1].

For ER, at the regional scale, there is the influence of topography and landform. Different topographies and landforms determine the main use modes of regional land and determine its main ecological function [7,8]. For example, agricultural land is well-distributed in flat terrain, which enhances ecosystem food production services. Meanwhile, slopes above 25 degrees are more suitable for planting trees than for farming, which can enhance ecosystem water and soil conservation services [9]. Land-use types can produce trade-offs or synergies between different ESs [10]. Jia et al. studied the relationship between "one increase and one decrease" and "co-increase and co-decrease" among ESs in the loess hilly region, and found that afforestation enhanced the "one increase and one decrease" interaction between regulating services and supplying services, and the increase in shrubs improved the soil and water conservation services of the ecosystem [11]. Therefore, it is necessary to study the impact of regional ER in detail according to geomorphic zoning. It is of great significance to find the problems in ER in order to guide and adjust the employed strategies.

The Loess Plateau, located in Northwest China, is one of the areas most seriously affected by soil and water erosion in the world [12]. The poor condition of the ecological environment in the area significantly restricts the social and economic development of the region [13]. In order to improve this situation, China has long governed the environment of the Loess Plateau. From 1950 to 1970, the environmental governance measures of the Loess Plateau included afforestation, the terracing of farmlands on slopes, and building sediment-trapping dams. From 1970 to the late 1990s, the government mainly carried out the comprehensive management of small watersheds and the construction of Three-North shelterbelts. After 2000, environmental governance focused on the projects of "Grain for Green" [14,15]. After years of environmental restoration, the overall vegetation cover area has increased, the sediment entering the Yellow River has reduced, and the land-use structure has also changed dramatically [13,16–19]. However, owing to the interaction between the ecological environment and the socio-economic relationship, the sustainability of ER can be affected by managers, land-lost farmers, and other stakeholders. It is thus of great significance to study the relationship between LUTs and ESV, find out the deficiencies of ER, and sum up the experience for system governance and precise policy formulation.

The Beiluo River Basin contains four typical landforms of the Loess Plateau, and is one of the key regions in the control of soil and water erosion and "Grain for Green". The four geomorphic regions are the loess hilly and gully (LHG) region, with the most severe soil erosion; the loess plateau gully (LPG) region; the forested rocky mountain (RM) region; and the terrace plain (TP) region. There have been many studies exploring ER from different perspectives. SUN Zexing et al. evaluated the comprehensive benefits of ER in Shaanxi province [20]. NIU Linan et al. evaluated the extent and potential of ER in the Loess Plateau [21]. XUE Fan et al. studied the evolution of water and sediment characteristics in different landforms and vegetation types in the Beiluo River Basin over the past 70 years [22]. Wang Zhuangzhuang et al. established an evaluation index system of ER benefits for five key fragile ecological zones in China [23]. Finally, Zhao Yonghua et al. explored the changes in the ecosystem service value in Shaanxi Province from 2001 to 2006 and the differences among different regions [24]. These studies provide a reference for the adjustment of ER policies, but few researchers have paid attention to the research on the differentiated effects of LUTs on the main ecosystem service functions in different landforms.

Aiming to address the above problems, this paper takes the Beiluo River Basin as the representative area to study the impact of ER on the Loess Plateau and determine its existing problems, so as to provide a basis for the adjustment of ER strategies. Specifically, this includes (1) using geo-informatic Tupu (GT) to analyze the spatio-temporal characteristics of LUTs, (2) quantifying the spatiotemporal variation in ESV, and (3) analyzing the impact of LUTs on the main services of ecosystems in different landform types and discussing the existing problems.

2. Materials and Methods

2.1. Study Area

The Beiluo River (34°40′–37°19′ N, 107°35′–110°20′ E) originates in the Baiyu Mountains, Dingbian County, Shaanxi Province (Figure 1) [25]. Its main stream has a total length of 680 km, with an average ratio of 1.52‰ and a watershed area of 26,905 km^2. There are four typical landforms of the Loess Plateau in this basin: the loess hilly and gully (LHG) region, the Lloess plateau gully (LPG) region, the rocky mountainous (RM) region, and the terrace plain (TP) region. The LHG region is located in the upper reaches of the Beiluo River Basin and exhibits severe gully erosion. The LPG region is located in the middle reaches with high and flat lands and gullies around these lands. The RM region has high vegetation coverage and a relatively complete natural forest [26]. The TP region is located in the lower reaches of the basin and is its alluvial plain. The largest area, covering 10,542 km^2 and accounting for 39.2% of the total watershed area, is the loess hilly forest area, which includes the famous Ziwuling forest region and the Huanglong Mountain forest region. The area of the LHG region is 6755 km^2, accounting for 25.1% of the watershed area. It serves as the main sediment-producing area, with a high soil erosion modulus of more than 10,000 t/km·a. The LPG region experiences moderate soil erosion, while the other areas have less pronounced erosion. The Beiluo River Basin has an annual average precipitation (1954–1996) of 514.2 mm, a runoff of 8.652 × 10^8 m^3, and a sediment transport of 8.65 × 10^7 t [27–29]. Over time, the region has undergone ER, with the vegetated area increasing from 41.12% in 1987 to 63.43% in 2007 [25,30]. During this period, there were also significant land-use changes, affecting 56.5% of the entire basin area between 1975 and 2015 [29].

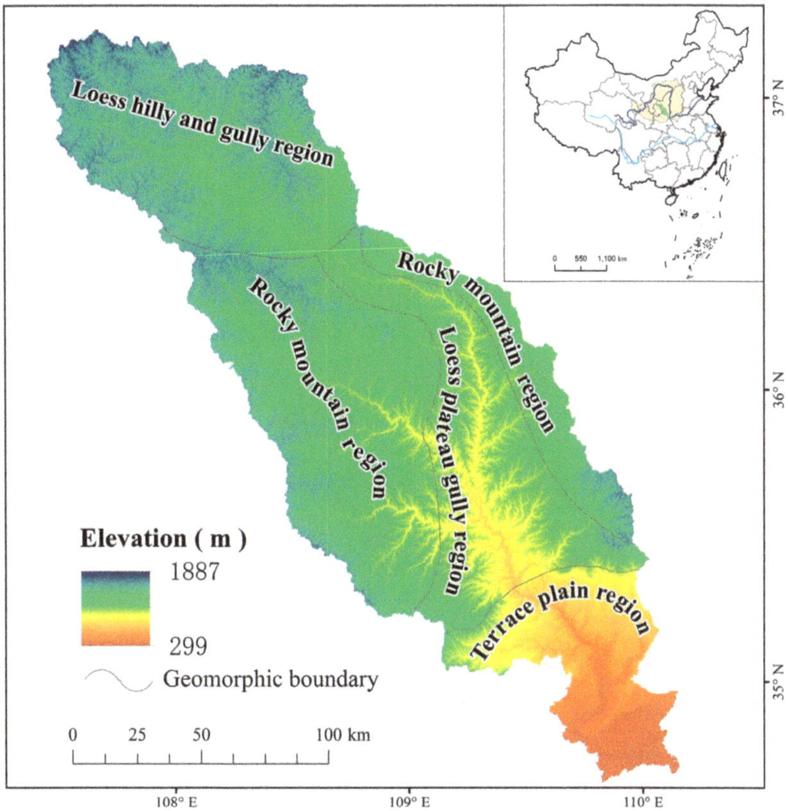

Figure 1. Location of the Beiluo River Basin. Note: The LHG region is developed from the high and flat loess land in the course of long-term soil erosion. There is a large amount of vegetation cover in the RM region, and the problem of soil erosion is not significant. The soil texture of the LPG region is the same as that of the LHG region, and there is still some flat land that has not completely eroded into gullies. The classification of the four landforms is based on data from *Research Theory and practice in Hydrology and Water Resources* by Changming LIU [31], which has been used many times by other people studying the Beiluo River Basin [32,33].

2.2. Data Sources

The Chinese administrative boundaries and typical geomorphology vector data for 1975, 1990, 2000, 2010, and 2015, along with the 30 × 30 m land-use database, were obtained from the RESDC (Resource and Environmental Science Data Center of the Chinese Academy of Sciences, http://www.resdc.cn (accessed on 2 January 2023)) [34–37]. These datasets were generated using Landsat TM/ETM (thematic mapper/enhanced thematic mapper) remote sensing images from different periods, which were then visually interpreted based on national field surveys. The land-use types were classified as forestland, scrub, grassland, cultivated land, water, construction land, and desert. Furthermore, the data for the national net primary productivity (NPP) of vegetation in 1975, 1990, 2000, 2010, and 2015 were derived from the Global Change Research Data Publishing and Repository [38,39].

2.3. Methodology

2.3.1. Geo-Informatic Tupu Method

This paper uses the geo-informatic Tupu (GT) method to establish a mathematical model. GT is a method for geographical spatiotemporal analysis that can project multidimensional spatial information onto two-dimensional maps. By doing so, it greatly reduces the complexity of model simulation. In addition, Tupu helps the model builder to understand the spatial information and its processes [40,41]. We used ArcGIS (10.8.1, Esri, Redlands, CA, USA) software to assign codes 1, 2, 3, 4, 5, 6, and 7 to seven land-use types, namely, forestland, shrub, grassland, cultivated land, construction land, water areas, and desert, respectively. The raster calculator was employed to generate the raster land-use data with a 30 m resolution over five periods. Based on this, a series of GT patterns were constructed.

GT can assign a color to each land-use unit on the map, enabling a more intuitive observation of its distribution. Similarly, the Tupu unit of land-use change can be assigned a color, facilitating the analysis of the spatial characteristics of LUTs. To integrate the value and spatial information of LUTs into the Tupu coding, two adjacent Tupu units can be selected for algebraic superposition. By screening the results and selecting the Tupu related to the LUTs of cultivated land, forest, and grassland, the Tupu can be created. The specific calculation formula is as follows [42]:

$$T = A \times 10 + B \quad (1)$$

where T is the Tupu code of LUTs in a given period, B is the code of the land-use unit at the beginning of the study, and A is the code of the land-use unit at the end of the study.

This study focuses on land-use change in the five periods of 1975, 1990, 2000, 2010, and 2015. The Tupu pattern of LUTs can be divided into five types: the prophase transition type, the middle transition type, the anaphase transition type, the repeated transition type, and the continuous transition type. The operation equation of the land-use unit code in each period is as follows:

$$T = 10000 \times Q + 1000 \times W + 100 \times E + 10 \times R + D \quad (2)$$

where T is the land-use change Tupu code in 1975, 1990, 2000, 2010, and 2015; Q is the land-use unit code in 1975; W is the land-use unit code in 1990; E is the land-use unit code in 2000; R is the land-use unit code in 2010; and D is the land-use unit code in 2015 (Table 1).

Table 1. Pattern classification of land-use change Tupu.

Tupu Pattern	Definition	Examples	Meaning
Prophase transition type	Only one change took place between 1975 and 2000.	12222, 22444	Reflects the LUTs for the time scale.
Middle transition type	Only changed between 2000 and 2010.	22233, 5522	
Anaphase transition type	Only changed between 2010 and 2015.	11112, 44443	
Repeated transition type	There have been two or more changes, involving only two types of land use.	22112, 22121, 22122, 22212	After at least two changes, the early land-use types reappeared in the later stage, highlighting the contradiction between the two land-use types.
Continuous transition type	There have been more than two changes, involving three or more land-use types.	11213, 11231, 14211, 14241	The frequency of prominent change is high, and the type of change is diverse.

2.3.2. Statistics of the Changing Characteristics of Tupu

The land-use change Tupu was analyzed using spatial statistics, and the degree of spatial separation and LUT rate were calculated. The formulae for calculating the rate of change and the degree of spatial separation are as follows:

$$A_{ij} = \frac{N_{ij}}{\sum_{i=1}^{n}\sum_{j=1}^{n} N_{ij}} \times 100\% (i \neq j) \quad (3)$$

$$S_{ij} = \frac{1}{2} \times \frac{\sqrt{F_{ij}/\sum_{i=1}^{n}\sum_{j=1}^{n} N_{ij}}}{N_{ij}/\sum_{i=1}^{n}\sum_{j=1}^{n} N_{ij}} (i \neq j) \tag{4}$$

where A_{ij} is the rate of change, which indicates the ratio of the area of a certain LUTs to the total area of LUTs in the study area over a period of time. S_{ij} is the degree of spatial separation, which reflects the degree of spatial dispersion of land-use Tupu units. The greater the absolute value, the greater the degree of dispersion of Tupu units. F_{ij} and N_{ij} indicate the number and area, respectively, of Tupu units in which the land-use type is transformed from land-use type i at the initial time t into j at the end ($t + \Delta t$), and n is the number of land-use types.

2.3.3. ESV Calculation

The concept of ESs was first proposed by King in 1966 [43]. Using an economics-based method to calculate the ESV can help human beings to realize the development, utilization, protection, and restoration of ecosystems more scientifically, and finally allow us to achieve the goal of sustainable development. The most common methods used to calculate the ESV are the energy evaluation method and the benefit transfer method [16]. However, regardless of which method is used, the subjective differences in the selection of parameters lead to differences in the calculation results. At present, there is no one standard recognized method for calculating the ESV [44,45]. Based on Burkhard et al.'s study on ESs in 2010 [46], this study classifies ESs into four types: supplying services, regulating services, supporting services, and cultural services. These four services are subdivided into nine services according to the specific conditions of the Chinese ecosystem, as shown in Table 2.

In this paper, the ESV of the Beiluo River Basin was calculated using the average vegetation net primary productivity (NPP) of each land-use ecosystem in China and the NPP of the Beiluo River Basin to modify the ecosystem value equivalent per unit area of ecosystems in China (Table 2), as proposed by Xie Gaodi (Formula (5)) [47]. This method sets the equivalent of food production service value of cultivated land ecosystem to 1.0, and the equivalent of various service values of other land-use ecosystems is derived from this. The ESV of construction land is set to 0. Because NPP cannot accurately reflect the biomass of water areas, this paper takes the average value equivalent of the national river and lake ecosystems proposed by Xie Gaodi as the value of the water area in the study area.

$$\eta_i = \frac{NPP_{bi}}{NPP_{ci}} \tag{5}$$

where η_i is the correction coefficient of the land-use type i, NPP_{ci} is the average net primary productivity of vegetation of the land-use type i in China, and NPP_{bi} is the net primary productivity of vegetation of the land-use type i in the Beiluo River Basin.

The ESV of food production in the cultivated land ecosystem per unit area is a standard equivalent factor, and the ESV is equivalent to 1/7 of the average grain yield market value in the study area. The calculation method is as follows:

$$E_c = \frac{1}{7}\sum_{i}^{n} \frac{s_i \times p_i \times q_i}{S} \tag{6}$$

where E_c represents the value of the food production service function per unit of cultivated land ecosystem (USD/hm^2), i is the crop type (the main food crops in the Beiluo River Basin are wheat, corn, millet, potatoes, and vegetables), S_i is the area of food crop i (hm^2), P_i is the average price of food crop i (USD/hm^2), and q_i is the per-unit yield of food crop i (t). S is the total area of food crops (hm^2).

As the market grain price is significantly affected by monetary value, we used the average grain price and grain output of Yan'an, Tongchuan, Yulin, Weinan, and Qingyang in 2015. According to data from the Municipal Bureau of Statistics, the average grain price in the Beiluo River Basin for 2015 was CNY 2.31/kg and the average grain output was CNY 3.11 (t/hm^2), which was substituted into Formula 5. From the average U.S. dollar exchange rate in the interbank foreign exchange market of CNY 1 to USD 0.1541 on 31 December 2015, the equivalent factor of the ESV in the Beiluo River Basin was calculated to be 158.71 USD/hm^2.

The ESV of the Beiluo River basin was calculated according to the area of various land-use types and the value of the unit equivalent factor in the study area. The specific formula is as follows:

$$VC_{ij} = E_c \times C_{ij} \times \eta_i \quad (7)$$

$$ESV = \sum_{i}^{n}\sum_{j}^{n} A_i \times VC_{ij} \quad (8)$$

where VC_{ij} is the unit ESV of service type j of the land-use type i, in USD/hm^2, and C_{ij} is the equivalent value of the ecosystem service type j of the land-use type i proposed by Xie Gaodi. A_i is the area of the land-use type i in the study area.

Table 2. The equivalent of various services in various land-use ecosystems (Xie et al., 2003, 2015 [45,48]).

First-Level Service	Second-Level Service	1 Forestland	2 Shrub	3 Grassland	4 Cultivated Land	5 Desert	6 Water Area
Supplying services (SuyS)	Food production	0.33	0.20	0.43	1.0	0.02	0.445
	Raw material	2.98	1.80	0.36	0.39	0.04	0.295
Regulating services (RegS)	Gas regulation	4.32	2.59	1.5	0.72	0.06	1.46
	Climate regulation	4.07	2.45	1.56	0.97	0.13	7.805
	Hydrological regulation	4.09	3.90	1.52	0.77	0.07	16.105
	Waste treatment	1.72	1.11	1.32	1.39	0.26	14.625
Supporting services (SutS)	Soil formation and retention	4.02	2.42	2.24	1.47	0.17	1.20
	Biodiversity protection	4.51	2.72	1.87	1.02	0.40	3.56
Cultural services (CulS)	Recreation and culture	2.08	3.44	0.87	0.17	0.24	4.565
Total		28.12	20.64	11.67	7.90	1.39	50.06

2.3.4. Hot Spot and Spatial Auto Correlation Analysis of ESV Change

Moran's I is a kind of index that can be used to test the autocorrelation of spatial data [49]. In this paper, GeoDa (1.20, Dr. Luc Anselin et al.) software was used to analyze the spatial autocorrelation of ESV changes in four geomorphic types in the Beiluo River Basin from 1975 to 2015 under the background of ER through the local Moran's I. The range of Moran's I is [−1, 1]. When it is greater than 0, it means that the region is a spatial aggregation of similar attributes; that is, the high (low) value of the region is also surrounded by high (low) values. When it is less than 0, it means that the region is a spatial aggregation of non-similar attributes; that is, the high (low) value of the region is surrounded by a low (high) value. When it is equal to 0, it means that there is no spatial association between the region and the neighborhood. A LISA (local indicators of spatial association) cluster map and z LISA significance map are drawn to determine whether the local correlation types of each region and their clustering regions are statistically significant.

3. Results

3.1. *Spatial and Temporal Evolution of Land-Use Types from 1975 to 2015*

From 1975 to 2015, the area of forest land, construction land, and desert in the Beiluo River Basin increased by 18.27%, 1.09%, and 0.32%, respectively, while the area of shrubland, grassland, cultivated land, and water areas decreased by 1.03%, 0.16%, 18.23%, and 0.26%, respectively. The main LUTs in different landform areas exhibit obvious differences. The area of different land-use types in each region is shown in Figure 2.

Grassland in the LHG region accounted for more than 50% at all five time points, reaching 76.96% in 2015, constituting the main land-use type in the region. From 2000 to 2010, the area of forestland increased by 7.31%, whereas cultivated land decreased by 17.27%. Furthermore, from 2010 to 2015, cultivated land experienced a further decrease of 17.52%, while scrub increased by 6.82%.

Before 2000, the main land-use types in the LPG region were forestland and cultivated land. However, there have been significant changes in land-use types since then. From 2000 to 2010, the area of forestland increased by 31.03%. This trend shifted again from 2010 to 2015, with forestland decreasing by 5.52%. During the same period, scrub and grassland saw increases of 6.69% and 10.12%, respectively, while cultivated land experienced a decrease of 11.95%.

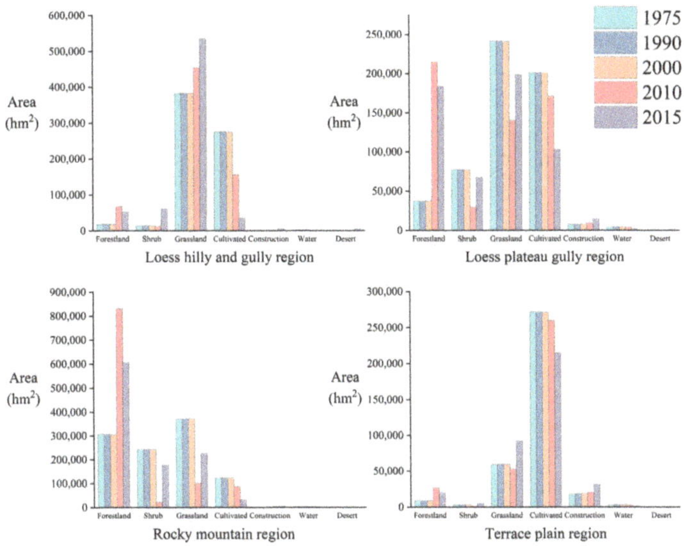

Figure 2. Areas of different land-use types in the Beiluo River Basin between 1975 and 2015.

Before 2000, the total area of forestland, scrub, and grassland accounted for about 85% of the total area in the RM region. However, by 2010, there had been an increase in the proportion of forestland to 79.29%, while scrub and grassland decreased to 2.26% and 9.63%, respectively. By 2015, there had been a decrease in forestland to 57.74%, while scrub and grassland increased to 16.9% and 21.6%, respectively. The area of cultivated land in this region is relatively small, and continued to decrease after 2000.

The TP region primarily consists of cultivated land, which accounted for more than 70% of land use between 1975 and 2010. However, in 2015, this proportion decreased to 58.59%. In contrast, the areas of forestland, grassland, and construction land all increased to some degree after 2000.

During the four periods after 1975, LUCC occurred in 0.07%, 2.47%, 36.46%, and 33.05% of the area of the whole basin, respectively. In short, there was no large-scale land-use change in the four geomorphological areas before 2000, but drastic changes occurred after 2000.

3.2. Analysis of Land-Use Change Tupu Process between 1975 and 2015

The most common forms of land-use change in the Beiluo River Basin are the interconversion between forestland and scrub and forestland and grassland, as well as the reduction in cultivated land area. The Tupu analysis revealed that the largest area of conversion occurs between forestland (12, 13) and scrub or grassland, totaling 6480.88 km^2. Additionally, a substantial area, totaling 8433.12 km^2, was converted from scrub or grassland to forestland (21, 31), which exceeds the conversions of other Tupu units. The conversion from cultivated land to forestland, scrub, or grassland (41, 42, 43) amounts to 5017.46 km^2, with 3766.44 km^2 converted to grassland. Conversely, a total of 361.73 km^2 was converted from forestland, scrub, and grassland to cultivated land (14, 24, 34), indicating that there was a larger area of cultivated land being transformed into other types of land use compared with the opposite direction. Figure 3 illustrates the direction of the land-use type conversion across the basin in each landform type area.

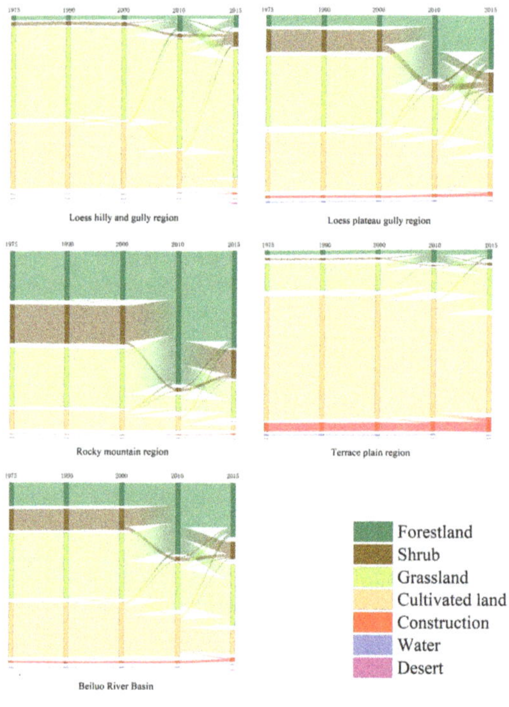

Figure 3. The direction of LUTs in the Beiluo River Basin between 1975 and 2015.

Land-Use Change Tupu Process between 1975 and 2015

Between 1975 and 1990, in the LHG region and in the LPG region, the Tupu unit of "water area → grassland" (63) is prominent, accounting for 73.15% and 21.65%, respectively. At the same time, the areas of 14 and 34 Tupu units in the LPG region and the RM region, respectively, are relatively large, and the inflow of cultivated land is larger than the outflow. In addition, there was a large proportion of "cultivated land → construction land" (45) Tupu units in the LHG region, RM region, and TP region. The number of LUT units in this period was relatively small, and they were mainly concentrated on the expansion of grassland, cultivated land, and construction land with low spatial isolation, indicating that these changes were more spatially concentrated.

From 1990 to 2000, the mutual transformation between grassland and cultivated land was the major type of land-use change, and their proportions are roughly the same. Their total proportions in the LHG region, LPG region, RM region, and TP region are 88.15%,

48.49%, 21.35%, and 50.29%, respectively. Units 34 and 43 showed low levels of spatial separation in the four zones (Figure 5).

The period from 2000 to 2010 was mainly characterized by the transformation of cultivated land into grassland, scrub, and woodland and grassland to woodland and shrub (43, 42, 41, 32, 31). The proportion of "cultivated land → grassland" in the LHG region was 63.87%. The proportion of "grassland → forestland" in the LPG region was 55.35%. The conversion of "grassland and scrub → forestland" was 51.70% and 40.37%, respectively, in the RM region. The Tupu units of "grassland → forestland" and "cultivated land → grassland" in the TP region accounted for 41.96% and 29.00%, respectively. The lowest levels of spatial isolation during this period were 43 in the LHG region; 31 and 41 in the LPG region; 21 and 31 in the RM region; and 43, 45, and 31 in the TP region.

From 2010 to 2015, the main types of Tupu were still the conversion of "cultivated land → grassland, scrub, and forestland" and "grassland → forestland and scrub" (43, 42, 41, 32, 31). Additionally, the conversion of "forestland → shrub"," grassland → shrub", and "forestland → grassland" continued to be present in the four geomorphological regions. In the LHG region, the largest proportion of the total area (41.86%) was "cultivated land → grassland", followed by "grassland → forestland and scrub" (25.15%) and "forestland → grassland" (18.34%). In the LPG region, "cultivated land → forest land" and "cultivated land → grassland" accounted for 17.11% and 9.97%, respectively, while "grassland → forestland" accounted for 13.99%. The conversion of forestland to grassland and scrub accounted for 27.98% and 14.51%, respectively. In the RM region, a significant amount of forestland was converted to scrub (42.45%) and grassland (32.41%). The conversion of "cultivated land → grassland and forestland" accounted for only 9.05% and 5.05%, respectively. In the TP region, "cultivated land → grassland" was still the main component, accounting for 40.89%. However, there was also some conversion of forestland to grassland, accounting for 19.93%. Similar to the previous two phases, the spatial isolation of the Tupu units involving 31, 32, and 43 was low in all four regions. Additionally, the spatial isolation of Tupu units of types 12 and 13 also showed low levels for this period.

3.3. *Land-Use Change Tupu Patterns between 1975 and 2015*

Land-use change activity was infrequent between 1975 and 2000, as shown in Figures 3 and 4. However, it increased rapidly after 2000. Accordingly, the period between 1975 and 2000 can be referred to as the prophase, the period between 2000 and 2010 as the middle phase, and the period between 2010 and 2015 as the anaphase. To examine the patterns of land-use change Tupu from 1975 to 2015, the top 90% of land-use change pathways were statistically analyzed, as shown in Table 3. The spatial distribution of various Tupu patterns is shown in Figures 5 and 6.

Table 3. Top 90% of land-use Tupu patterns in each region.

	Prophase Transition Type	Middle Transition Type	Anaphase Transition Type	Repeated Transition Type	Continuous Transition Type
LHG region	44333 (0.90%)	44433 (23.51%) 33311 (2.21%)	44443 (27.51%) 33332 (6.83%) 33331 (5.88%) 11113 (3.60%) 44442 (2.58%) 44441 (1.62%)	33313 (5.75%) 33433 (2.04%) 44343 (1.03%)	44413 (2.24%) 44431 (2.14%) 44432 (1.80%) 33312 (0.95%)
LPG region	—	33311 (18.53%) 22211 (10.01%) 44433 (4.10%) 44422 (0.84%)	44443 (10.13%) 33331 (7.23%) 44441 (5.90%) 11113 (3.19%) 33332 (2.37%) 22221 (2.02%) 44442 (1.84%) 11112 (1.27%) 44445 (0.97%) 66663 (0.82%)	33313 (10.08%) 22212 (1.78%)	33312 (5.50%) 22213 (2.79%) 44431 (1.10%)

Table 3. *Cont.*

	Prophase Transition Type	Middle Transition Type	Anaphase Transition Type	Repeated Transition Type	Continuous Transition Type
RM region	——	33311 (22.56%) 22211 (22.55%) 44433 (1.56%)	11112 (5.96%) 11113 (4.86%) 44443 (4.36%) 44441 (2.39%) 33331 (1.94%)	33313 (7.64%) 22212 (5.50%)	33312 (9.00%) 22213 (2.61%)
TP region		44433 (8.69%) 33311 (5.60%) 44455 (1.67%) 22211 (1.39%)	44443 (29.98%) 44445 (9.25%) 44441 (5.01%) 11113 (4.72%) 33331 (3.76%) 66665 (1.79%) 44446 (1.08%) 33332 (0.88%) 33335 (0.88%) 44442 (0.74%)	33313 (8.69%) 33433 (1.58%) 55455 (0.83%) 44343 (0.75%) 44344 (0.63%)	33312 (1.46%) 44413 (0.69%)

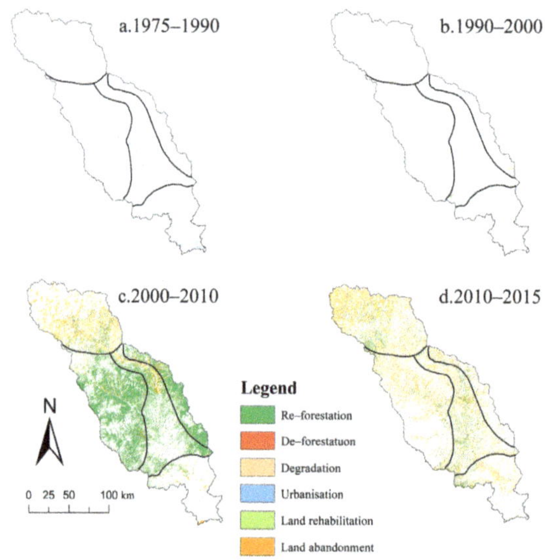

Figure 4. Spatial distribution of LUT features after classification in the Beiluo River Basin. Note: This figure is grouped according to the LUT features in four stages in the Beiluo River Basin, in which "Re–forestation" includes (21, 31, 31, 41, 51, 52, 61, 62, 71, 72, 73); "De–forestation" includes (14, 17); "Degradation" includes (12, 13, 16, 23, 26, 27, 36, 37, 46, 47, 57, 63, 67); "Urbanisation" includes (15, 25, 35, 45, 65, 75); "Land rehabilitation" includes (24, 34, 54, 64, 74); and "Land abandonment" includes (42, 43, 53).

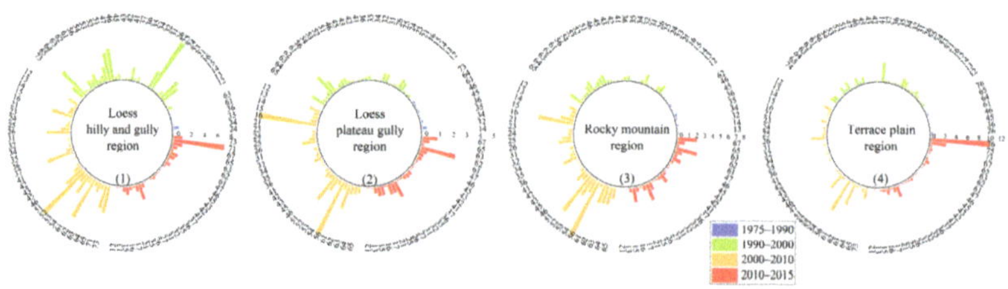

Figure 5. Degree of spatial separation of LUTs in the Beiluo River Basin from 1975 to 2015. Note: The unit of spatial separation is "100%"; because most of the data exceeded 100%, in order to simplify the picture, we simplified 100%, 200%, and so on, to 1, 2, and so on.

3.3.1. Prophase Transition Type

From 1975 to 2015, the prophase transition type accounted for the smallest proportion among the five transition types, with a total area of 9.02 km^2. This type was predominantly characterized by the transformation of "cultivated land → construction land" (45555), accounting for 53.13%, and it was mainly concentrated in the TP region. This indicated that, from 1975 to 2015, the land-use type did not often change after the transformation of "cultivated land → construction land" with the progress of urbanization.

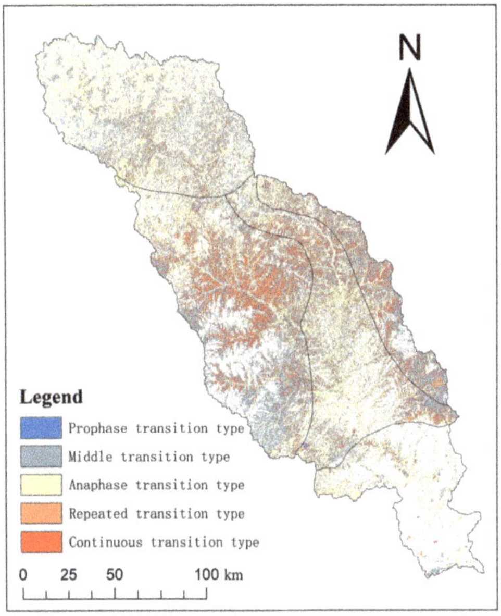

Figure 6. Land-use change Tupu patterns in the Beiluo River Basin from 1975 to 2015.

3.3.2. Middle Transition Type

The area of this transition type was larger than the previous type, with a wider spatial distribution, especially for "cultivated land → grassland" (44433), which not only occupied 23.51% and 8.69% of the LPG region and TP region, respectively, but also appeared in the top 90% of the other two landform types. "Grassland → forestland" (33311) is also a major component of this type, accounting for 18.53% and 22.56% of the RM region and the LPG region, respectively. In addition, there is a high proportion of "scrub → forestland" (22211) within these two landform types, with proportions of 10.01% and 22.55%, respectively.

3.3.3. Anaphase Transition Type

The anaphase accounts for the largest proportion and greatest diversity in the top 90% of the transition types in the other three areas, in addition to the RM region. The anaphase transition type is mainly composed of "cultivated land → forest, shrub, and grassland" and "grassland → forest and shrub". In the LHG region, "cultivated land → grassland" (44,443) accounts for 27.51%. Similarly, in the TP region, "cultivated land → grassland" (44,443) accounts for 29.98%. It is noteworthy that, although "cultivated land (grassland) → forest and shrub (grassland)" is the predominant type in all districts, there is also the existence of "forest (shrub) → grassland", particularly in the RM region. Here, the proportion of forest converted to shrub and grassland (11112, 11113) is 5.96% and 4.96%, respectively.

3.3.4. Repeated and Continuous Transition Types

There are 87 and 425 change paths in the repeated transition type and the continuous transition type, respectively. This is far more than for the other three transition types; the areas of these two types account for 13.25% and 16.09% of the five transition types, respectively. For the repeated transition type, the transformation from "grassland → forestland → grassland" (33313) was predominant in all four areas, followed by the repeated transformation of cultivated land and grassland (33433, 44343, 44344). The transition from "scrub → forestland → scrub" (22212) was observed in both the LPG region and the RM region, accounting for 1.78% and 5.50%, respectively. The continuous transition type is dominated by the transition of "cultivated land → grassland → forestland, cultivated land → grassland → scrub, cultivated land → forestland → grassland, and grassland → forestland → scrub" (44431, 44432, 44413, and 33312). Additionally, within the RM region, there was also the occurrence of "scrub → forestland → grassland" (22213), accounting for 5.50% of the pattern.

3.4. The Influence of LUCC on ESV

3.4.1. Changes in ESV

As shown in Figures 7 and 8, overall, the ESV in the Beiluo River Basin shows an upward trend year by year, increasing by USD 3.209 billion, or by 54.16%, from 1975 to 2015. For first-level ecosystem services, the largest proportion of the ESV was provided by regulating services, followed by supporting services and supplying services, with cultural services accounting for the smallest proportion.

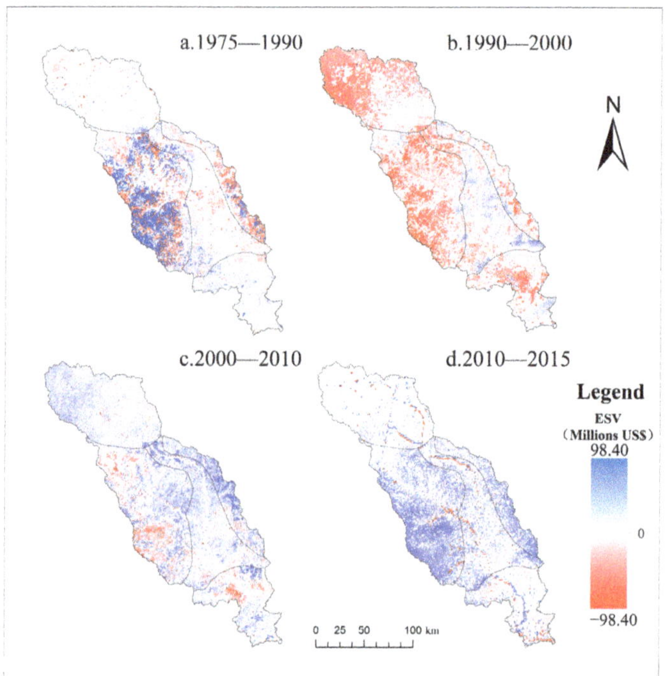

Figure 7. Changes in the ESV in the Beiluo River Basin from 1975 to 2015.

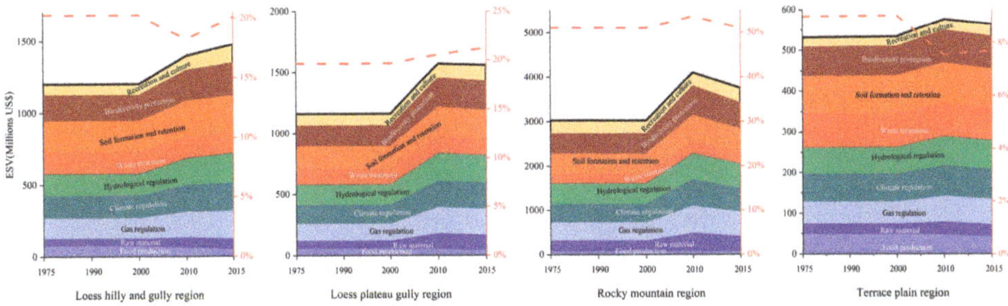

Figure 8. ESV of secondary services in various geomorphological types of the Beiluo River Basin from 1975 to 2015.

The ESV in the four geomorphic regions showed an upward trend, and the LHG region's ESV increased by USD 274.5 million from 1975 to 2015. The ESVs of regulating services, supporting services, and cultural services all increased by 23.1%, 21.86%, and 25.47%, respectively. The ESV of supplying services decreased by USD 12 million, or 1.50%. Specific to the secondary services, the ESVs of soil formation and retention increased by 16.89%; the ESVs of raw material supply, gas regulation, climate regulation, hydrological regulation, and biodiversity protection increased by more than 20%; and the ESVs of recreation and culture increased by 66.44%. The ESV for food production decreased by USD 24.79 million, or 34.63%.

Among the four geomorphological areas, only the LHG region showed a downward trend in ESV, with a total decrease of USD 1.23 million from 1975 to 2015. Among the nine services, there was only a small increase in the following ESV types: hydrological regulation, gas regulation, recreation and culture, and the supply of raw materials. The value of other services has decreased. The value of waste treatment services decreased the most, followed by food production and soil formation and retention. The reduction in the value of other services was not significant. Among them, food production services decreased by 54.83% and waste treatment services decreased by 31.38%.

The values of the nine services in the LPG region have increased significantly since 2000. Among them, the value of biodiversity protection increased the most, reaching USD 175 million, an increase of 67.22%. The second-largest increase is seen in soil formation and retention and gas regulation, with increases of 112.74% and 133.17%, respectively. The value of food production services also increased by USD 12 million, but only by 27.85%.

The ESV in the RM region increased the most, reaching USD 2.079 billion, or 61.54%. The value of all kinds of services has increased significantly. As with the gully region of the plateau, the growth is most obvious for biodiversity conservation, gas regulation, and hydrological regulation, with increases of 67.22%, 73.27%, and 59.64%, respectively. The value of food production services also increased by USD 4.32 million, but only by 5.38%.

The overall ESV of the TP region increased by USD 159 million, with the largest increases in value seen for soil formation and retention and biodiversity protection, with increases of 36.45% and 45.55%, respectively. The value of food production services also rose by USD 4 million, but only by 12.53%.

3.4.2. Response of ESV to LUCC

Figure 9 shows the transfer matrix of ESV changes caused by the transformation of land use types in different geomorphic regions in different periods. In the LHG region, grassland is the main contributor to ESV, accounting for 74.84% in 2015. The expansion of the grassland area has led to a 39.70% increase in the ESV in this region. This is mainly due to the conversion of 2126.83 km^2 of "cultivated land → grassland" between 2000 and 2015, resulting in an increase of USD 127 million in the ESV, as shown in Figure 8. The share of cultivated land remained at about 28.3% before 2010, but fell to 3.3% in 2015. The changes in 43, 41, 31, 13, and 32 are the main LUTs in the LHG region.

Before 2000, grassland was the biggest contributor to the ESV in the LPG region, accounting for 38.5%. The contributions to ESV are similar for shrub and cultivated land. Since 2000, forestland has become the biggest contributor to the ESV, accounting for 61.22%. This situation is mainly due to the emergence of a large number of type 31, 21, and 41 Tupu patterns in 2000 and 2010, resulting in a net increase in the forestland area of 1772.31 km^2. Specifically, the transfer of cultivated land led to a net increase of USD 47.23 million in ESV, while converting other land types to forestland increased the ESV by USD 40 million. Types 21, 31, 41, 13, and 12 are the biggest driving forces for the change in the overall ESV in the LPG region.

The most obvious changes in the Tupu patterns of the RM region from 2000 to 2010 were observed in 21 and 31, with changes in areas of 2241.25 km^2 and 2870.29 km^2, respectively. These changes accounted for 92.07% of the total modified area. Moreover, a significant amount of other land was converted into forestland, resulting in a net increase of USD 163 million in the ESV. This net growth accounted for 98.52% of the total net growth of the ESV. It is noteworthy that the biggest driving forces for the change in ESV in the RM region were 12, 13, 21, 31, and 41.

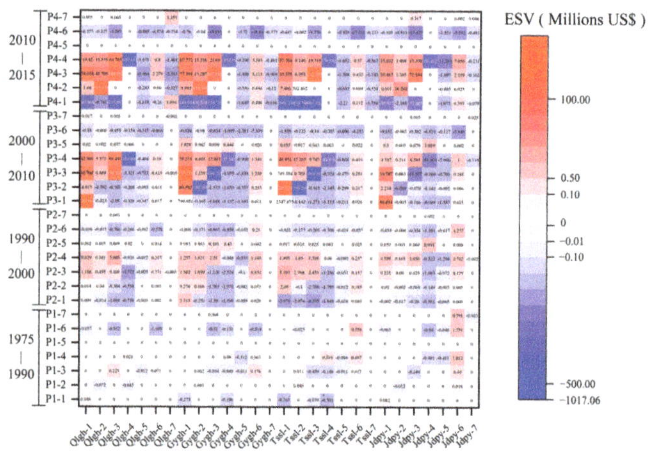

Figure 9. Ecosystem service value transfer matrix. Note: The abbreviations of "Qlgh, Gygh, Tssl, and Jdpy" in the figure are the abbreviations for the LHG region, the LPG region, the RM region, and the TP region, respectively. The numbers 1, 2, 3, 4, 5, 6, and 7 are the codes for forestland, shrub, grassland, cultivated land, construction land, water areas, and desert, respectively. P1, P2, P3, and P4 represent the four periods, in chronological order. The vertical axis represents the initial land-use type, the horizontal axis represents the land-use type at the telophase of the study period for the different regions, and the number corresponding to the horizontal and vertical coordinates is the change in ESV. For example, (Qlgh-4,P3-1) means that, after the grassland in the LHG region was turned into forestland in 2010, the ESV decreased by USD 3280.

The main contributor to the ESV in the TP area is cultivated land. However, its proportion decreased from 63.92% in 1975 to 47.75% in 2015. Significant changes were observed in the Tupu unit in this area from 2000 to 2010, with a shift leading to a net increase of USD 40 million in ESV. The most significant change from 2010 to 2015 was 43, leading to a net increase of USD 17 million in the ESV. Additionally, there has been a gradual intensification in the conversion of "cultivated land $\rightarrow$ construction land", resulting in a total of 89.88 km^2 of cultivated land being transformed into construction land within a span of 5 years. The driving forces behind the ESV change in the TP are units 31, 43, 41, and 13.

The main contributor to ESV in the LHG region is grassland, as a large amount of cultivated land has been transformed into grassland and forestland, which has led to an increase in ESV. In the anaphase, the conversion of some of the forestland into shrub and grassland has resulted in a decrease in the ESV. However, the increase in the grassland and shrub area has significantly improved the value of soil formation and retention services in this region. In the LPG region and the RM region, grassland and forestland are the main contributors to the ESV, with reasons 13 and 31 accounting for the changes in the ESV. In the RM region, the increase in the area of forestland, shrubland, and grassland has greatly improved the value of soil formation and retention, as well as the hydrological regulation service. Moreover, in the LPG region and the TP region, a significant amount of cultivated land has been converted into grassland and forestland, increasing the overall ESV of the land without any loss in the value of the food production service.

3.4.3. Identifying the Hot and Cold Spots of Changes in ESV

The results of the spatial autocorrelation analysis of ESV changes using GeoDa software are shown in Figure 10. The Moran's I (a) of ESV change in the Beiluo River Basin from 1975 to 2015 were all greater than 0, indicating the existence of autocorrelation of regional ESV changes, and the distribution of spatial attribute values is a positive correlation. There are four cluster types in the LISA cluster diagram: High–High, Low–Low, Low–High, and High–Low. According to the LISA significance map, the confidence of the High–High and Low–Low core is 99.9%, and the confidence of the edge is 99%. It can be considered that High–High and Low–Low are the hot and cold spots of ESV changes.

In the first stage, the distribution range of hot spots and cold spots is small, in which the hot spots were mainly concentrated in the RM region on the west side of the basin and the cold spots were distributed in the RM region on the east side of the basin. From 1990 to 2000, the hot spots were mainly concentrated in the whole LPG region and the southern part of the LHG region and the TP region, while the cold spots were distributed in the RM region on the west side of the basin and the northern part of the LHG region. From 2000 to 2010, the hot spots of the ESV were concentrated in the east of the basin and the north of the LHG region, while the cold spots were distributed in the west of the basin and the north of the TP region. From 2010 to 2015, hot spots were concentrated in the RM region and cold spots were concentrated in the LHG region and TP region. The distribution trend of hot and cold spots of ESV changes from 1975 to 2015 is similar to that from 2010 to 2015.

On the whole, the development sequence of hot spots is the west RM region → the LPG area and the south LHG region → the east RM region and the north LHG region → the east and west RM region.

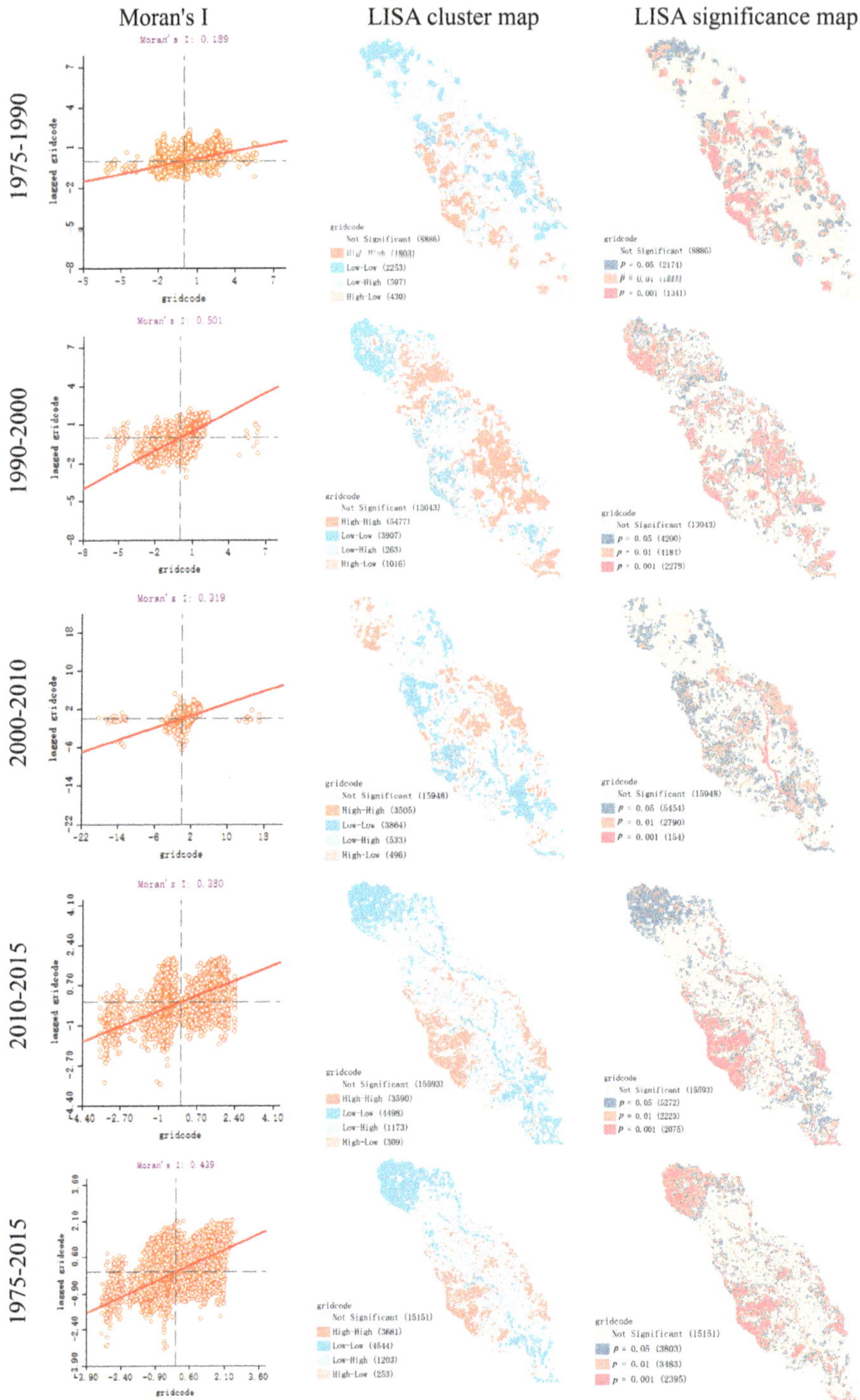

Figure 10. Moran's I and its LISA cluster map and LISA significance map of ESV changes.

4. Discussion

The quantitative expression and evaluation of LUTs and ecosystem service functions is an important research topic that guides ecological protection planning. From 1975 to 2015, the ESV in the Beiluo River Basin showed an overall upward trend, which is consistent with the findings of other researchers [11,21,50,51]. From 1975 to 1990, the ESV of the Beiluo River Basin increased by USD 1.082 billion, indicating that small-scale soil and water conservation engineering measures, such as small-scale check dams and the dry terraces built in villages and towns in this area, played a positive role in the ER of the Loess Plateau [52,53]. From 1990 to 2000, the ER measures of the Loess Plateau mainly focused on the comprehensive management of small watersheds, which were more systematic and scientific [54,55]. From 2000 to 2015, the Loess Plateau began a large-scale "Grain for Green" project. A large number of sloped cultivated lands greater than 25° were converted into forestland or grassland [56]. Most of the ESVs in the Beiluo River Basin increased during this period. The project of returning cultivated land to forest and grassland is currently the largest ER measure in developing countries. While improving the regional ecological environment, such projects have also caused drastic changes in the land-use structure.

In the Tupu pattern, the proportion of the prophase transition type is relatively low in the four regions. That is, engineering interventions are secondary and natural recovery is the main focus [57,58]. Since 2000, land use has undergone significant changes that are mainly attributable to the Tupu patterns of 43, 41, 32, and 31. Since 2000, significant changes have taken place in land use, mainly caused by 44433, 33311, 22211, and 44443 in the Tupu patterns of middle and anaphase transition types. It is also these types of Tupu patterns that have become the backbone of improving the environment of the Beiluo River Basin and improving the main ESs of the four typical geomorphological types.

The ER of the Beiluo River Basin provides basic conditions for its ESs. However, in order to achieve sustainable ER, the coupling of social, economic, and environmental processes needs to be taken into account, as well as the attitudes and strategies of stakeholders [59]. In some of the Tupu patterns, such as the repeated transition type and continuous transition type, we found that there are resource-wasting Tupu units, such as 11112, 33313, 33433, 22212, 44413, and 33312. The emergence of these frequent types of LUCC and ecological degradation indicates the existence of policy discontinuity in the region and the game phenomenon between small-scale stakeholders and policies. After the conversion of cultivated land to forest land, the income of farmers suffered losses. In order to maintain the original living standards, some farmers redeveloped the forest land in the restoration area, and the ER benefit could not achieve sustainable development [60]. The conversion of "cultivated land → grassland and forestland" leads to the loss of supplying services and the enhancement of regulating services, which represents a trade-off between supplying services and regulating services, which can be directly confirmed in the LHG region and RM region in this study. Although the supplying services and regulating services in the LPG region and TP region increased at the same time, there was a big difference in the growth rate of the two regions.

The TP region is the main source of the food production service in the Beiluo River Basin, followed by the LPG region. The service value of food production in the LPG region and the TP region still maintains a positive growth trend, despite the decreasing area of cultivated land [61]. There are several reasons for this: (1) The average elevation and slope of cultivated land have decreased, and most of the cultivated land converted to forest land is at a slope above 25°, so the situation of soil erosion is serious, soil fertility is insufficient, and food production is not high [30]. (2) After the "Grain for Green" project, the implementation of agricultural engineering improved the sustainability of cultivated land ecosystems. In conclusion, the optimization of cultivated land's spatial pattern improves the environment, and ecological security plays a positive role in promoting food security [20,61,62]. Therefore, in the process of formulating and implementing land-use policy and environmental protection policy, it is necessary to deal with the trade-off between various services. Integrating an ecosystem service assessment system into the formulation and implementation of policies and plans, quantifying the impact of policies and plans, and adopting multiple compensation mechanisms can help to coordinate the relationship between various services and various stakeholders, so as to make land use more reasonable and social development more sustainable [63–66].

In order to improve the comprehensive benefits of ER in different landforms, it is necessary to strengthen the control of grassland and shrub in the LHG region, strengthen the control of forests in the RM region, strengthen the control of cultivated land in the TP region and the LPG region, and establish a management system to avoid repeated changes or destruction of land use. In addition, the work of the "Grain for Green" project in different landforms requires a scientific assessment of the losses of landless farmers and the implementation of compensation for farmers. Xie et al. once suggested that China should increase its investment in ecological compensation, enrich and diversify ecological compensation tools, and establish and improve the institutional system of ecological compensation [48].

It should be noted that there are still some shortcomings to this research, detailed as follows. (1) This paper uses the average grain output and average grain market price of the Beiluo River Basin in 2015 to calculate the value of the unit equivalent factor. The ESV in 1975, 1990, 2000, and 2010 was found to be on the high side owing to the influence of currency depreciation and the rising output per unit of cultivated land. (2) Similar to many other studies, this paper assumes that the ESV of construction land is zero [67–69]. However, it is noteworthy that, today, many cities and villages are increasingly prioritizing green environments and human settlement experiences. Hence, this assumption may no longer accurately reflect the actual situation. For instance, the construction land ecosystem also serves the functions of absorbing carbon dioxide, purifying the air, and providing entertainment opportunities, to some extent [70–73]. (3) As this study focuses solely on LUCC analysis, it does not consider the impact of hidden land-use changes on ESs, such as quality, property rights, management modes, input, output, and function [74]. Hence, it is impossible to comprehensively analyze the changes in ESs. Consequently, exploring the influence of hidden LUTs on ESs constitutes a valuable future research direction.

5. Conclusions

In this study, the spatio-temporal characteristics of LUTs in the Beiluo River Basin from 1975 to 2015 were analyzed using the GT method, the ESV was assessed using the revised benefit transfer method, and the experience of ER measures applied in four typical geomorphic regions of the Loess Plateau was summarized.

The results indicate that, from 1975 to 2015, the proportion of forestland in the Beiluo River Basin increased by 18.27%, while the areas of shrub, grassland, cultivated land, and water decreased by 1.03%, 0.16%, 18.23%, and 0.26%, respectively. In the past 40 years, the overall ESV of the basin increased by USD 3.209 billion (54.16%).

In the Tupu pattern of middle transition and anaphase transition types, the transformations of "cultivated land → forestland, shrub, and grassland" and "shrub and grassland → forestland" (44433, 44443, 44441, 33311, 33331, 33332) are key to enhancing the main services of the four typical geomorphic regions. The transition of "cultivated land → grassland" enhanced the water and soil conservation services in the LHG region. The transition of "grassland/shrub → forestland" improved the regulating and support services in the LPG region and RM region. Since 2000, a large amount of cultivated land has been converted to forestland, shrub, or grassland, which has significantly enhanced the regulating and support services of the ecosystem, but has resulted in losses in the supplying services. However, in the TP region and LPG region, the value of the supplying service showed a slight increase, owing to the increase in grain yield per unit area.

The emergence of repeated transition and continuous transition Tupu patterns, such as "grassland → forestland → grassland" and "cultivated land → forestland → grassland", indicates that there was a discontinuity between policy formulation and implementation, and there were game problems between farmers and policy. To achieve sustainable development, it is necessary to strengthen the ecological protection of the areas that have completed ER, put ecological compensation schemes in place, prevent repeated damage, evaluate the unrepaired areas, and identify reasonable areas that need ecological restoration to promote targeted precise ecological restoration.

Author Contributions: Conceptualization, J.X. and J.Z.; Data curation, J.X. and J.Z.; Formal analysis, J.X. and J.Z.; Funding acquisition, J.Z.; Investigation, J.X. and J.Z.; Methodology, J.Z.; Project administration, J.Z.; Resources, J.Z.; Software, J.X., J.W. and S.N.; Supervision, J.Z.; Validation, J.Z.; Visualization, J.X.; Writing—original draft, J.X.; Writing—review and editing, J.X., M.Q., M.L. and J.Z. All authors have read and agreed to the published version of the manuscript.

Funding: This research was funded by the China University of Geosciences (Beijing) University Student Innovation and Entrepreneurship Training Program, grant number S202311415124; the Inner Mongolia Science and Technology Major Project, grant number 2020ZD0020; and the National Natural Science Foundation of China, grant number 41701207.

Data Availability Statement: The data presented in this study are available upon request from the corresponding author.

Acknowledgments: We would like to express our respect and gratitude to the anonymous reviewers and editors for their professional comments and suggestions.

Conflicts of Interest: The authors declare no conflict of interest.

References

1. Gann, G.D.; McDonald, T.; Walder, B.; Aronson, J.; Nelson, C.R.; Jonson, J.; Hallett, J.G.; Eisenberg, C.; Guariguata, M.R.; Liu, J.; et al. International principles and standards for the practice of ecological restoration. *Restor. Ecol.* **2019**, *27*, S1–S46. [CrossRef]
2. Harris, J.; Hobbs, R.; Higgs, E.; Aronson, J. Ecological Restoration and Global Climate Change. *Restor. Ecol.* **2006**, *14*, 170–172. [CrossRef]
3. Nations, U. United Nations Decade on Ecosystem Restoration (2021~2030) [EB/OL]. Available online: https://undocs.org/A/RES/73/284 (accessed on 10 July 2023).
4. Congress, I.W.C. The Marseille Manifesto [EB/OL]. Available online: https://www.iucncongress2020.org/programme/marseille-manifesto (accessed on 10 July 2023).
5. Benayas, J.M.R.; Newton, A.C.; Diaz, A.; Bullock, J.M. Enhancement of Biodiversity and Ecosystem Services by Ecological Restoration: A Meta-Analysis. *Sci. Rep.* **2009**, *325*, 1121–1124. [CrossRef] [PubMed]
6. Feng, X.; Fu, B.; Lu, N.; Zeng, Y.; Wu, B. How ecological restoration alters ecosystem services: An analysis of carbon sequestration in China's Loess Plateau. *Sci. Rep.* **2013**, *3*, 2846. [CrossRef] [PubMed]
7. Carlucci, M.B.; Brancalion, P.H.S.; Rodrigues, R.R.; Loyola, R.; Cianciaruso, M.V. Functional traits and ecosystem services in ecological restoration. *Restor. Ecol.* **2020**, *28*, 1372–1383. [CrossRef]
8. Lawler, J.J.; Lewis, D.J.; Nelson, E.; Plantinga, A.J.; Polasky, S.; Withey, J.C.; Helmers, D.P.; Martinuzzi, S.; Pennington, D.; Radeloff, V.C. Projected land-use change impacts on ecosystem services in the United States. *Proc. Natl. Acad. Sci. USA* **2014**, *111*, 7492–7497. [CrossRef]
9. Kalogirou, S. Expert systems and GIS: An application of land suitability evaluation. *Comput. Environ. Urban Syst.* **2002**, *26*, 89–112. [CrossRef]
10. Seppelt, R.; Lautenbach, S.; Volk, M. Identifying trade-offs between ecosystem services, land use, and biodiversity: A plea for combining scenario analysis and optimization on different spatial scales. *Curr. Opin. Environ. Sustain.* **2013**, *5*, 458–463. [CrossRef]
11. Jia, X.; Fu, B.; Feng, X.; Hou, G.; Liu, Y.; Wang, X. The tradeoff and synergy between ecosystem services in the Grain-for-Green areas in Northern Shaanxi, China. *Ecol. Indic.* **2014**, *43*, 103–113. [CrossRef]
12. Zhang, J.; Gao, G.; Fu, B.-J.; Gupta, H. Investigation of the relationship between precipitation extremes and sediment discharge production under extensive land cover change in the Chinese Loess Plateau. *Geomorphology* **2020**, *361*, 107176. [CrossRef]
13. Liu, L.; Zhang, H.; Li, F.; Chen, X. Research progress and prospect of soil and water conservation measures in the Loess Plateau. In Proceedings of the 5th International Conference on Traffic Engineering and Transportation System, ICTETS 2021, Chongqing, China, 24–26 September 2021; Academic Exchange Information Center (AEIC): Guangzhou, China, 2021.
14. Ning, Z.; Gao, G.Y.; Fu, B.J. Changes in streamflow and sediment load in the catchments of the Loess Plateau, China: A review. *Acta Ecol. Sin.* **2020**, *40*, 2–9. [CrossRef]
15. Lü, Y.; Fu, B.; Feng, X.; Zeng, Y.; Liu, Y.; Chang, R.; Sun, G.; Wu, B. A Policy-Driven Large Scale Ecological Restoration: Quantifying Ecosystem Services Changes in the Loess Plateau of China. *PLoS ONE* **2012**, *7*, e31782. [CrossRef] [PubMed]
16. Chen, Y.; Bai, L.; Jiao, J.; Wang, N.; Li, J.; Xu, Q.; Yan, X.; Tang, B. Recognition of suitable small watersheds for check dam construction on the Loess Plateau. *Land Degrad. Dev.* **2023**, *34*, 4441–4455. [CrossRef]

17. Li, Y.; Li, Y.; Fang, B.; Wang, Q.; Chen, Z. Impacts of ecological programs on land use and ecosystem services since the 1980s: A case-study of a typical catchment on the Loess Plateau, China. *Land Degrad. Dev.* **2022**, *33*, 3271–3282. [CrossRef]
18. Liu, B.; Jia, X.; Shao, M.a.; Jia, Y. Assessing soil water recovery after converting planted shrubs and grass to natural grass in the northern Loess Plateau of China. *Agric. Water Manag.* **2022**, *264*, 0378–3774. [CrossRef]
19. Yunlong, W.D.C. Evaluation of Ecological Restoration Effects in China: A Review. *Prog. Geogr.* **2009**, *28*, 622–628.
20. Zexing, S.; Wenyi, L.; Jiamin, L. Evaluation of comprehensive benefit for ecological restoration in Shannxi Province. *Acta Ecol. Sin.* **2022**, *42*, 2718–2729.
21. Niu, L.; Shao, Q.; Ning, J.; Yang, X.; Liu, S.; Liu, G.; Zhang, X.; Huang, H. Evaluation on the degree and potential of ecological restoration in Loess Plateau. *J. Nat. Resour.* **2023**, *38*, 779–794. [CrossRef]
22. Fan, C.Z., Liang, H.F.; Haijie, Y.I.; Xiaoming, X.U.; Zou, Y.; He, J.; Lu, D.; Zhang, X. Hydrologic and sediment responses to ecological restoration in different geomorphological and vegetation type areas in past 70 years. *Acta Ecol. Sin.* **2023**, *43*, 3247–3260.
23. Wang, Z.Z.; Wang, H.; Feng, X.M.; Wang, X.; Zhang, L.; Fu, B. Evaluation index system of comprehensive benefits of ecological restoration in key ecologically vulnerable regions. *Acta Ecol. Sin.* **2019**, *39*, 7356–7366. [CrossRef]
24. Zhao, Y.-H.; Zhang, L.-L.; Wang, X.-F. Assessment and spatiotemporal difference of ecosystem services value in Shaanxi Province. *Minist. Educ. Key Lab.* **2011**, *22*, 2662–2672.
25. Tian, Q.L.; Xu, X.M.; Lyu, D.; Wang, H.J.; Lei, S.Y.; Yi, H.J.; He, J.; He, L.; Xue, F.; Zhou, Y.D. Relationship between geographical pattern of plant diversity and environmental factors in Beiluo River Basin. *China Environ. Sci.* **2021**, *41*, 4378–4387. [CrossRef]
26. Ran, D.C. Water and Sediment Variation and Ecological Protection Measures in the Middle Reach of the Yellow River. *Resour. Sci.* **2006**, *28*, 93–101.
27. Qin, W.; Zhu, Q.-K.; Liu, G.-Q.; Zhang, Y. Regulation effects of runoff and sediment of ecological conservation in the upper reaches of Beiluo River. *Shuili Xuebao/J. Hydraul. Eng.* **2010**, *41*, 1325–1332.
28. Zhang, J.; Gao, G.; Li, Z.; Fu, B.; Gupta, H.V. Identification of climate variables dominating streamflow generation and quantification of streamflow decline in the Loess Plateau, China. *Sci. Total Environ.* **2020**, *722*, 137935. [CrossRef] [PubMed]
29. Xue, F.; Zhang, X.; Zhang, L.; Liu, B.; Yang, Q.; Yi, H.; He, L.; Zou, Y.; He, J.; Xu, X.; et al. Attribution recognition of streamflow and sediment changes based on the Budyko hypothesis and fractal theory: A case study in the Beiluo River Basin. *Dili Xuebao/Acta Geogr. Sin.* **2022**, *77*, 79–92. [CrossRef]
30. Yan, R.; Zhang, X.-P.; Yan, S.-J.; Zhao, W.-H. Topographical distribution characteristics of vegetation restoration in the Beiluo River Basin from 1995 to 2014. *Dongbei Daxue Xuebao/J. Northeast. Univ.* **2016**, *37*, 1598–1603. [CrossRef]
31. Liu, C. *Research Theory and Practice in Hydology and Water Resources*; Science Press: Beijing, China, 2004; p. 716.
32. Chen, N.; Ma, T.; Zhang, X. Responses of soil erosion processes to land cover changes in the Loess Plateau of China: A case study on the Beiluo River basin. *Catena* **2016**, *136*, 118–127. [CrossRef]
33. Zhang, J.; Zhang, X.; Li, R.; Chen, L.; Lin, P. Did streamflow or suspended sediment concentration changes reduce sediment load in the middle reaches of the Yellow River? *J. Hydrol.* **2017**, *546*, 357–369. [CrossRef]
34. Liu, J.; Kuang, W.; Zhang, Z.; Xu, X.; Qin, Y.; Ning, J.; Zhou, W.; Zhang, S.; Li, R.; Yan, C. Spatiotemporal characteristics, patterns, and causes of land-use changes in China since the late 1980s. *J. Geogr. Sci.* **2014**, *24*, 195–210. [CrossRef]
35. Liu, J.Y.; Deng, X.Z.; Liu, M.L.; Zhang, S.W. Study on the spatial patterns of land-use change and analyses of driving forces in northeastern China during 1990–2000. *Chin. Geogr. Sci.* **2002**, *12*, 299–308. [CrossRef]
36. Liu, J.; Zhang, Z.; Xu, X.; Kuang, W.; Zhou, W.; Zhang, S.; Li, R.; Yan, C.; Yu, D.; Wu, S.; et al. Spatial patterns and driving forces of land use change in China during the early 21st century. *J. Geogr. Sci.* **2010**, *20*, 483–494. [CrossRef]
37. Ning, J.; Liu, J.; Kuang, W.; Xu, X.; Zhang, S.; Yan, C.; Li, R.; Wu, S.; Hu, Y.; Du, G.; et al. Spatiotemporal patterns and characteristics of land-use change in China during 2010–2015. *J. Geogr. Sci.* **2018**, *28*, 547–562. [CrossRef]
38. Chen, P.F. Monthly NPP Dataset Covering China's Terrestrial Ecosystems at North of 18° N (1985–2015). *J. Glob. Chang. Data Discov.* **2019**, *3*, 34–41. [CrossRef]
39. Zhu, S.H.; Yan, Y.; Shao, H.; Li, C.F. The Responses of the Net Primary Productivity of the Dryland Ecosystems in Central Asia to the CO_2 and Climate Changes during the Past 35 Years. *J. Nat. Resour.* **2017**, *32*, 1844–1856. [CrossRef]
40. Donghai, L.; Bin, A.; Xia, L.; Qiusheng, W.; Xiaoguang, X. Influence of land-use changes on soil erosion based on geo-information Tupu theory in Zhujiang Delta. In Proceedings of the Geoinformatics 2008 and Joint Conference on GIS and Built Environment: Monitoring and Assessment of Natural Resources and Environments, Guangzhou, China, 28–29 June 2008.
41. Yu, H.; Li, J.; Zhang, C.; Yuan, X.; Huang, Y.; Liu, X. Analysis on spatial and temporal change of land use based on geo-information Tupu. *Sci. Soil Water Conserv.* **2022**, *20*, 109–117. [CrossRef]
42. Wang, J.; Shao, J.; Li, Y. Geo-Spectrum Based Analysis of Crop and Forest Land Use Change in the Recent 20 Years in the Three Gorges Reservoir Area. *J. Nat. Resour.* **2015**, *30*, 235–247. [CrossRef]
43. King, R.T. Wildlife and Man. *NY Conserv.* **1966**, *20*, 8–11.
44. Costanza, R.; d'Arge, R.; De Groot, R.; Farber, S.; Grasso, M.; Hannon, B.; Limburg, K.; Naeem, S.; O'neill, R.V.; Paruelo, J.; et al. The value of the world's ecosystem services and natural capital. *Nature* **1997**, *387*, 253–260. [CrossRef]
45. Xie, G.D.; Lu, C.X.; Leng, Y.F.; Zheng, D.U.; Li, S.C. Ecological assets valuation of the Tibetan Plateau. *J. Nat. Resour.* **2003**, *18*, 189–196.
46. Vihervaara, P.; Kumpula, T.; Tanskanen, A.; Burkhard, B. Ecosystem services–A tool for sustainable management of human–environment systems. Case study Finnish Forest Lapland. *Ecol. Complex.* **2010**, *7*, 410–420. [CrossRef]

47. Zeng, C.; Li, Y.; Duan, X.; Xu, Y. Assessment and Driving Force Analysis of Ecosystem Service 2022.04.014Value in the Urban Agglomeration Along the Middle Reaches of the Yangtze River. *Res. Soil Water Conserv.* **2022**, *29*, 362–371. [CrossRef]
48. Ord, J.K.; Getis, A. Local spatial autocorrelation statistics: Distributional issues and an application. *Geogr. Anal.* **1995**, *27*, 286–306. [CrossRef]
49. Wu, Z.; Zhao, S.; Luo, Z.; Chai, B.; Wang, W.; Nie, K. Land Use Change and Its Ecosystem Service Value in Zunyi City. *Saf. Environ. Eng.* **2023**, *30*, 252–260. [CrossRef]
50. Zhao, G.; Mu, X.; Wen, Z.; Wang, F.; Gao, P. Soil erosion, conservation, and eco-environment changes in the loess plateau of china. *Land Degrad. Dev.* **2013**, *24*, 499–510. [CrossRef]
51. Liu, C.; Jia, X.; Ren, L.; Duan, G.; Shao, M.a. A preliminary investigation of water storage in check-dams across China's Loess Plateau using electrical resistivity tomography. *Hydrol. Process.* **2023**, *37*, 14826. [CrossRef]
52. Dong, H.; Song, Y.; Chen, L.; Liu, H.; Fu, X.; Xie, M. Soil erosion and human activities over the last 60 years revealed by magnetism, particle size and minerals of check dams sediments on the Chinese Loess Plateau. *Environ. Earth Sci.* **2022**, *81*, 162. [CrossRef]
53. Li, Z.S.; Yang, L.; Wang, G.L.; Hou, J.; Xin, Z.; Liu, G.; Fu, B. The management of soil and water conservation in the Loess Plateau of China: Present situations, problems, and counter-solutions. *Acta Ecol. Sin.* **2019**, *39*, 7398–7409. [CrossRef]
54. Gao, P.; Lei, T. Dynamic process simulation model for soil erosion of small-scale watershed system. *Nongye Gongcheng Xuebao/Trans. Chin. Soc. Agric. Eng.* **2010**, *26*, 45–50. [CrossRef]
55. National Forestry and Grassland Administration. Circular of the National Forestry and Grassland Administration on issuing the Technical regulations on the Design of the Operation of grain for green and the measures for Archives Management of grain for green. *Natl. Land Resour. Inf.* **2020**, *40–45*, 47. [CrossRef]
56. Yang, K.; Liu, G.; Wu, F.; Sun, B. Hydrological and environmental responses to comprehensive control of soil loss in a typical watershed of hill and gully region of the Loess Plateau. *Acta Ecol. Sin.* **2008**, *28*, 2042–2051.
57. Yuan, H.D.; Xin, Z.B.H.J. Models of soil and water conservation in the loess hilly region of China. *Acta Ecol. Sin.* **2021**, *41*, 6398–6416. [CrossRef]
58. Zhang, K.; Lv, Y.; Fu, B. Ecosystem service evolution in ecological restoration: Trend, process, and evaluation. *Acta Ecol. Sin.* **2016**, *36*, 6337–6344.
59. Seppelt, R.; Dormann, C.F.; Eppink, F.V.; Lautenbach, S.; Schmidt, S. A quantitative review of ecosystem service studies: Approaches, shortcomings and the road ahead. *J. Appl. Ecol.* **2011**, *48*, 630–636. [CrossRef]
60. Zhang, J.; Zhang, X.; Hao, M.; Lei, Y.; Liu, J.; Liu, E.; Sun, Y. Potentiality Analysis of Grrain Productivity and Its Security in Changwu County in the Loess Plateau Gully Region against the Background of "Grain for Green" Projection. *J. Soil Water Conserv.* **2011**, *25*, 231–235.
61. Liu, X.; Qin, Z.; Li, F. Spatiotemporal changes of the sustainability of grain production system in the Loess Plateau based on Grain for Green Project. *Trans. Chin. Soc. Agric. Eng.* **2022**, *38*, 249–257.
62. Goldstein, J.H.; Caldarone, G.; Duarte, T.K.; Ennaanay, D.; Hannahs, N.; Mendoza, G.; Polasky, S.; Wolny, S.; Daily, G.C. Integrating ecosystem-service tradeoffs into land-use decisions. *Proc. Natl. Acad. Sci. USA* **2012**, *109*, 7565–7570. [CrossRef]
63. Bateman, I.J.; Harwood, A.R.; Mace, G.M.; Watson, R.T.; Abson, D.J.; Andrews, B.; Binner, A.; Crowe, A.; Day, B.H.; Dugdale, S.; et al. Bringing Ecosystem Services into Economic Decision-Making: Land Use in the United Kingdom. *Science* **2013**, *341*, 45–50. [CrossRef]
64. Schmidt, J.P.; Moore, R.; Alber, M. Integrating ecosystem services and local government finances into land use planning: A case study from coastal Georgia. *Landsc. Urban Plan.* **2014**, *122*, 56–67. [CrossRef]
65. Chen, W.X.; Chi, G.Q.; Li, J.F. The spatial association of ecosystem services with land use and land cover change at the county level in China, 1995–2015. *Sci. Total Environ.* **2019**, *669*, 459–470. [CrossRef]
66. Gaodi, X.; Shuyan, C.; Chunxia, L.; Changshun, Z.; Yu, X. Current status and future trends for eco-compensation in China. *J. Resour. Ecol.* **2015**, *6*, 355–362. [CrossRef]
67. Wang, W.; Guo, H.; Chuai, X.; Dai, C.; Lai, L.; Zhang, M. The impact of land use change on the temporospatial variations of ecosystems services value in China and an optimized land use solution. *Environ. Sci. Policy* **2014**, *44*, 62–72. [CrossRef]
68. Yue, S.; Zhang, S.; Yan, Y. Impacts of Land Use Change on Ecosystem Services Value in the Northeast China Transect (NECT). *Acta Geogr. Sin.* **2007**, *62*, 879–886. [CrossRef]
69. Wang, Z.; Zhang, B.; Zhang, S.; Song, K.; Duan, H. Estimates of loss in ecosystem service values of Songnen plain from 1980 to 2000. *J. Geogr. Sci.* **2005**, *15*, 80–86. [CrossRef]
70. Chuai, X.; Huang, X.; Wu, C.; Li, J.; Lu, Q.; Qi, X.; Zhang, M.; Zuo, T.; Lu, J. Land use and ecosystems services value changes and ecological land management in coastal Jiangsu, China. *Habitat Int.* **2016**, *57*, 164–174. [CrossRef]
71. Kienast, F.; Bolliger, J.; Potschin, M.; De Groot, R.S.; Verburg, P.H.; Heller, I.; Wascher, D.; Haines-Young, R. Assessing Landscape Functions with Broad-Scale Environmental Data: Insights Gained from a Prototype Development for Europe. *Environ. Manag.* **2009**, *44*, 1099–1120. [CrossRef]
72. Yi, H.; Güneralp, B.; Filippi, A.M.; Kreuter, U.P.; Güneralp, İ. Impacts of Land Change on Ecosystem Services in the San Antonio River Basin, Texas, from 1984 to 2010. *Ecol. Econ.* **2017**, *135*, 125–135. [CrossRef]

73. Costanza, R.; de Groot, R.; Sutton, P.; van der Ploeg, S.; Anderson, S.J.; Kubiszewski, I.; Farber, S.; Turner, R.K. Changes in the global value of ecosystem services. *Glob. Environ. Chang. Part A Hum. Policy Dimens.* **2014**, *26*, 152–158. [CrossRef]
74. Long, H. Explanation of Land Use Transitions. *China Land Sci.* **2022**, *36*, 1–7.

forests

MDPI

Article

Landscape Dynamics in a Poverty-Stricken Mountainous City: Land-Use Change, Urban Growth Patterns, and Forest Fragmentation

Chen Wen [1] and Luqi Wang [2,*]

[1] School of Architecture and Urban Planning, Huazhong University of Science and Technology, No.1037, Luoyu Road, Wuhan 430074, China
[2] School of Civil Engineering and Architecture, Wuhan University of Technology, No. 122, Luoshi Road, Wuhan 430070, China
* Correspondence: luqiwang@whut.edu.cn

Abstract: For poverty-stricken mountainous cities in China, both poverty alleviation and ecological restoration projects are sources of land-use change in urban development. However, the patterns in changes are understudied in light of sustainable forest management. The study aims to explore the characteristics of land-use change in a poverty-stricken mountainous city with a focus on forests. This research proposed a three-step approach to explore the multi-aspect dynamics of land change, including the differences among land-use categories, spatial characteristics of urban expansion, and forest fragmentation. This study investigated Enshi City, China, based on land-use data from 2000, 2010, and 2020. Throughout the two intervals, the gain of water bodies and the loss of grassland were active. Artificial surfaces increased most intensively from 2010 to 2020, with transitions from grassland and cultivated land. Edge-expansion was the dominant type of artificial surface growth. Furthermore, forests had the largest size of gain across the two intervals, and there was a substantial reduction in forest fragmentation in the western part of the city. The findings confirm that recent planning measures are effective in restoring the natural environment. The identified key areas can support sustainable forest management in urban growth.

Keywords: forest fragmentation; land-use change; landscape pattern; mountainous city; urban growth

Citation: Wen, C.; Wang, L. Landscape Dynamics in a Poverty-Stricken Mountainous City: Land-Use Change, Urban Growth Patterns, and Forest Fragmentation. *Forests* **2022**, *13*, 1756. https://doi.org/10.3390/f13111756

Academic Editors: Chao Wang, Fan Zhang and Wei Liu

Received: 22 September 2022
Accepted: 19 October 2022
Published: 25 October 2022

1. Introduction

With a unique topography and climate, mountains provide a number of crucial benefits and services, such as fresh water supply and biodiversity maintenance [1–6]. However, mountain ecosystems are fragile. For example, a mountain's steep slope might facilitate erosion, thus leading to poor soil quality [7]. They are sensitive to anthropogenetic impacts, natural disasters, and climate change [2,3,7]. In recent decades, many mountainous regions have undergone changes in their socioecological systems as a result of human activities [6,8].

Mountains are home to 12% of the world's population, and this proportion is even higher in developing countries [4,9]. In China, many inhabitants of mountainous areas live in poverty, partly because of transportation difficulties and limited cultivated land [10]. Subsequently, poverty alleviation through urban development is an important task for local authorities [11]. While human settlement expansion and infrastructure construction can make people's lives more convenient, some activities might lead to the deterioration of the local environment [12]. For instance, some of the forest and grassland in sloping areas has been converted to farmland, leading to soil erosion, sandstorms, and other disasters [13,14].

In order to alleviate the anthropogenetic damages to natural environments, ecological restoration has been playing a more significant role in mountainous areas in recent decades. Typical ecological restoration projects include forestation, water source protection, restoration of rocky desertification, and treatment of mine areas [15]. Since 1999, China has widely implemented a "Grain for Green" policy, which aims to convert farmland back to forest and grassland [13]. At the same time, built-up areas are being expanded in mountainous regions because of urban and rural development. Both urban growth and ecological restoration projects would lead to land-use changes.

In such contexts, a better understanding of land-use dynamics would contribute to sustainable regional development and harmonized human–nature relationships [16]. It would benefit both current land-use management and future planning in mountainous areas. To understand the detailed land-use and land cover changes for decision making, many researchers have monitored and investigated the changes spatiotemporally [1–3,9,17–19]. For instance, one study in Peruvian Jalca has identified threatened areas based on monitoring landscape change [1]. Researchers also explored land-use transitions in peri-urban areas and discussed the associations with human activities [17]. Spatiotemporal analysis can help investigate where and when land-use transitions occur, providing additional information about how much has changed and how fast [20]. Since land-use change is a continuous process, spatiotemporal analysis at several time points can help build a conceptual model to quantify landscape patterns and understand trends. These insights provide a detailed perspective to investigate human–nature relationships and the interactions between biophysical structures and socioeconomic changes [21]. When focusing on a single land-use category—for example, artificial surfaces—the analysis is able to identify the trends and direction of urban expansion [22]. When focusing on the differences between rates of change in various land-use categories, it helps researchers to understand how overall landscape pattern change and which ecosystem might be the leading source of change [20]. In mountainous areas, a growing number of studies have explored urban development spatiotemporally, and the increase in built-up areas has been confirmed globally [3,6,8,9,17,23]. Because of topographical restrictions, the urban growth and land-use change of cities in mountainous areas might show different characteristics compared to cities in plain areas [3]. A recent study modeling and predicting land-use change in a mountainous region of Oman found that, in spite of the complex terrain, the built-up area will continue to grow and occupy more local arable land in the future [9]. A study in the mountainous area of Brazil found significant transitions between natural vegetation and agriculture. Several studies have focused on the impact of land-use change on ecosystems in mountainous regions [2,18,19], and one study in Zhangjiakou City found that the changes in forestland, arable land, and grassland greatly impacted the local ecosystem [18].

Despite the increasing number of studies that have explored land-use change spatiotemporally, studies in mountainous cities remain insufficient [3], particularly in poverty-stricken mountainous areas [11]. For poverty-stricken mountainous cities in China, both poverty alleviation and ecological restoration projects are key tasks. Many cities are experiencing rapid urban growth in recent years. Different land-use changes can be associated with landscape patterns and ecological processes [24]. Monitoring land-use and landscape pattern change helps design efficient forest management and nature conservation strategies [1,8,25], which are critical for poverty-stricken mountainous areas [11,18]. In China, many mountainous cities are in the western or central regions, which are economically disadvantaged. The reasons include not only location, transport, and regional economic geography, but also the mountainous environment that may hinder industrial development [26,27]. Therefore, practices related to poverty alleviation in the city often focus on agriculture, urban development, and harvesting value from the forests. These practices can impact the sustainable management of forests [28]. However, an insufficient number of studies have spatial-explicitly evaluated the change of ecological conditions from the perspective of landscape pattern. Existing studies of monitoring land-use change in mountainous areas often neglected the socio-historical background and therefore did not integrate the related key points into the analysis. Studies on land-use change should investigate the change intensity of different land cover categories for exploring potential sources and driving forces for forest change in poverty alleviation. Moreover, close attention should be paid to areas that are undergoing the rapid process of urban expansion, which is often considered a challenge to sustainable forest management.

For supporting sustainable forest management, this study aims to explore the land use and landscape dynamics in Enshi City, applying a framework that considers different aspects of dynamics in the poverty-stricken mountainous area. Enshi City is located in one such area in China. Based on the land-use data from Enshi City in 2000, 2010, and 2020, this study integrates the methods of intensity analysis, landscape metrics, and mapping. The overarching objective of this study is to explore the characteristics of land-use change in a poverty-stricken mountainous city. The specific research questions are: (1) To what extent have different types of land use changed in Enshi city from 2000 to 2020? (2) What are the characteristics of the expansion of urban areas in the study area? (3) What are the spatiotemporal changes in the fragmentation of forests in the study area from 2000 to 2020?

2. Methodology

2.1. Study Area

Enshi City is the seat of Enshi Tujia and Miao Autonomous Prefecture, located in the southwest mountainous region of Hubei Province, spanning latitudes 29°50′33″–30°39′30″ N and longitudes 109°4′48″–109°58′42″ E. The total area is 3972 km^2. Enshi City has a subtropical climate. The mean annual temperature is 16 °C and the mean annual precipitation is between 1400 and 1600 mm [29]. As part of the Wuling Mountain area, Enshi City has a varying elevation (Figure 1). The city center is in the middle of the region, where the terrain is relatively flat. The elevation is much higher in the northwestern and southeastern parts of the city. The Wuling Mountain area is one of the key regions in China for implementing the "Grain for Green" policy [30]. With a large area covered by forest, Enshi City is rich in natural resources and biodiversity. Since many areas are ecologically fragile zones, the conservation of the natural environment is one of the city's most important tasks. The northeastern part of the city belongs to a national nature reserve. Enshi City is also considered a suitable place for recreation and tourism because of its diverse landscape, fresh air, and cultural heritage.

The Wuling Mountain area is a typical poverty-stricken area in China. This is also the case for Enshi City [31]. It was not until April 2020 that Enshi City was excluded from the National Poverty-Stricken County list [32]. As a transport hub, Enshi City is the political, economic, and cultural center of the prefecture, so its development is crucial to the region's economic growth and for alleviating poverty. The population of Enshi City has risen over the last two decades, increasing from 0.764 million in 2000 to 0.795 in 2010, and to 0.813 million in 2019 [33]. Ethnic minorities consist of 41.12% of the total population [29]. A recent study analyzing the urban expansion of 275 Chinese cities found that Enshi Prefecture is among the top ten cities with regard to the development level of outlying expansion patches [34]. Therefore, a specific investigation of Enshi City can improve our understanding of land use and landscape change in mountainous regions, where rapid urban growth is taking place against the background of nature conservation and poverty alleviation.

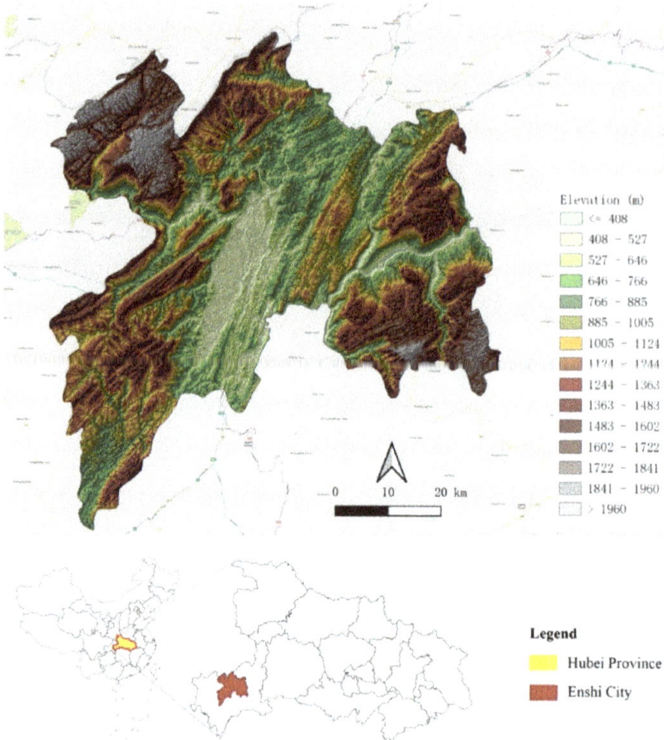

Figure 1. Location and elevation of Enshi City. Elevation data was obtained from the ASTER global digital elevation model [35].

2.2. Data Sources and Analysis

2.2.1. Data Sources and Analytical Framework

In order to explore the land-use and landscape changes in Enshi City in recent decades, this study used the GlobeLand30 dataset [36]. The dataset includes land use data from 2000, 2010, and 2020, with a resolution of 30 m. Ten types of land use are classified: cultivated land, forest, grassland, shrubland, wetland, water bodies, tundra, artificial surfaces, bare land, and permanent snow and ice [37]. This dataset has been proven by previous studies to have sufficient accuracy in analyzing land-use change [38,39].

In the early stage of the study, different dataset options are considered and tested based on a review of the data. For the research objectives, the choice of input data should meet the following requirements simultaneously: (1) The time span of the spatial data must be over 20 years, during which Enshi City has undergone rapid urbanization and development. (2) The spatial data must be fine-scaled so that the land-use change can be investigated at the city scale. (3) The dataset is validated and standardized in the study region. This is for the purpose of generalizing the proposed methods in the future to a larger scale or other mountainous study sites with similar situations. Therefore, we did not consider classifying our own land-use data based on historical satellite images. For these criteria, we tested but decided not to use some other commonly used land cover datasets. For example, the datasets FROM-GLC10 and WorldCover-10 m may have a finer resolution, but they do not offer historical data over twenty years [40–43]. The LC-CCI data have long time spans, but their spatial resolution is relatively scarce. After comparison, GlobeLand30 is by far the most suitable dataset for our study purpose, and it meets all three above criteria and has relatively high accuracy.

For exploring the complex nature of mountainous urban areas, this research proposed a three-step approach to explore the dynamics of land-use and landscape change (Figure 2). Each step addressed one research question, linking different facets of the changes. The first was to investigate the differences in changes among land-use categories by using the method of intensity analysis. The second was to explore the types and locations of urban expansion with the Landscape Expansion Index (LEI). The third was to explore forest fragmentation by calculating the spatiotemporal change of effective mesh size with cross-boundary connections (CBC) methods. All of the analyses were conducted using software (ArcGIS 10.5, QGIS 3.16, and R 4.0) through packages [44–48].

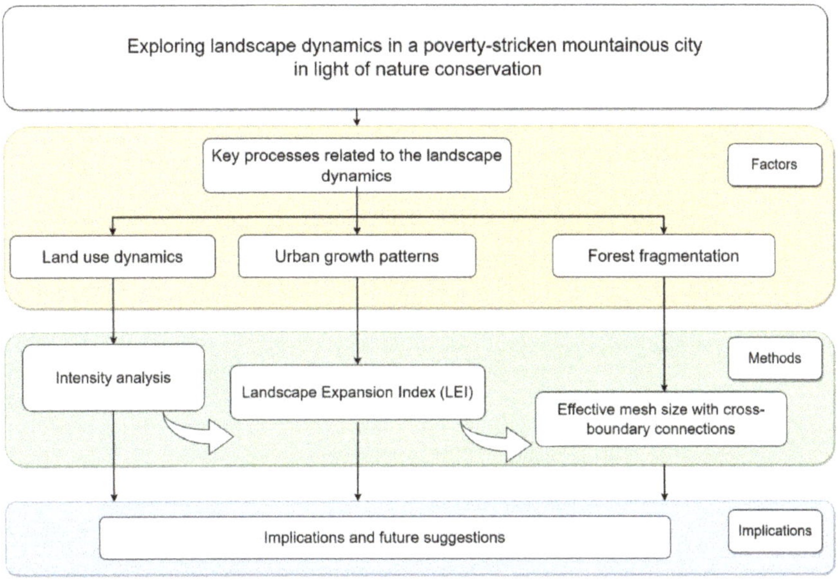

Figure 2. Analytic framework of this study.

2.2.2. Intensity Analysis

Intensity analysis provides a mathematic framework to examine land-use change for three or more time points (i.e., two time intervals or more). It is a unified approach to draw insights into different aspects of land-use change, and it has been widely used by scholars around the world (for example: [24,39,49–51]).

Intensity analysis concerns three levels: interval, category, and transition [16,52]. The interval level focuses on the speed of change across intervals. It compares the speed of change of each time interval with the speed of change of the overall interval. If any interval's speed of change is larger than the overall interval's, the change of this interval would be considered fast, and vice versa [16,52]. At the categorical level, this method compares the annual intensity of loss and gain of each category with the uniform intensity of annual change. The uniform intensity supposes that each category was to gain or lose with the same intensity. If one category's intensity of change (gain or loss) is larger than the uniform intensity, then the change is active. If one category's intensity of change is smaller than the uniform intensity, then the change of this category is dormant [16,52]. The transition level helps identify whether a specific category's loss or gain is targeting or avoiding other categories. It assumes that the selected category gains or loses uniformly across the spatial extent. If the intensity of transition (gain or loss) of the selected category to another category is larger than the uniform intensity, then the transition is targeted. If the transition is smaller than the uniform intensity, then the gain or loss of the selected category avoids the another category [16,52]. Furthermore, at the categorical and transition level, if a given category's change is always on one side of the uniform intensity (either larger or smaller than the uniform intensity) for all time intervals, the change or transition of this category would be considered stationary [16]. One previous study has successfully applied intensity analysis to compare the land-use transformation of two mountainous areas in Athens, and a variety of systematic stationary transitions were detected [17].

2.2.3. Urban Growth Pattern Analysis

The second step was to investigate the characteristics of urban growth patterns in Enshi City. We adopted a widely used index called the Landscape Expansion Index (LEI) invented by Liu et al. [53]. This index is calculated based on buffer analysis of the geographic information system (GIS) using the following formula:

$$\mathrm{LEI} = 100 \times \frac{A_o}{A_o + A_v} \tag{1}$$

where A_o refers to the intersection area between the buffer zone of the newly grown patch and the old urban patch, and A_v is the intersection between the buffer zone of the newly grown patch and the vacant land (non-urban patch). It differentiates urban growth patterns according to the results of the LEI, which ranges from 0 to 100. Three urban growth types are defined, namely edge-expansion ($0 < \mathrm{LEI} \leq 50$), infilling ($50 < \mathrm{LEI} \leq 100$), and outlying ($\mathrm{LEI} = 0$). Infilling is when the buffer zone of a newly grown patch primarily belongs to an old patch. Edge-expansion is when a newly grown patch's buffer zone is a combination of vacant land and an old patch. Outlying is when the buffer zone of a newly grown patch belongs only to vacant land [53]. The LEI is confirmed by studies in different contexts to characterize urban growth patterns [3,34,54–56]. For example, a study in the large mountainous city, Chongqing, found that the city showed different dominant modes of urban growth during different periods of development [3].

2.2.4. Fragmentation Analysis

The third step was to explore the fragmentation of the forest landscape by applying landscape metrics. Landscape metrics are widely used in quantifying and assessing landscape patterns [1,57–62]. While intensity analysis provides details of the interval, size, and intensity of the land-use change in the study area, calculating the landscape metrics for different time points helps provide further information about the changes in landscape patterns. Regarding fragmentation, several metrics are frequently used, such as patch density, number of patches, landscape division index, and effective mesh size [63–67]. This study chose an effective mesh size for its high reliability in measuring fragmentation [67,68]. Effective mesh size measures the landscape fragmentation based on the probability that two randomly chosen animals in different areas of the same region can find each other. Mathematically, it calculates the size of patches when the investigated region is divided into S areas (each of the same size) with the same degree of landscape division as obtained for the observed cumulative area distribution [67,69]. It can be calculated using the following formula:

$$\mathrm{m} = A_t \sum_{i=1}^{n} \left(\frac{A_i}{A_t}\right)^2 = \frac{1}{A_t} \sum_{i=1}^{n} A_i^2 \tag{2}$$

where A_t is the total area, n is the number of patches in the study area, and A_i is the size of each patch ($i = 1, \ldots, n$). While effective mesh size is widely used in quantifying landscape fragmentation, it suffers from "boundary problems" [70]. Because boundaries are also considered to be fragmenting the landscape, the results can be biased. To address this problem, the study proposed a new method called cross-boundary connections (CBC) [70].

This method improves the accuracy of comparing the fragmentation of different areas within a region. Mathematically, it is calculated using the following formula:

$$m_{eff}^{CBC} = \frac{1}{A_t} \sum_{i=1}^{n} A_i \times A_i^{cmpl} \tag{3}$$

where A_t is the total area, n is the number of patches in the study area, A_i is the size of the n patches ($i = 1, \ldots, n$), and A_i^{cmpl} is the size of the complete patch (including the area outside the boundaries) [70]. Previous studies have used the CBC method to calculate effective mesh size in different contexts for a more accurate measurement of fragmentation [64,71]. In order to help visualize the spatiotemporal dynamics of landscape fragmentation, we used an R package called *zonebuilder* [48]. This package can uniformly break the analytical unit (e.g., the city) into zones of different rings and segments so that we can better investigate the direction of change and compare the differences across time periods [48].

3. Results

3.1. Land-Use Change

3.1.1. Land Use of Enshi City in 2000, 2010, and 2020

Figure 3 presents the land use of Enshi City in 2000, 2010, and 2020. The land use consists of forest, grassland, cultivated land, water bodies, and artificial surfaces. The primary land-use categories are forest and cultivated land. From 2000 to 2010, grassland decreased noticeably, and water bodies became more visible. From 2010 to 2020, artificial surfaces increased notably. Figure 3d exhibits the detailed locations where land-use changes occurred. While 72.26% of the area did not experience any changes in land use across the two time intervals, 22.52% of the area changed once (either from 2000 to 2010 or from 2010 to 2020), and 5.23% of the area changed twice.

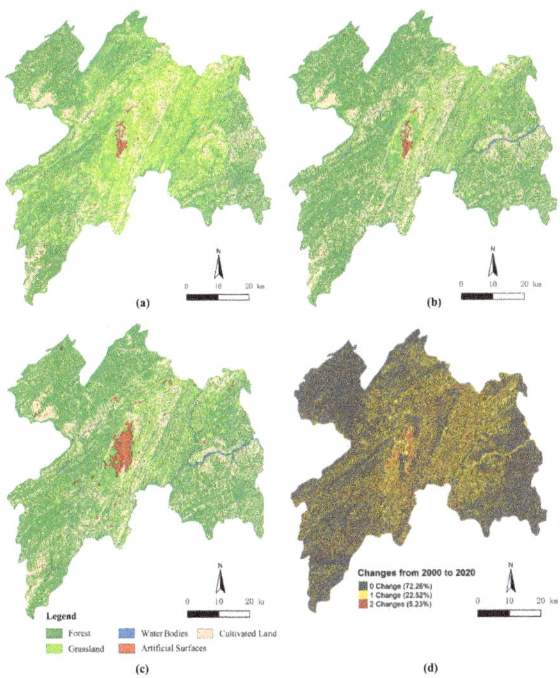

Figure 3. Land use in Enshi City. (**a**) Land use in Enshi City in 2000; (**b**) Land use in Enshi City in 2010; (**c**) Land use in Enshi City in 2020; (**d**) Accumulated changes from 2000 to 2020.

An overview of the changes in the five categories is presented in Figure 4. From 2000 to 2020, while artificial surfaces, forest, and water bodies experienced a net gain in land-use area, cultivated land and grassland experienced a net loss. Forest and grassland had relatively larger changes in size. Cultivated land experienced the smallest net change size; however, its gross change area is much larger than those of artificial surfaces and water bodies. This suggests that, despite the small change in the total area of cultivated land from 2000 to 2020, there has been intensive land-use exchange between cultivated land and other land categories. Tables 1–3 demonstrate the transition matrix of land-use types in Enshi City from 2000 to 2020.

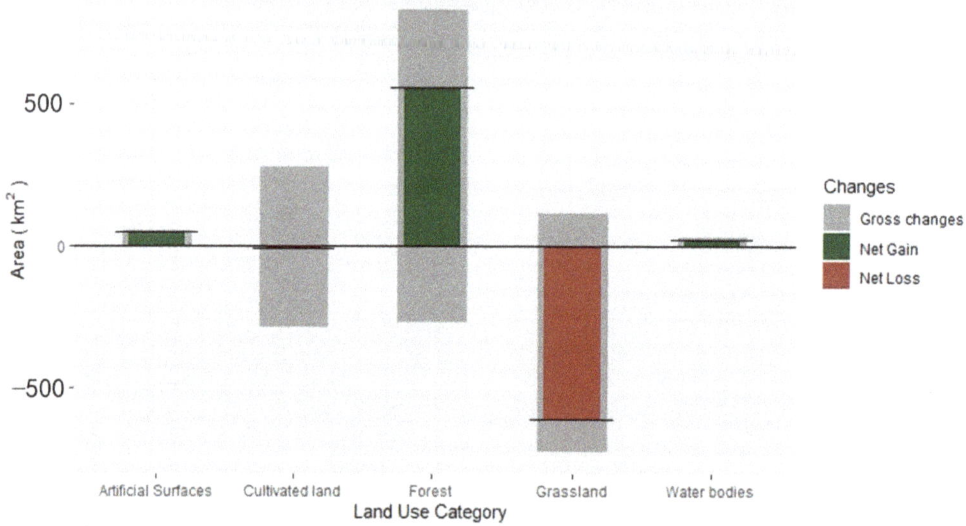

Figure 4. The area of land-use change in Enshi City from 2000 to 2020. The bar plot combines both gross changes and net changes. Bars extending to the upper side of zero refer to gross gain (shown in grey) and net gain (shown in green). Bars extending below zero refer to gross loss (grey) and net loss (red).

Table 1. Transition matrix of land-use types in Enshi City from 2000 to 2010 (km^2).

	2000–2010				
	Cultivated Land	**Forest**	**Grassland**	**Water Bodies**	**Artificial Surfaces**
Cultivated land	1167.0030	90.9099	24.1812	5.4252	0.5931
Forest	59.787	1756.0611	3.0996	2.6145	0.0216
Grassland	70.2594	549.6012	184.7556	8.4069	0.2412
Water bodies	3.0717	1.7901	0.4329	7.3557	0.0225
Artificial surfaces	0.7551	0.0891	0.2196	0.2061	18.9495

Table 2. Transition matrix of land-use types in Enshi City from 2010 to 2020 (km^2).

	2010–2020				
	Cultivated Land	**Forest**	**Grassland**	**Water Bodies**	**Artificial Surfaces**
Cultivated land	1132.5762	129.1041	12.3498	2.4021	24.444
Forest	111.1356	2191.0716	75.2436	9.0846	11.916
Grassland	31.1373	56.313	109.0953	1.5453	14.598
Water bodies	0.8046	1.0908	0.0162	21.8385	0.2583
Artificial surfaces	0.6093	0.1341	0.0738	0.1476	18.8631

Table 3. Transition matrix of land-use types in Enshi City from 2000 to 2020 (km^2).

	2000–2020				
	Cultivated Land	**Forest**	**Grassland**	**Water Bodies**	**Artificial Surfaces**
Cultivated land	1136.4705	107.6472	14.8239	6.4845	22.7358
Forest	74.0268	1691.9811	45.8109	7.8462	2.0853
Grassland	63.1512	577.2888	135.8649	11.3265	25.659
Water bodies	2.1555	0.8217	0.225	9.1089	0.369
Artificial surfaces	0.5112	0.1467	0.0639	0.2673	19.2303

3.1.2. Results of Intensity Analysis

This section presents the results of the intensity analysis, which includes three levels: interval, categorical, and transition. The interval level helps identify which time interval changes fast or slow compared with the overall annual rate of change. As shown in Figure 5, compared with the uniform rate (1.65%), the first time interval changed fast (2.08%), and the second interval changed slowly (1.22%). It also shows that a greater area underwent change in the first interval.

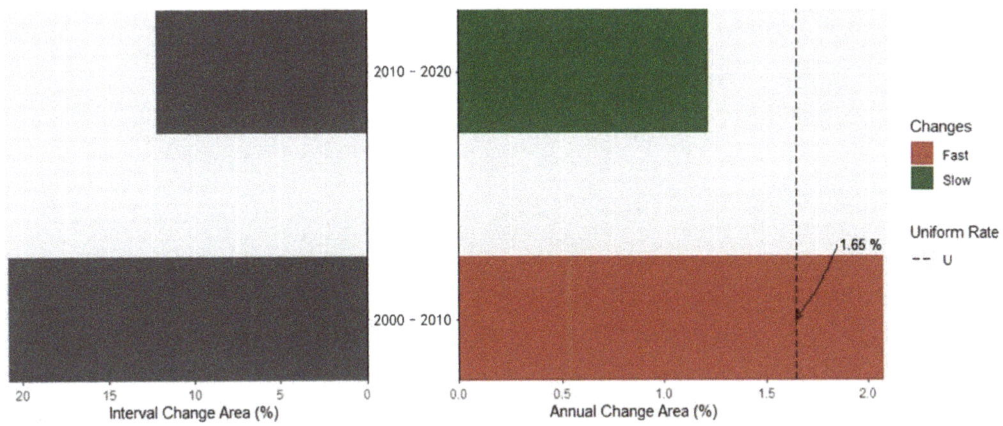

Figure 5. Time intensity analysis for two time intervals: 2000 to 2010 and 2010 to 2020. The left side represents gross area of overall change in each interval, and the right side represents the intensity of annual area of change within each interval. The dash line represents the uniform annual intensity of change from 2000 to 2020. If a category's bar extends beyond the uniform line, then the change of intensity is fast. If a category's bar ends before the uniform line, then the change of intensity is slow.

The category level further explores the detailed change of each land-use category and helps detect which categories are dormant or active during different time intervals. Figure 6a shows the detailed situation of area gain and intensity of each category over the last two decades. The dash lines represent the uniform annual intensity of change in the respective time interval. The interval from 2000 to 2010 (2.08%) had a larger annual rate of intensity gain than the interval from 2010 to 2020 (1.22%). From 2000 to 2010, the gain of artificial surfaces, cultivated land, and grassland was dormant, and the gain of forest and water bodies was active. Although forest had the largest size of gain, the gain of water bodies was more intensive. From 2010 to 2020, artificial surfaces experienced the largest gain intensity, followed by grassland and water bodies. The gain of cultivated land and forest was dormant.

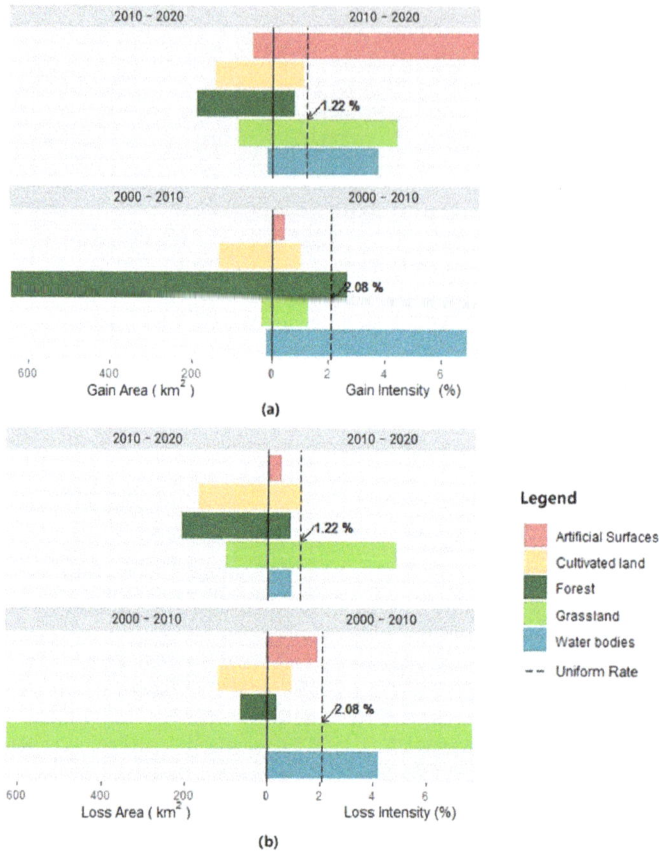

Figure 6. Intensity of change at the categorical level. Each bar plot combines the area of change (bars that extend to the left of zero) and intensity of change (bars that extend to the right of zero). (**a**) Gain intensity analysis for two time intervals: 2000–2010 and 2010–2020. (**b**) Loss intensity analysis for two time intervals: 2000–2010 and 2010–2020. If a category's bar extends beyond the uniform line (the dash line), then the gain or loss of intensity is active. If a category's bar ends before the uniform line, then the gain or loss of intensity is dormant.

As regards the loss intensity and size (Figure 6b), grassland lost most intensively across the two time intervals. Water bodies lost actively in the first interval and became dormant in the second interval. Artificial surfaces and forest show stationarity across the two intervals as they stayed on the left side of the uniform rate line. The loss of cultivated land was dormant in the first interval and active in the second interval.

The transition level analyzes the gain of artificial surfaces and loss of grassland. According to the research objectives and site background, the study pays close attention to the risks and challenges to forests and the ecological environment under the pressure of urban expansion. As known from the category level, artificial surfaces gained intensively from 2010 to 2020. Figure 7a indicates that the gain of artificial surfaces from 2010 to 2020 mostly targeted grassland and cultivated land. Although cultivated land contributed the largest area, the gain from grassland was the most intensive, meaning that the gain from grassland had the largest rate of transition. The gain of artificial surfaces avoided forest and water bodies, because the transition rate is lower than the annual transition rate (0.13%).

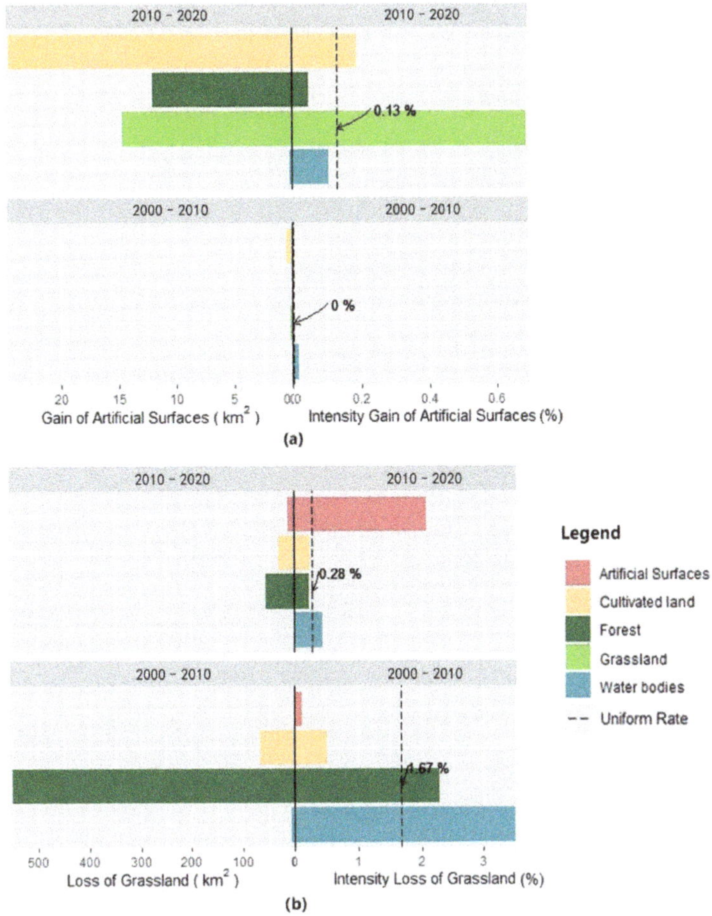

Figure 7. Intensity of change at the transition level. Each bar plot combines the area of change (bars that extend to the left of zero) and intensity of change (bars that extend to the right of zero). (**a**) Transition intensity analysis to artificial surfaces for two time intervals: 2000–2010 and 2010–2020. (**b**) Transition intensity analysis from grassland for two time intervals: 2000–2010 and 2010–2020. If a category's bar extends beyond the uniform line (the dash line), then the gain or loss of the selected category targets this category. If a category's bar ends before the uniform line, then the gain or loss of the selected category avoids this category.

The loss of grassland was analyzed because grassland is the only category that is active in loss in both time intervals. From 2000 to 2010, water bodies and forest targeted the loss of grassland, and artificial surfaces and cultivated land avoided it (Figure 7b). Although the size of the loss to forest is the largest, the loss to water bodies is most intensive because water bodies have the largest rate of transition. From 2010 to 2020, artificial surfaces and water bodies targeted the loss of grassland. The loss of grassland to forest remains the largest in terms of size, but it avoided forest as the transition intensity is less than the uniform intensity. The loss to artificial surfaces is the most intensive.

3.2. Urban Growth Pattern

The urban growth pattern of Enshi City from 2000 to 2020 is presented in Figure 8. The main location of growth is in the central part of the city, with a large newly developed area being detected as the edge-expansion type. As shown in Table 4, edge-expansion is the dominant mode of urban growth in Enshi City, and this type accounted for 83.05% of the new artificial area. Most edge-expansion areas are found in the north of the old urban center, and they are now shaping the urban forms of the Enshi City center into a north–south strip of land. Infilling and outlying accounted for 4.09% and12.86%, respectively. Overall, the artificial land use of Enshi City increased from 20.22 km^2 in 2000 to 70.08 km^2 in 2020.

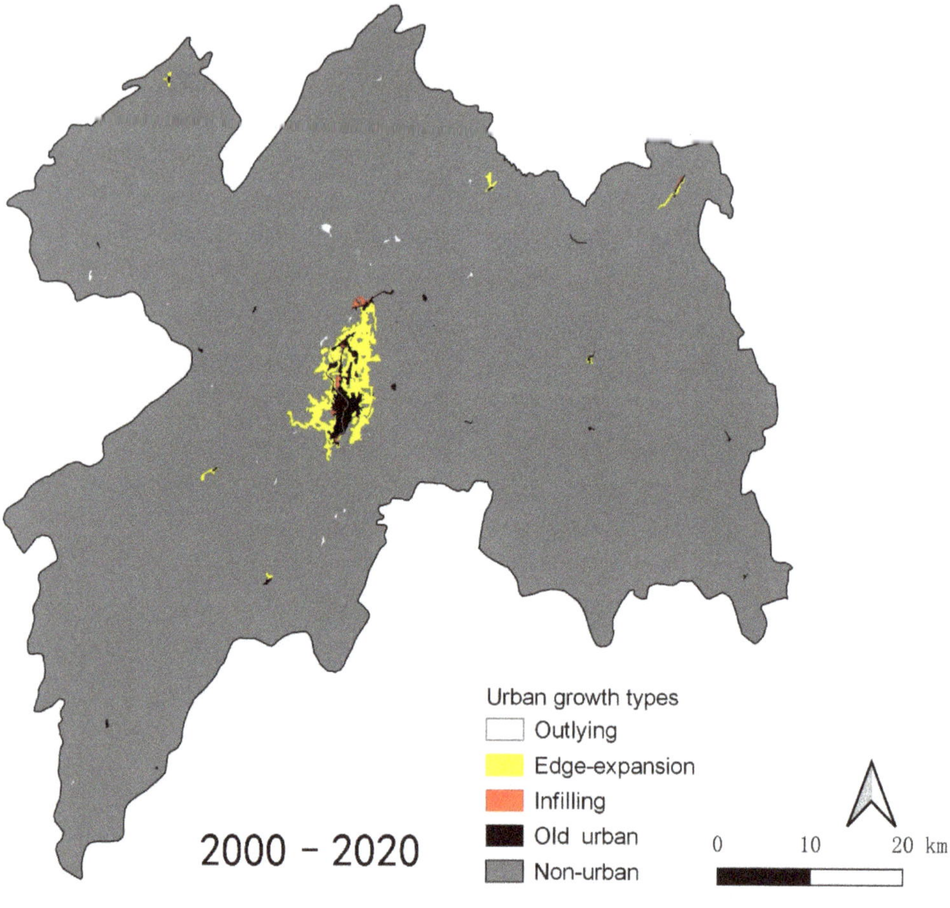

Figure 8. Urban growth pattern of Enshi City.

Table 4. Results of urban growth types of Enshi City from 2000 to 2020.

	Urban Growth Type (2000–2020)		
	Outlying	**Edge-Expansion**	**Infilling**
LEI interval	0	$0 < LEI \leq 50$	$50 < LEI \leq 100$
Area (km^2)	6.54	42.23	2.08
Area (%)	12.86	83.05	4.09

3.3. Forest Fragmentation

We used effective mesh size with cross-boundary connections to quantify the fragmentation of forests. Figure 9 indicates that the central part of the city was the most fragmented from 2000 to 2020, as the value of effective mesh size was relatively smaller. The degree of fragmentation reduced substantially in the western part of the area from 2000 to 2010, particularly in the northwestern part. From 2010 to 2020, although there was a significant increase in the artificial area of the city, the fragmentation of the forest only increased slightly. The southwestern part became a little bit more fragmented. However, the eastern part of the city did not show noticeable improvement regarding forest fragmentation. This region is fragmented and densely covered by small agricultural land slots.

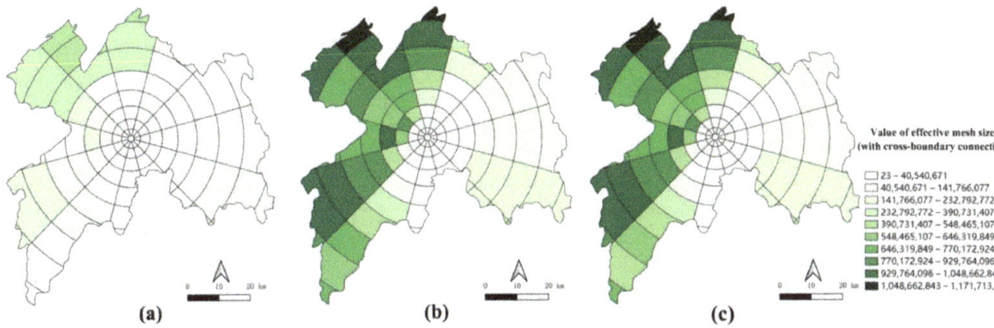

Figure 9. Results of the value of effective mesh size of forest land use in Enshi City. (**a**) Results from 2000; (**b**) Results from 2010; (**c**) Results from 2020.

4. Discussion

4.1. Insights from Analysis of Land-Use Change

In summary, this study explored the dynamics of land use and landscape pattern change in Enshi City from 2000 to 2020 by combining different methods. According to the intensity analysis, the speed of land-use change was much faster from 2000 to 2010 than from 2010 and 2020. At the categorical level, throughout the two intervals, water bodies and cultivated land showed stationarity in terms of the intensity of gain. Artificial surfaces, forest, and grassland showed stationarity in terms of the intensity of loss. At the transition level, we focused on two specific types of land: artificial surfaces and grassland. Artificial surfaces gained the most intensively from 2010 to 2020, with transitions from grassland and cultivated land. Grassland showed stationarity across the two intervals, as it continued an active trend of area loss. We confirmed that the losses were mostly transitioned to water bodies and forest from 2000 to 2010 and artificial surfaces and water bodies from 2010 to 2020.

The intensive loss of grassland is in line with several studies [1,11,39,66]. For example, a recent study on land-use transformation in Sangzhi County in the Wuling Mountain area found that the area of grassland decreased significantly from 2010 to 2018 [11]. They also confirmed a significant increase in cultivated land, driven by the implementation of policies relating to the protection of cultivated land [11]. This finding differs from ours because the total area of cultivated land in Enshi City is relatively stable. Our findings also reflect the implementation of policies regarding the remediation of water bodies. The area of water bodies continued an active trend of gain from 2000 to 2020. Local authorities have efficiently protected and remediated the main rivers in Enshi City. It also showed an active trend of loss from 2000 to 2010. It is possible that some scattered distributed water patches have vanished. Attention should be paid to small and scattered water bodies to maintain biodiversity and landscape connectivity.

Since the artificial area of Enshi City has increased rapidly, we further explored the urban growth pattern from 2000 to 2020 by calculating the LEI. The results showed that the expansion of artificial surfaces in Enshi City is primarily occurring in flat areas close to the urban center. This echoes previous findings that, in mountainous regions, flat areas are more inclined to be transitioned to built-up areas [9]. While previous studies in larger mountainous cities (e.g., Chongqing) found that the dominant urban growth mode is outlying in recent years [3], we found that the dominant mode in Enshi City is edge-expansion, which is similarly occurring in other cities in plain areas. This is partly because Enshi City is underdeveloped, with plain (vacant) space surrounding old urban areas. For Enshi City, the intensive gain of artificial surfaces began after 2010, which is relatively late compared to many other cities and counties in China. Since the terrain is relatively flat in the middle of the city, it is possible that the built-up area of Enshi City will continue to grow in this locale in the future. Local decision-makers should carefully control the boundaries of artificial surfaces and consider the possible negative impacts of land transitions, for instance, the loss of biodiversity, the creation of urban heat islands, poor air quality, and the decrease in landscape connectivity [3,9].

4.2. Impacts of Land-Use Change on Landscape Fragmentation

We confirmed a substantial reduction in the degree of forest fragmentation from 2000 to 2010, which primarily occurred in the western part of the city. For ecologically fragile mountainous areas, our findings clearly reflect the successful implementation of forestation projects aiming to restore the natural environment. The notable increase in forest between 2000 and 2010 echoes many Chinese studies and is a result of the nationwide "Grain for Green" policy [2,18,39]. While urbanization frequently leads to landscape fragmentation [72], the rapid growth of urban areas in Enshi City from 2010 to 2020 did not seem to influence the fragmentation of local forest areas. Nevertheless, it is necessary to identify and fortify urban forests that are located near the boundary of current built-up environments. Some of them are vital for maintaining ecological integrity but are also in potential danger of land-use transition. It is crucial to continuously monitor land-use change and spatially model different scenarios for future development.

For mountainous cities to be lifted out of poverty, it is vital to monitor how poverty alleviation practices can be reflected in the landscape dynamics and how they can impact sustainable forest management. In Enshi City's case, recent urban development has been focused on developing agriculture and expanding urban areas for housing and industry. Therefore, social contexts are meaningful in analyzing decades of change centering around forests via potential driving forces and impacts. This study proposed a framework of three analytical modules. First, it investigates the land-use change intensity of different categories. It explores potential sources and driving forces for forest change in poverty alleviation practices. Second, it investigates areas that are undergoing the process of urban expansion. Third, it evaluates the ecological conditions of forests under those driving forces, using the indicators of landscape fragmentation.

4.3. Limitations

The data have an issue that Qingjiang River was not clearly displayed in 2000. This inconsistency could impact the analysis of land cover in some river segments. This is an issue related to the original data of GlobeLand30. However, the central objective of this study is to serve sustainable forest management by analyzing land-use change, urban development and expansions, and forest fragmentation. The impacted analysis is limited in the transition between water bodies and agriculture, but less in forests. Using our pixel-based evaluation, the influenced area covers less than 0.3% of the study area. The influenced area is relatively small.

5. Conclusions

According to the intensity analysis, the speed of land-use change was much faster between 2000 and 2010 than between 2010 and 2020. The gain of water bodies and loss of grassland were active throughout the two intervals. Artificial surfaces increased most intensively from 2010 to 2020 with transitions from grassland and cultivated land, indicating a rapid urbanization process. Edge-expansion was the dominant type of artificial surface growth. Furthermore, forests had the largest size of gain across the two intervals, and there was a substantial reduction in forest fragmentation in the western part of the city.

Overall, this research contributes to our understanding of land-use change and urbanization in poverty-stricken mountainous cities. The proposed analytical framework helps synthesize and understand various characteristics of land-use dynamics, ranging from the overall trends, differences among land-use categories, types and locations of urban expansion, forest fragmentation, and spatially explicit evaluation. The research findings reflect that some recent planning measures are effectively restoring the natural environment. For sustainable land management in mountainous areas, it is necessary to further refer to comprehensive spatial analysis to identify key areas to protect ecological integrity and harmonize the relationship between nature conservation and urban growth.

Author Contributions: All authors contributed to the study conception and design. C.W.: Conceptualization, Methodology, Software, Writing—Original draft preparation, Writing—Reviewing and Editing. L.W.: Conceptualization, Methodology, Software, Writing—Original draft preparation, Writing-Reviewing and Editing. All authors have read and agreed to the published version of the manuscript.

Funding: This work was supported by the Fundamental Research Funds for the Central Universities (No. 2020kfyXJJS105, 2022IVA036) and the National Natural Science Foundation of China (No. 52208060).

Data Availability Statement: The source of data used in this study is described in Section 2.2.

Acknowledgments: The authors appreciate the constructive comments and suggestions of anonymous reviewers.

Conflicts of Interest: The authors declare no conflict of interest.

References

1. Tovar, C.; Seijmonsbergen, A.C.; Duivenvoorden, J.F. Monitoring Land Use and Land Cover Change in Mountain Regions: An Example in the Jalca Grasslands of the Peruvian Andes. *Landsc. Urban Plan.* **2013**, *112*, 40–49. [CrossRef]
2. Wang, Y.; Dai, E.; Yin, L.; Ma, L. Land Use/Land Cover Change and the Effects on Ecosystem Services in the Hengduan Mountain Region, China. *Ecosyst. Serv.* **2018**, *34*, 55–67. [CrossRef]
3. Jia, L.; Ma, Q.; Du, C.; Hu, G.; Shang, C. Rapid Urbanization in a Mountainous Landscape: Patterns, Drivers, and Planning Implications. *Landsc. Ecol.* **2020**, *35*, 2449–2469. [CrossRef]
4. FAO (Food and Agriculture Organization of the United Nations). Mountains as the Water Towers of the World: A Call for Action on the Sustainable Development Goals (SDGs). Available online: http://www.fao.org/fileadmin/templates/mountain_partnership/doc/POLICY_BRIEFS/SDGs_and_mountains_water_EN.pdf (accessed on 6 November 2020).
5. Price, M.F. Mountains: Globally Important Ecosystems. *Unasylva-FAO* **1998**, *49*, 3–12.
6. Bicudo da Silva, R.F.; Millington, J.D.A.; Moran, E.F.; Batistella, M.; Liu, J. Three Decades of Land-Use and Land-Cover Change in Mountain Regions of the Brazilian Atlantic Forest. *Landsc. Urban Plan.* **2020**, *204*, 103948. [CrossRef]
7. FAO (Food and Agriculture Organization of the United Nations). The Vulnerability of Mountain Environments and Mountain People. Available online: https://www.fao.org/3/Y7352E/y7352e04.htm (accessed on 6 November 2020).
8. Vannier, C.; Lefebvre, J.; Longaretti, P.-Y.; Lavorel, S. Patterns of Landscape Change in a Rapidly Urbanizing Mountain Region. *Cybergeo* **2016**. UMR 8504 Géographie-cités. [CrossRef]
9. Mansour, S.; Al-Belushi, M.; Al-Awadhi, T. Monitoring Land Use and Land Cover Changes in the Mountainous Cities of Oman Using GIS and CA-Markov Modelling Techniques. *Land Use Policy* **2020**, *91*, 104414. [CrossRef]
10. Liang, X.; Li, Y. Identification of Spatial Coupling between Cultivated Land Functional Transformation and Settlements in Three Gorges Reservoir Area, China. *Habitat Int.* **2020**, *104*, 102236. [CrossRef]
11. Xie, W.; Jin, W.; Chen, K.; Wu, J.; Zhou, C. Land Use Transition and Its Influencing Factors in Poverty-Stricken Mountainous Areas of Sangzhi County, China. *Sustainability* **2019**, *11*, 4915. [CrossRef]
12. Liu, Y.; Yang, R.; Long, H.; Gao, J.; Wang, J. Implications of Land-Use Change in Rural China: A Case Study of Yucheng, Shandong Province. *Land Use Policy* **2014**, *40*, 111–118. [CrossRef]
13. National Foresty and Grassland Administration. "Grain for Green" Regulation. Available online: http://www.forestry.gov.cn/main/3950/20170314/459878.html (accessed on 17 May 2021).
14. Long, H.; Wu, X.; Wang, W.; Dong, G. Analysis of Urban-Rural Land-Use Change during 1995–2006 and Its Policy Dimensional Driving Forces in Chongqing, China. *Sensors* **2008**, *8*, 681–699. [CrossRef]

15. Development and Reform Commission of Hunan Province The 13th Five-Year Plan on Hunan Wuling Mountian Area Development and Poverty Alleviation. Available online: http://fgw.hunan.gov.cn/xxgk_70899/tzgg/201710/t20171031_4628536.html (accessed on 20 May 2021).
16. Aldwaik, S.Z.; Pontius, R.G. Intensity Analysis to Unify Measurements of Size and Stationarity of Land Changes by Interval, Category, and Transition. *Landsc. Urban Plan.* **2012**, *106*, 103–114. [CrossRef]
17. Mallinis, G.; Koutsias, N.; Arianoutsou, M. Monitoring Land Use/Land Cover Transformations from 1945 to 2007 in Two Peri-Urban Mountainous Areas of Athens Metropolitan Area, Greece. *Sci. Total Environ.* **2014**, *490*, 262–278. [CrossRef]
18. Huang, A.; Xu, Y.; Sun, P.; Zhou, G.; Liu, C.; Lu, L.; Xiang, Y.; Wang, H. Land Use/Land Cover Changes and Its Impact on Ecosystem Services in Ecologically Fragile Zone: A Case Study of Zhangjiakou City, Hebei Province, China. *Ecol. Indic.* **2019**, *104*, 604–614. [CrossRef]
19. Peng, J.; Liu, Y.; Li, T.; Wu, J. Regional Ecosystem Health Response to Rural Land Use Change: A Case Study in Lijiang City, China. *Ecol. Indic.* **2017**, *72*, 399–410. [CrossRef]
20. Pijanowski, B.C.; Robinson, K.D. Rates and Patterns of Land Use Change in the Upper Great Lakes States, USA: A Framework for Spatial Temporal Analysis. *Landsc. Urban Plan.* **2011**, *102*, 102–116. [CrossRef]
21. Evans, T.P.; Moran, E.F. Spatial Integration of Social and Biophysical Factors Related to Landcover Change. *Popul. Dev. Rev.* **2002**, *28*, 165–186.
22. Nechyba, T.J.; Walsh, R.P. Urban Sprawl. *J. Econ. Perspect.* **2004**, *18*, 177–200. [CrossRef]
23. Liu, H.; Yuan, X.; Gan, S.; Yan, H.; Zhang, X.; He, Q. The Urban Space Expansion Analysis in Mountain Areas Based on GIS and RS. *Appl. Mech. Mater.* **2014**, *444–445*, 1255–1259. [CrossRef]
24. Tang, J.; Li, Y.; Cui, S.; Xu, L.; Ding, S.; Nie, W. Linking Land-Use Change, Landscape Patterns, and Ecosystem Services in a Coastal Watershed of Southeastern China. *Glob. Ecol. Conserv.* **2020**, *23*, e01177. [CrossRef]
25. Ferrara, A.; Salvati, L.; Sateriano, A.; Carlucci, M.; Gitas, I.; Biasi, R. Unraveling the 'Stable' Landscape: A Multi-Factor Analysis of Unchanged Agricultural and Forest Land (1987–2007) in a Rapidly-Expanding Urban Region. *Urban Ecosyst.* **2016**, *19*, 835–848. [CrossRef]
26. Liu, Y.; Xu, Y. A Geographic Identification of Multidimensional Poverty in Rural China under the Framework of Sustainable Livelihoods Analysis. *Appl. Geogr.* **2016**, *73*, 62–76. [CrossRef]
27. Jiao, W. Analyzing Multidimensional Measures of Poverty and Their Influences in China's Qinba Mountains. *Chin. J. Popul. Resour. Environ.* **2020**, *18*, 214–221. [CrossRef]
28. Zhou, L.; Xiong, L.Y. Natural Topographic Controls on the Spatial Distribution of Poverty-Stricken Counties in China. *Appl. Geogr.* **2018**, *90*, 282–292. [CrossRef]
29. The Government of Enshi City. About Enshi City. Available online: http://www.es.gov.cn/zjes/ (accessed on 19 February 2021).
30. Li, H.; Niu, X.; Wang, B.; Zhao, Z. Coupled Coordination of Ecosystem Services and Landscape Patterns: Take the Grain for Green Project in the Wuling Mountain Area as an Example. *Acta Ecol. Sin.* **2020**, *40*, 4316–4326.
31. Zhang, J. Research on Ecological Security Assessment of Enshi. Doctoral thesis, Central China Normal University, Wuhan, China, 2015.
32. Hubei Provincial People's Government. All the 37 Poverty-Stricken Counties in Hubei Province Eliminated Poverty. Available online: http://www.hubei.gov.cn/hbfb/bmdt/202004/t20200423_2239712.shtml (accessed on 19 December 2020).
33. Enshi Prefecture Statistics Bureau. *Enshi Statistical Yearbook*; Enshi Prefecture Statistics Bureau: Enshi, China, 2019.
34. He, Q.; Zhou, J.; Tan, S.; Song, Y.; Zhang, L.; Mou, Y.; Wu, J. What Is the Developmental Level of Outlying Expansion Patches? A Study of 275 Chinese Cities Using Geographical Big Data. *Cities* **2020**, *105*, 102395. [CrossRef]
35. Tachikawa, T.; Kaku, M.; Iwasaki, A.; Gesch, D.B.; Oimoen, M.J.; Zhang, Z.; Danielson, J.J.; Krieger, T.; Curtis, B.; Haase, J. ASTER Global Digital Elevation Model Version 2-Summary of Validation Results. Available online: https://pubs.er.usgs.gov/publication/70005960 (accessed on 10 November 2020).
36. National Geomatics Center of China. GlobeLand30 Product Introduction. Available online: http://www.globallandcover.com/Page/EN_sysFrame/dataIntroduce.html?columnID=81&head=product&para=product&type=data (accessed on 8 November 2020).
37. Jun, C.; Ban, Y.; Li, S. Open Access to Earth Land-Cover Map. *Nature* **2014**, *514*, 434. [CrossRef]
38. Feng, Y.; Lei, Z.; Tong, X.; Gao, C.; Chen, S.; Wang, J.; Wang, S. Spatially-Explicit Modeling and Intensity Analysis of China's Land Use Change 2000–2050. *J. Environ. Manag.* **2020**, *263*, 110407. [CrossRef]
39. Sun, X.; Li, G.; Wang, J.; Wang, M. Quantifying the Land Use and Land Cover Changes in the Yellow River Basin While Accounting for Data Errors Based on GlobeLand30 Maps. *Land* **2021**, *10*, 31. [CrossRef]
40. Dong, S.; Gao, B.; Pan, Y.; Li, R.; Chen, Z. Assessing the Suitability of FROM-GLC10 Data for Understanding Agricultural Ecosystems in China: Beijing as a Case Study. *Remote Sens. Lett.* **2020**, *11*, 11–18. [CrossRef]
41. Gong, P.; Liu, H.; Zhang, M.; Li, C.; Wang, J.; Huang, H.; Clinton, N.; Ji, L.; Li, W.; Bai, Y.; et al. Stable Classification with Limited Sample: Transferring a 30-m Resolution Sample Set Collected in 2015 to Mapping 10-m Resolution Global Land Cover in 2017. *Sci. Bull.* **2019**, *64*, 370–373. [CrossRef]
42. Venter, Z.S.; Barton, D.N.; Chakraborty, T.; Simensen, T.; Singh, G. Global 10 m Land Use Land Cover Datasets: A Comparison of Dynamic World, World Cover and Esri Land Cover. *Remote Sens.* **2022**, *14*, 4101. [CrossRef]
43. Wang, H.; Wen, X.; Wang, Y.; Cai, L.; Peng, D.; Liu, Y. China's Land Cover Fraction Change during 2001–2015 Based on Remote Sensed Data Fusion between Mcd12 and Cci-Lc. *Remote Sens.* **2021**, *13*, 341. [CrossRef]

44. Exavier, R.; Zeilhofer, P. OpenLand: Quantitative Analysis and Visualization of LUCC. Available online: https://cran.r-project.org/package=OpenLand (accessed on 10 November 2020).
45. Hijmans, R.J. Terra: Spatial Data Analysis. Available online: https://cran.r-project.org/package=terra (accessed on 6 November 2020).
46. Pebesma, E. Simple Features for R: Standardized Support for Spatial Vector Data. *R J.* **2018**, *10*, 439. [CrossRef]
47. Hesselbarth, M.H.K.; Sciaini, M.; With, K.A.; Wiegand, K.; Nowosad, J. Landscapemetrics: An Open-Source R Tool to Calculate Landscape Metrics. *Ecography* **2019**, *42*, 1648–1657. [CrossRef]
48. Lovelace, R.; Martijn, T. ClockBoard: A Zoning System for Urban Analysis. Available online: https://zonebuilders.github.io/zonebuilder/articles/paper.html (accessed on 1 May 2021).
49. Estoque, R.C.; Murayama, Y. Intensity and Spatial Pattern of Urban Land Changes in the Megacities of Southeast Asia. *Land Use Policy* **2015**, *48*, 213–222. [CrossRef]
50. Teixeira, Z.; Marques, J.C.; Pontius, R.G. Evidence for Deviations from Uniform Changes in a Portuguese Watershed Illustrated by CORINE Maps: An Intensity Analysis Approach. *Ecol. Indic.* **2016**, *66*, 282–290. [CrossRef]
51. Akinyemi, F.O.; Pontius, R.G.; Braimoh, A.K. Land Change Dynamics: Insights from Intensity Analysis Applied to an African Emerging City. *J. Spat. Sci.* **2017**, *62*, 69–83. [CrossRef]
52. Pontius, R.G.; Gao, Y.; Giner, N.M.; Kohyama, T.; Osaki, M.; Hirose, K. Design and Interpretation of Intensity Analysis Illustrated by Land Change in Central Kalimantan, Indonesia. *Land* **2013**, *2*, 351–369. [CrossRef]
53. Liu, X.; Li, X.; Chen, Y.; Tan, Z.; Li, S.; Ai, B. A New Landscape Index for Quantifying Urban Expansion Using Multi-Temporal Remotely Sensed Data. *Landsc. Ecol.* **2010**, *25*, 671–682. [CrossRef]
54. Bosch, M.; Jaligot, R.; Chenal, J. Spatiotemporal Patterns of Urbanization in Three Swiss Urban Agglomerations: Insights from Landscape Metrics, Growth Modes and Fractal Analysis. *Landsc. Ecol.* **2020**, *35*, 879–891. [CrossRef]
55. He, Q.; Zeng, C.; Xie, P.; Tan, S.; Wu, J. Comparison of Urban Growth Patterns and Changes between Three Urban Agglomerations in China and Three Metropolises in the USA from 1995 to 2015. *Sustain. Cities Soc.* **2019**, *50*, 101649. [CrossRef]
56. Pili, S.; Serra, P.; Salvati, L. Landscape and the City: Agro-Forest Systems, Land Fragmentation and the Ecological Network in Rome, Italy. *Urban For. Urban Green.* **2019**, *41*, 230–237. [CrossRef]
57. Peng, J.; Wang, Y.; Zhang, Y.; Wu, J.; Li, W.; Li, Y. Evaluating the Effectiveness of Landscape Metrics in Quantifying Spatial Patterns. *Ecol. Indic.* **2010**, *10*, 217–223. [CrossRef]
58. Frank, S.; Fürst, C.; Koschke, L.; Witt, A.; Makeschin, F. Assessment of Landscape Aesthetics—Validation of a Landscape Metrics-Based Assessment by Visual Estimation of the Scenic Beauty. *Ecol. Indic.* **2013**, *32*, 222–231. [CrossRef]
59. Wan, L.; Zhang, Y.; Zhang, X.; Qi, S.; Na, X. Comparison of Land Use/Land Cover Change and Landscape Patterns in Honghe National Nature Reserve and the Surrounding Jiansanjiang Region, China. *Ecol. Indic.* **2015**, *51*, 205–214. [CrossRef]
60. Xiao, H.; Liu, Y.; Li, L.; Yu, Z.; Zhang, X. Spatial Variability of Local Rural Landscape Change under Rapid Urbanization in Eastern China. *ISPRS Int. J. Geo-Inf.* **2018**, *7*, 231. [CrossRef]
61. Sakieh, Y.; Salmanmahiny, A. Performance Assessment of Geospatial Simulation Models of Land-Use Change—A Landscape Metric-Based Approach. *Environ. Monit. Assess.* **2016**, *188*, 169. [CrossRef]
62. Li, C.; Li, J.; Wu, J. Quantifying the Speed, Growth Modes, and Landscape Pattern Changes of Urbanization: A Hierarchical Patch Dynamics Approach. *Landsc. Ecol.* **2013**, *28*, 1875–1888. [CrossRef]
63. Fan, C.; Myint, S. A Comparison of Spatial Autocorrelation Indices and Landscape Metrics in Measuring Urban Landscape Fragmentation. *Landsc. Urban Plan.* **2014**, *121*, 117–128. [CrossRef]
64. Girvetz, E.H.; Thorne, J.H.; Berry, A.M.; Jaeger, J.A.G. Integration of Landscape Fragmentation Analysis into Regional Planning: A Statewide Multi-Scale Case Study from California, USA. *Landsc. Urban Plan.* **2008**, *86*, 205–218. [CrossRef]
65. Liu, S.; Dong, Y.; Deng, L.; Liu, Q.; Zhao, H.; Dong, S. Forest Fragmentation and Landscape Connectivity Change Associated with Road Network Extension and City Expansion: A Case Study in the Lancang River Valley. *Ecol. Indic.* **2014**, *36*, 160–168. [CrossRef]
66. Wang, L.; Wen, C. Traditional Villages in Forest Areas: Exploring the Spatiotemporal Dynamics of Land Use and Landscape Patterns in Enshi Prefecture, China. *Forests* **2021**, *12*, 65. [CrossRef]
67. Jaeger, J.A.G. Landscape Division, Splitting Index, and Effective Mesh Size: New Measures of Landscape Fragmentation. *Landsc. Ecol.* **2000**, *15*, 115–130. [CrossRef]
68. Schmiedel, I.; Culmsee, H. The Influence of Landscape Fragmentation, Expressed by the 'Effective Mesh Size Index', on Regional Patterns of Vascular Plant Species Richness in Lower Saxony, Germany. *Landsc. Urban Plan.* **2016**, *153*, 209–220. [CrossRef]
69. Mcgarigal, K. *FRAGSTATS Help*; University of Massachusetts: Amherst, MA, USA, 2015.
70. Moser, B.; Jaeger, J.A.G.; Tappeiner, U.; Tasser, E.; Eiselt, B. Modification of the Effective Mesh Size for Measuring Landscape Fragmentation to Solve the Boundary Problem. *Landsc. Ecol.* **2007**, *22*, 447–459. [CrossRef]
71. Luo, T.; Zhang, T.; Wang, Z.; Gan, Y. Driving Forces of Landscape Fragmentation Due to Urban Transportation Networks: Lessons from Fujian, China. *J. Urban Plan. Dev.* **2016**, *142*, 04015013. [CrossRef]
72. Tarabon, S.; Calvet, C.; Delbar, V.; Dutoit, T.; Isselin-Nondedeu, F. Integrating a Landscape Connectivity Approach into Mitigation Hierarchy Planning by Anticipating Urban Dynamics. *Landsc. Urban Plan.* **2020**, *202*, 103871. [CrossRef]

forests

Article

Research on Vegetation Cover Changes in Arid and Semi-Arid Region Based on a Spatio-Temporal Fusion Model

Zhihong Liu [1], Donghua Chen [2,3,*], Saisai Liu [2], Wutao Feng [1], Fengbing Lai [1], Hu Li [3,*], Chen Zou [3], Naiming Zhang [2] and Mei Zan [1]

1 School of Geography and Tourism, Xinjiang Normal University, Urumqi 830054, China
2 School of Computer and Information Engineering, Chuzhou University, Chuzhou 239000, China
3 School of Geography and Tourism, Anhui Normal University, Wuhu 241003, China
* Correspondence: chendonghua@chzu.edu.cn (D.C.); lihu2881@ahnu.edu.cn (H.L.); Tel.: +86-055-0351-0251 (D.C.); +86-183-2537-6353 (H.L.)

Abstract: Vegetation dynamics in arid and semi-arid regions have an important impact on carbon cycle, water cycle, and energy exchange at local, regional, and global scales. Therefore, it is of great significance for scientists to grasp the changes of vegetation cover in arid and semi-arid regions timely and accurately. Based on this, the applicability of ESTARFM model in the complex terrain area of arid and semi-arid Xinjiang was explored using Landsat and MODIS data fusion, and the overall change characteristics of vegetation cover (FVC) and the distribution and change patterns of different terrains in the study area in the past 15 years were analyzed by combining the dimidiate pixel model, unary linear regression and digital elevation model. The results show that: (1) the *NDVI* data fused by ESTARFM Model has high consistency with the real *NDVI* data, and it can be used for subsequent *FVC* estimation. (2) From 2006 to 2020, the inter *FVC* was at a high level as a whole, and the average annual *FVC* showed a weak increasing trend in fluctuation; there are obvious differences in spatial distribution, which is characterized by high distribution in the north and low in the south. (3) The improved area of vegetation cover in the study area is greater than the degraded area, accounting for 52.3% and 47.7% respectively; (4) In the elevation range of 2000 to 3500 m, the *FVC* showed a slight degradation trend on 25° to 45° slopes and south and southeast slopes, and the rest showed a slight improvement trend. ESTARFM-based model enables monitoring of vegetation cover changes in complex terrain areas of the arid and semi-arid regions in Xinjiang over a long time series. The overall *FVC* level in the study area is high, and there both are serious degradation and improvement phenomena.

Keywords: remote sensing; ESTARFM; arid and semi-arid region; FVC; time-series

Citation: Liu, Z.; Chen, D.; Liu, S.; Feng, W.; Lai, F.; Li, H.; Zou, C.; Zhang, N.; Zan, M. Research on Vegetation Cover Changes in Arid and Semi-Arid Region Based on a Spatio-Temporal Fusion Model. *Forests* **2022**, *13*, 2066. https://doi.org/10.3390/f13122066

Academic Editors: Chao Wang, Fan Zhang and Wei Liu

Received: 1 November 2022
Accepted: 2 December 2022
Published: 4 December 2022

1. Introduction

Vegetation is an important component of terrestrial ecosystems and plays an irreplaceable role in global material and energy cycling, regulating carbon balance, and maintaining climate stability [1–4], as well as an "indicator" of global ecosystem changes [5,6]. The monitoring of dynamic changes of vegetation has long been important and studied by many scholars at home and abroad [7–10]. Vegetation cover (FVC) is the percentage of the vertical projected area of the above-ground part of vegetation (including leaves, stems, and branches) on the ground to the total area of the statistical area [11], which is a comprehensive quantitative indicator describing the growth of vegetation on the ground and an important indicator of the regional ecosystem environment [12,13].

Two methods are mainly used to study *FVC*: ground survey method and remote sensing monitoring method [14,15]. The traditional ground survey method has problems such as large workload, high labor intensity, and difficulty in achieving monitoring of long time series. With the continuous development of remote sensing technology, the use of remote sensing technology to monitor vegetation changes has become a feasible means and has been widely used [16,17]. Some scholars have estimated and analyzed the *FVC* change trends in Tianshan or North Tianshan for the past 16 years based on 3–5 Landsat TM images, respectively [18–20]. Some scholars have also studied the changes in vegetation cover or grassland cover and their influencing factors in Xinjiang and even in a larger region based on MODIS data [21,22]. Another group of scholars studied the vegetation growth trends in China (Northwest) based on the GIMMS dataset and analyzed to expose the influencing factors [23–25].

From the above studies, it can be found that MODIS and GIMMS data are time-sensitive and can better express the time-series changes of *FVC*, especially for the monitoring of abrupt change years; by contrast, Landsat data have higher spatial resolution and can achieve finer monitoring of vegetation change information, but it is difficult to find abrupt change information in time-series change monitoring due to the limitations of data re-simulation period and cloud volume [18,21]. Gao et al. [26] first proposed the spatial and temporal adaptive reflection fusion model (STARFM) in 2006 to achieve the fusion of Landsat images and MODIS images. This model can accurately predict reflectance values with high spatial and temporal resolution. However, the STARFM model also has limitations, and the method does not fuse well in areas with high heterogeneity. In view of this limitation, Zhu et al. [27] proposed the enhanced spatio-temporal adaptive reflection fusion model (ESTARFM) in 2010. This model improved the con-version coefficient based on the STARFM model to improve the accuracy of image fusion in heterogeneous regions. It was shown that the model has higher fusion accuracy in complex surface mountainous areas.

Xinjiang is a typical arid and semi-arid region with an extremely fragile internal ecological environment, and the West Tianshan forest region within its territory is one of the areas rich in vegetation resources and plays the role of an ecological barrier in western China [28]. The topography in this region is complex and undulating, and each topographic factor has a complex and integrated influence on the vegetation growth environment, so it is important to carry out vegetation change research with high spatial resolution for ecological environmental protection in complex topographic areas. On the other hand, the ecological environment in this area is complex and extremely sensitive to climate change and human activities, so monitoring vegetation change in time series can help to discover the year of sudden vegetation change, and then analyze the change factors to achieve ecological environmental protection more accurately. However, there is no research on the analysis of vegetation change in this region based on long time series remote sensing data with high spatial resolution. In this paper, based on Landsat data and MODIS data, the ESTARFM model was used to fuse the high spatial resolution time series normalized vegetation index (NDVI) data of each growing season in the study area from 2006 to 2020, and the image dichotomy model was used to estimate the *FVC*. The distribution and change characteristics of *FVC* under different topographic factors were revealed by combining with digital elevation model (DEM), in order to provide reference for ecological environment improvement in semi-arid areas of Xinjiang.

2. Materials and Methods

2.1. Study Area

Tekes Forest Farm, Qapqal Forest Farm and Zhaosu Forest Farm of the State-owned Forest Administration Bureau in the Western Tianshan Mountains of Xinjiang Uygur Autonomous Region are selected as the study areas in this study. They are mainly located in Ili Region, with a geographical location of 42°15′26″–43°33′38″ N, 80°12′19″–82°37′52″ E, with a total area of about 14,656 km^2 (Figure 1). The overall terrain is high in the north and south and low in the middle with an altitude of about 919~6170 m, an average sea wave of 2725 m, and undulating terrain. The study area is deeply inland and belongs to a temperate continental semi-arid climate. The annual and daily temperature range is large. The average annual temperature in the mountain area is 2.8 °C, and the average annual temperature in the river valley is 5.3 °C; on the other hand, it has more precipitation than surrounding areas due to its location in the mid latitude westerly belt and the influence of mountains as well as water vapor from the Atlantic Ocean. It is the precipitation center in the arid region of Central Asia, with an annual precipitation of 550~700 mm. The soil vertical zones in the study area are relatively complete, including mountain chestnut soil, mountain chernozem, mountain leaching taupe forest soil, mountain common taupe forest soil, and mountain grass taupe forest soil.

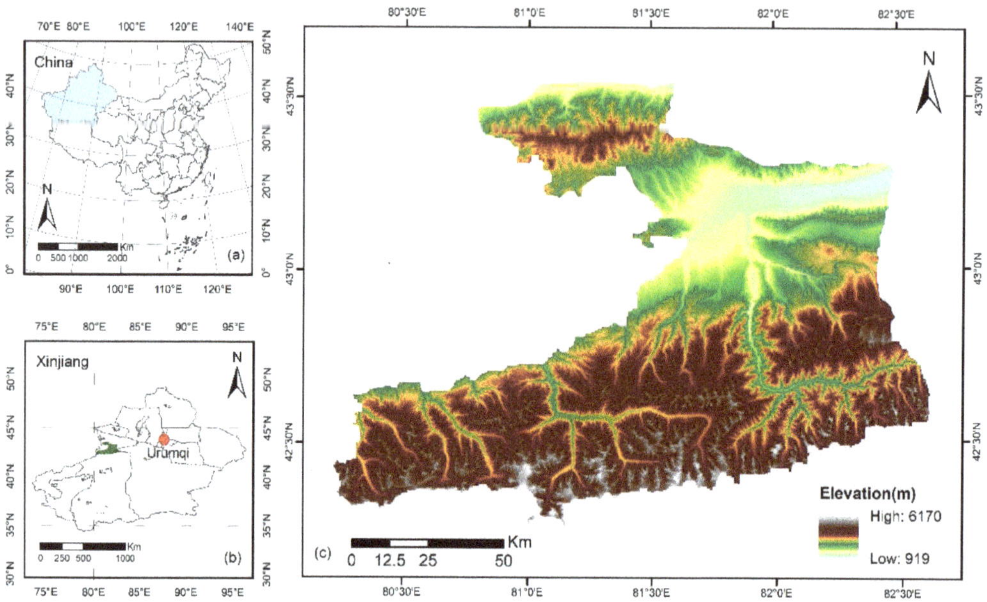

Figure 1. The location and topographic features of the study area, (**a**) in China, (**b**) in Xinjiang Uygur Autonomous Region and (**c**) are topographic features of the study area, and elevation data were obtained from ASTER Global Digital Model.

2.2. Data Source and Pre-Processing

To avoid errors in monitoring results caused by seasonal changes in cloud cover and vegetation, the selected remote sensing data are Landsat series data and MOD09A1 data from July-August, the peak vegetation growth season each year. The cloud amount is less than 10% and the data source is the United States Geological Survey (USGS) (https://earthexplorer.usgs.gov/, accessed on 28 May 2021). The interval between MOD09A1 data collection time and Landsat data is not more than 4 days. Please refer to Appendix A Tables A1 and A2 for specific information of remote sensing data. There are 11 years of directly available Landsat data from 2006 to 2020, and the years to be predicted are shown in Table 1, where the predicted data for 2007 and 2019 are correlated with Landsat TM data for 2007 and Landsat OLI data for 2019 to verify the feasibility of the ESTARFM algorithm. The auxiliary data include the administrative map of the Tianxi Bureau and the DEM data with a spatial resolution of 30 m from the USGS.

Table 1. Study Area Demand Forecast Data 2006–2020.

Year	Forecast Date	Year	Forecast Date
2007 (Experiment)	08.13	2015	08.13
2009	08.13	2017	08.13
2010	08.13	2019 (Experiment)	08.13

The Landsat data are radiometric calibrated, atmospheric corrected, mosaic, and cropped by ENVI software to obtain the image map of the study area. MRT (MODIS Reconstruction Tool) Tool is used for splicing MOD09A1 data, converting the projection into the same projection as that of Landsat. The infrared band and near-infrared band were selected as the bands, and the nearest neighbor sampling method was used for re-projecting the data to the same 30 m spatial resolution as Landsat data, and was outputted in GEOTIFF format. ENVI software is used for cutting the processed MOD09A1 data to ensure that the boundary range and pixel size are consistent with the Landsat data after cutting. Pre-processing was carried out, such as splicing, re-projection and clipping of DEM data, and altitude, slope and slope direction were extracted using ArcGIS software and spatial analysis tools.

2.3. Research Methods

2.3.1. ESTARFM Spatiotemporal Fusion MODEL

ESTARFM Model requires two pairs of high-resolution and low-resolution images acquired in the same period, and a set of low-resolution images used for predicting dates [27]. The Model sets a sliding window of a certain size around the predicted pixel, and convolutes the pixels in the window with the weight function to obtain the predicted value of the central pixel. The sliding window moves one by one on the whole image to obtain the predicted image. The calculation Formula is as follows:

$$L\left(x_{\frac{w}{2}},y_{\frac{w}{2}},t_p,B\right) = L\left(x_{\frac{w}{2}},y_{\frac{w}{2}},t_0,B\right) + \sum_{i=1}^{N} W_i \times V_i \times \left(M\left(x_i,y_i,t_p,B\right) - M(x_i,y_i,t_0,B)\right) \quad (1)$$

$$\beta_t = \frac{\frac{1}{\left|\sum_{j=1}^{w}\sum_{i=1}^{w} M\left(x_i,y_j,t_\beta,B\right) - \sum_{j=1}^{w}\sum_{i=1}^{w} M\left(x_i,y_j,t_p,B\right)\right|}}{\sum_{t=b,e}\left(\frac{1}{\left|\sum_{j=1}^{w}\sum_{i=1}^{w} M\left(x_i,y_j,t_\beta,B\right) - \sum_{j=1}^{w}\sum_{i=1}^{w} M\left(x_i,y_j,t_p,B\right)\right|}\right)}, (t = b,e) \quad (2)$$

$$L\left(x_{\frac{w}{2}},y_{\frac{w}{2}},t_p,B\right) = \beta_b \times L\left(x_{\frac{w}{2}},y_{\frac{w}{2}},t_b,B\right) + \beta_e \times L\left(x_{\frac{w}{2}},y_{\frac{w}{2}},t_e,B\right) \quad (3)$$

where, L represents Landsat data; M represents MODIS data; w is the size of the sliding window; $(x_{w/2},y_{w/2})$ represents the center position of the pixel to be measured; B represents the band of the image; $L(x_{w/2},y_{w/2},t_p,B)$ is the predicted high-resolution pixel value at time t_p; $L(x_{w/2},y_{w/2},t_0,B)$ is the high-resolution pixel value at t_0; N is the number of similar pixels including predicted pixels; (x_i,y_i) is the position of the ith similar pixel; W_i is the weight of the ith similar pixel jointly determined by the distance of space, time and spectrum; V_i is the conversion coefficient of similar pixels.

In Equation (1), the following aspects need to be noted:

(a) Determining the window size. The window size w is set to 25 according to the recommended parameters used by Zhu [27].

(b) Searching for similar image elements, which are obtained based on the spectral similarity of Landsat *NDVI* data to the central image element $(x_{w/2},y_{w/2})$ using a sliding window w search.

(c) Calculation of weights. To calculate the weights, we first calculate the correlation coefficient Ri between Landsat *NDVI* and MODIS *NDVI* data, and then find the geographic distance di between the similar image element and the central image element, and finally find the weight Wi of the similar image element to the central image element based on the correlation coefficient and geographic distance.

$$R_i = \frac{E[(L_i - E(L_i))(M_i - E(M_i))]}{\sqrt{D(L_i)} \times \sqrt{D(M_i)}} \quad (4)$$

$$L_i = \{L(x_i,y_i,T_b,B_1),\cdots L(x_i,y_i,T_b,B_n),L(x_i,y_i,T_e,B_1),\ldots,L(x_i,y_i,T_e,B_n)\} \quad (5)$$

$$M_i = \{M(x_i,y_i,T_b,B_1),\cdots M(x_i,y_i,T_b,B_n),M(x_i,y_i,T_e,B_1),\ldots,M(x_i,y_i,T_e,B_n)\} \quad (6)$$

In Equations (4)–(6), the correlation coefficient is found for each similar image element, i represents the ith similar image element. L_i/M_i is the ith similar image element data set of Landsat/MODIS image, respectively. b is the number of bands, e is the mean value, and d is the variance. The value of correlation coefficient R is in the range of (−1, 1), and the value of R is positively correlated with the spectral similarity.

$$d_i = 1 + \frac{\sqrt{(x_{w/2} - x_i)^2 + (y_{w/2} - y_i)^2}}{\frac{w}{2}} \tag{7}$$

where d_i is the geographic distance between the similar image element and the central image element, and the larger d_i is, the farther the geographic distance is.

Based on the above, the weighting formula is as follows.

$$D_i = (1 - R_i) \times d_i \tag{8}$$

$$w_i = \frac{1/D_i}{\sum_{i=1}^{N}(1/D_i)} \tag{9}$$

where the range of the weight w_i values is (0, 1) and the sum of the weights of all similar image elements is 1.0.

(d) Calculate the conversion coefficients.

$$V_i = \frac{L_{ib} - L_{ie}}{M_{ib} - M_{ie}} \tag{10}$$

where L_{ia}, L_{ie}, M_{ia}, and M_{ie} are the image element values of Landsat and MODIS data corresponding to different periods (T_b and T_e) respectively, i.e., different image elements correspond to different conversion factors.

Through Formula (1), MODIS data of two different periods (t_b and t_e) are selected for calculating the high-resolution image reflectance of the prediction date T_p, which is recorded as L_b ($x_{w/2}$,$y_{w/2}$,t_p,B) and L_e ($x_{w/2}$,$y_{w/2}$,t_p,B). Combining the two prediction results, the predicted central pixel reflectance is more accurate. The weight is calculated as Formula (2) with higher weight closer to the prediction period as the criterion. The high-resolution reflectance data of the final prediction time is calculated by Formula (3) [27].

Based on different fusion sequences, the ESTARFM model has two schemes. Fusion then Index (BI) and Index then Fusion (IB). BI first simulates the reflectance of the predicted date image and then calculates the vegetation index based on the pre-predicted reflectance; IB calculates the vegetation index first and then uses these vegetation index data to simulate the vegetation index of the predicted date. According to Jarihani et al. [29], the IB scheme has lower computational cost and less error compared with the BI scheme. On this basis, in this paper, continuous normalized vegetation index data with high spatial resolution are obtained by fusion of IB schemes.

2.3.2. *FVC* Estimation

Methods for measuring *FVC* based on remote sensing data chiefly include the regression model method, vegetation index method, and pixel decomposition model [30]. Among them, the pixel dichotomous model in the pixel decomposition model is commonly used. Its principle is that, assuming that the surface corresponding to each pixel only contains two components, vegetation and bare land, the proportion of vegetation in the pixel is the *FVC* of the pixel. The calculation formula is [31]:

$$FVC = \frac{S - S_{soil}}{S_{veg} - S_{soil}} \tag{11}$$

where *FVC* denotes the vegetation cover, S is the remote sensing information reflected by each image element, S_{soil} is the remote sensing information reflected by pure bare ground cover image element, and S_{veg} is the remote sensing information reflected by pure vegetation cover image element. Rundquist [32] showed that *NDVI* has a good correlation with *FVC*, which is in line with the conditions of pixel binary model. Li [33] established a *FVC* estimation model based on *NDVI*, and the calculation formula is:

$$FVC = \frac{NDVI - NDVI_{soil}}{NDVI_{veg} - NDVI_{soil}} \quad (12)$$

where $NDVI_{soil}$ is the *NDVI* value of pure bare ground cover image elements and $NDVI_{veg}$ is the *NDVI* value of pure vegetation cover image elements. Due to the influence of meteorological elements, vegetation types, and spatial and temporal differences, $NDVI_{soil}$ and $NDVI_{veg}$ will vary in different images. Based on the experience of previous studies [33], combined with the histogram of *NDVI* frequency distribution in this experiment, the cumulative frequency values of 5% and 95% were selected to represent different $NDVI_{soil}$ and $NDVI_{veg}$ in different years, respectively, and the *FVC* was classified into five levels with reference to the Soil Erosion Classification and Grading Standard and related studies [34]. The very low *FVC* is <10%, the low *FVC* is 10%–30%, the medium-low *FVC* is 30%–50%, the medium-high *FVC* is 50%–70%, and the high *FVC* is >70%.

2.3.3. Accuracy Verification

In this paper, we use the subimage comparison method based on the comparison of high-resolution images of the same period to validate the vegetation cover estimated based on Landsat data in the Western Tian Shan [35,36]. The pretreated 18 August 2019 Gaofen-1 Image (2 m) and 11 August 2019 Landsat Image were randomly selected for verification, 100 sample points on the Landsat Image were evenly selected and mapped to the Gaofen-1 Image, and the 225 Gaofen-1 pixels corresponding to each sample point were directly manually digitized to calculate the actual *FVC*. Then, the *FVC* estimated by the pixel binary model was compared and analyzed, and the model accuracy was judged by the Root Mean Square Error (*RMSE*). The *RMSE* Calculation Formula is:

$$RMSE = \left[\sum_{i=1}^{n} \frac{(X_i - Y_i)^2}{n}\right]^{\frac{1}{2}} \quad (13)$$

where X_i is the *FVC* estimated by the Model; Y_i is the actual *FVC*; n is the number of samples.

2.3.4. Annual Variation Trend of *FVC*

The univariate linear regression trend analysis method was used, whilst the least square method was used for fitting the slope of the annual average *FVC* pixel by pixel, simulating the change trend of each grid [37,38]. The Calculation Formula is as follows:

$$\theta_{slope} = \frac{n \times \sum_{i=1}^{n}(i \times FVC_i) - (\sum_{i=1}^{n} i)(\sum_{i=1}^{n} FVC_i)}{n \times \sum_{i=1}^{n} i^2 - (\sum_{i=1}^{n} i)^2} \quad (14)$$

where θ_{slope} is the slope of the change trend of the *FVC* in the multi-year time series, i is the serial number of the study year, n is the length of the study time series, and FVC_i is the *FVC* value in the ith year. If $\theta_{slope} > 0$ indicates that the change trend of *FVC* is increasing, $\theta_{slope} < 0$ indicates that the change trend of *FVC* is decreasing. The significance test of the trend adopts F test, and the significance indicates the reliability of the change trend. According to the test results, the change trend is divided into five grades: extremely significant decrease ($\theta_{slope} < 0$, $p \leq 0.01$); significantly reduced ($\theta_{slope} < 0$, $0.01 < p \leq 0.05$); there was no significant change ($p > 0.05$); significant increase ($\theta_{slope} > 0$, $0.01 < p \leq 0.05$); and extremely significant increase ($\theta_{slope} > 0$, $p \leq 0.01$).

3. Results

3.1. ESTARFM Model Fusion Results

In order to verify the fusion results of ESTARFM Model, TM *NDVI* was taken in this paper on 21 June 2007 and 25 September 2007, as well as OLI *NDVI* on 30 May 2019 and 14 October 2019 and the corresponding MODIS *NDVI* as an input data. The *NDVI* data was fused on 13 August 2007 and 13 August 2019, and then compared with the real *NDVI* data in the corresponding period. Figure 2 is a comparison of the fusion results of the ESTARFM Model. It can be found that compared with MODIS *NDVI* data through visual interpretation, the spatial resolution of the fused *NDVI* data is significantly improved; the two data have good consistency compared with TM / OLI *NDVI* data. Meanwhile, it is found from Figure 2b,c that the *NDVI* data fused by ESTARFM restores the original spectral information of the cloud cover area.

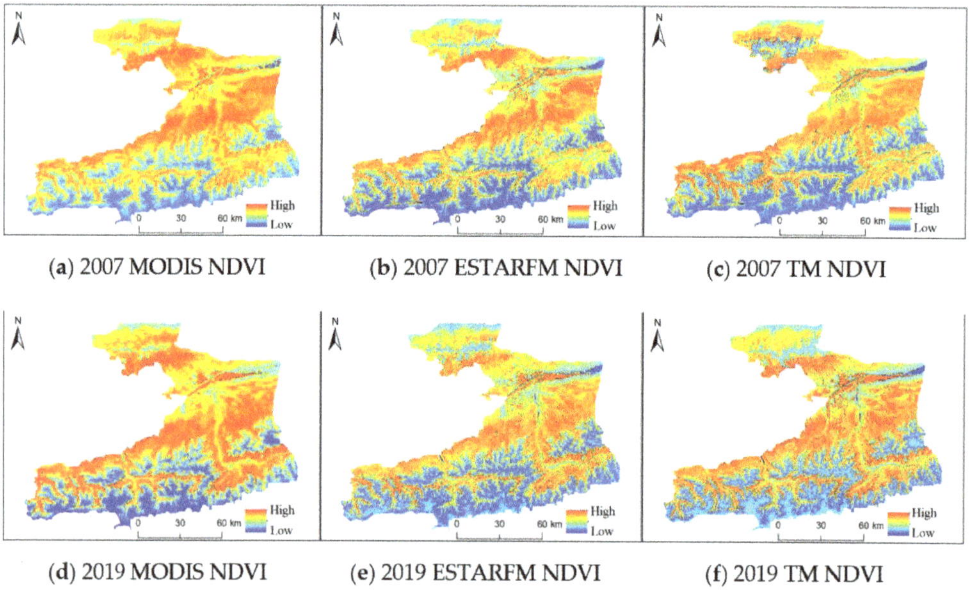

Figure 2. Comparison of ESTARFM model fusion results.

To further analyze the fusion effect, five typical 100 × 100 image subsets (Figure 3) were selected in this paper, and the fused *NDVI* data were compared with the real Landsat *NDVI* data, and a scatter plot was obtained (Figure 4). From Figure 4, it can be seen that the scatter points are basically distributed on both sides of the contour 1:1, indicating that the fused *NDVI* data are similar to the real *NDVI* data, and the determination coefficients of the fused *NDVI* data and the real *NDVI* data in the two scenes are 0.86 and 0.88, and the root mean square errors are 0.15 and 0.12, respectively, with high correlation. Therefore, the *NDVI* data obtained by fusion based on the ESTARFM model can be used for the estimation of secondary vegetation.

3.2. FVC Accuracy Verification

The accuracy verification results (Figure 5) show that the *RMSE* between the estimated value of the pixel binary model and the real value is 0.10, whilst the R^2 is 0.90, indicating that the *FVC* estimation results of Landsat data using the pixel binary model have a high accuracy and meet the requirements of this study.

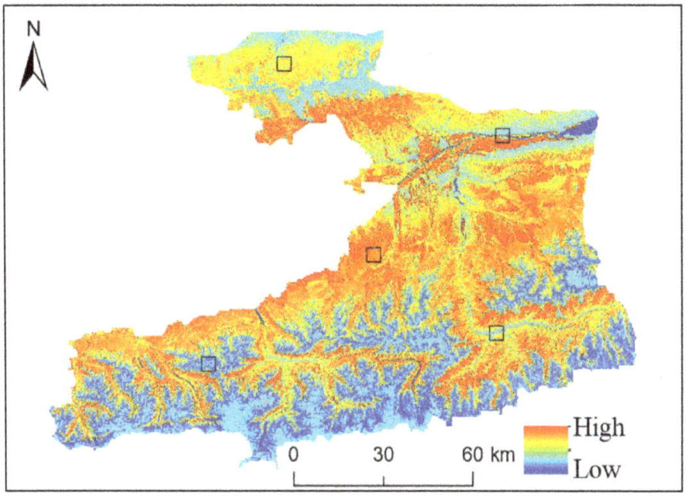

Figure 3. Distribution of 5 typical image subsets.

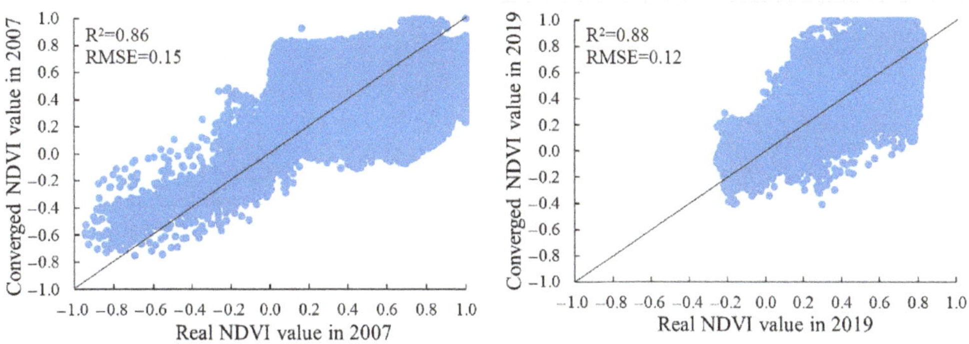

(**a**) 2007 (**b**) 2019

Figure 4. Scatter plot of fused *NDVI* and real *NDVI* and correlation coefficient.

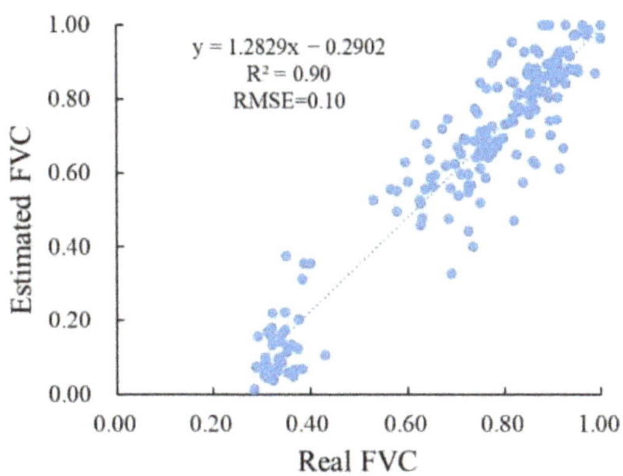

Figure 5. Comparison of estimated *FVC* and real *FVC*.

3.3. Temporal and Spatial Variation Characteristics of FVC

Figure 6 shows the inter annual difference of *FVC*, which is the average of the difference of *FVC* every two years in the study time series to analyze its inter annual change degree. As can be seen from Figure 6, *FVC* showed an increasing trend from 2006 to 2010; except for 2012, *FVC* showed a decreasing trend from 2011 to 2014; except for 2017, *FVC* showed an increasing trend from 2015 to 2019; and in 2020, it showed a decreasing trend again. Among them, the largest increases and decreases were in 2007 (0.028) and 2013 (−0.040), respectively.

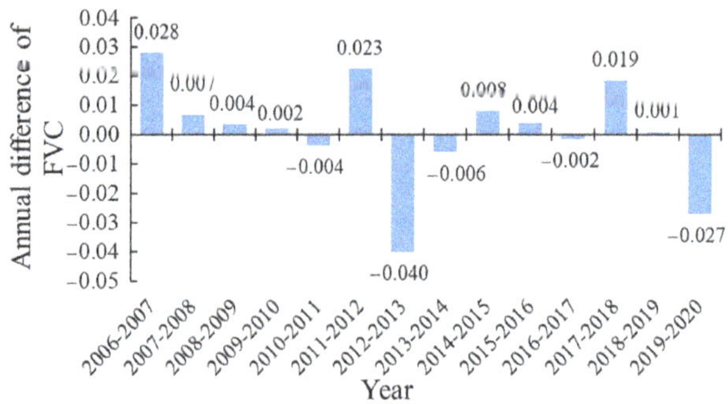

Figure 6. Annual *FVC* variance chart.

As shown in Figure 7, through statistical analysis of *FVC* at all levels in the study area, it can be found that from 2006 to 2020, the proportion of high *FVC* area in the study area showed a law of increase-decrease-increase, reaching a minimum of about 38% in 2006 and a maximum of about 49% in 2012; the proportion of medium high and medium *FVC* area showed a trend of decrease-increase- decrease, with the minimum proportions of 13% in 2013 and 9% in 2007, and the maximum proportion of 18% in 2019 and 14% in 2014; the area proportion of low and very low *FVC* changes in a wave pattern, and is stable between 14~16% and 12~14%, respectively. The high, medium high, medium, low and extremely low *FVC* accounts for 45%, 18%, 11%, 14% and 12% of the total area of the study area respectively according to the 15 years of average *FVC*. In general, the proportion of high *FVC* area is the largest and shows an increasing trend, the proportion of medium, high and medium *FVC* area shows a fluctuating and decreasing trend, and the proportion of low and very low *FVC* area shows a wavy change.

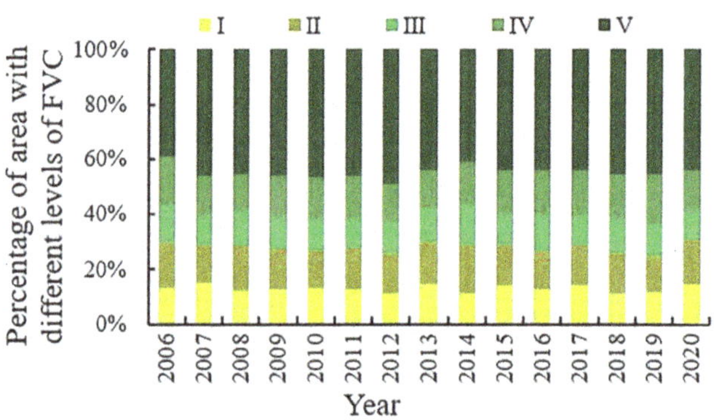

Figure 7. Interannual variation of different classes of vegetation, where I is Very Low *FVC*; II is Low *FVC*; III is Medium *FVC*; IV is Medium High *FVC*; V is High *FVC*. This is the same in Figure 8.

As shown in Figure 8, it can be found that the regional difference of average *FVC* in the study area is significant. The overall *FVC* in the north and central Tex River Valley is high, while the overall *FVC* in the southern mountainous area is low, and only the river valley has a high *FVC*. This is because the overall altitude of the southern mountainous area is high, and there are glaciers and snow, which are inhospitable to vegetation growth. The vegetation cover was counted in the forestry administrative district, and the results showed that: the best vegetation cover was in Tex, with an average *FVC* of 0.62; the second best vegetation cover was in Qapqal, with an average *FVC* of 0.6; and the vegetation cover in Zhaosu was relatively poor, with an average *FVC* of 0.47.

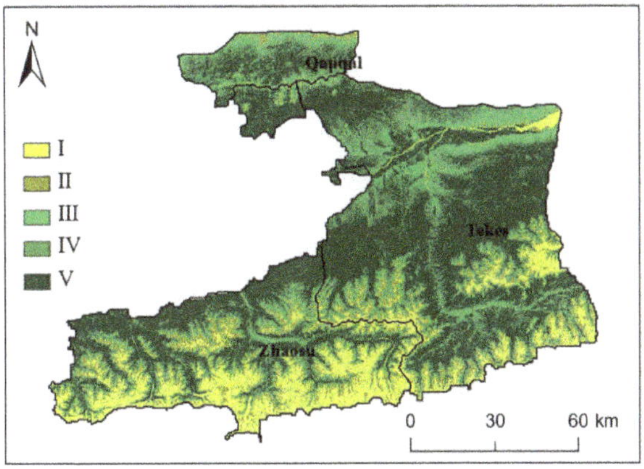

Figure 8. Spatial Distribution of Annual Average *FVC* in West Tianshan Mountains in the Past 15 Years.

3.4. Slope Change Trend of FVC

The change trend of *FVC* in the Western Tianshan Mountains from 2006 to 2020 is analyzed at the pixel scale, and the results are tested for significance (Figure 9, Table 2) based on the univariate linear regression model. The average change trend of *FVC* in the Western Tianshan region is 0.3%/10A, indicating that the overall *FVC* in the Western Tianshan Region is increasing, but the growth rate is relatively slow. The proportion of areas with positive and negative slope is 52.3% and 47.7%, respectively. Among the pixels with decreasing *FVC*, the proportion of pixels with larger decreasing trend (slope < -0.01) is 28%, and the proportion of pixels with larger increasing trend (slope > 0.01) is 27%. The results of significance test (Figure 9b) show that the *FVC* in the Western Tianshan Mountains shows extremely significant decrease, significant decrease, no significant change, significant increase and extremely significant increase, accounting for 43%, 2.1%, 6%, 3.3% and 45.6%, respectively. It can be found that there is a large area and a high degree of vegetation improvement and degradation in the Western Tianshan Region, and its area proportion and change degree are basically the same as the degradation phenomenon. This also makes the vegetation change in the Western Tianshan Region show an insignificant trend generally, but its internal changes are quite different.

Table 2. Vegetation cover change and significance statistics.

Slope	Percentage/%	Significance	Percentage/%
Slope < -0.02	3.7%	Extremely significant decrease	43%
$-0.02 <$ Slope < -0.01	9.7%	Significant decrease	2.1%
$-0.01 <$ Slope < 0	34.3%	Insignificant change	6%
$0 <$ Slope < 0.01	38%	Significant increase	3.3%
$0.01 <$ Slope < 0.02	10.4%	Extremely significant increase	45.6%
Slope > 0.02	3.9%		

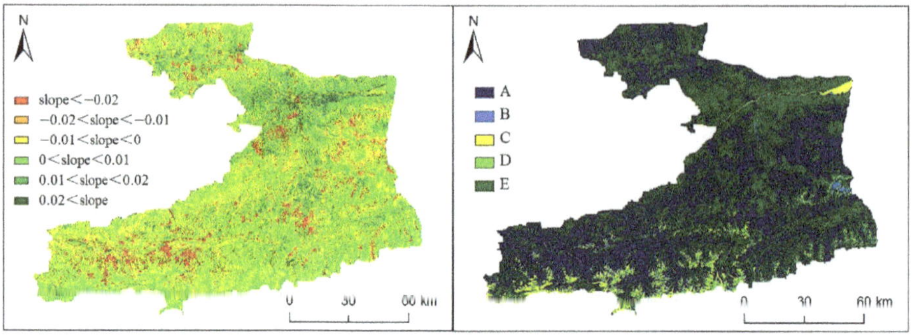

(**a**) Trends (**b**) Significance

Figure 9. Trends (**a**) and significance (**b**) of vegetation cover changes in the western Tian Shan region in the last 15 years, where A is a extremely significant decrease; B is a significant decrease; C is no significant change; D is a significant increase; and E is a significant increase. This is the same in Table 2.

Compared with Figure 8, it can be found that the insignificant changes are mainly in the extremely low *FVC* area, which is also the glacier snow coverage area, and the interference of human activities is small; the extremely significant increase mainly occurs in the low and medium *FVC* areas, and the extremely significant decrease mainly occurs in the medium high and high *FVC* areas, which changes are mainly affected by global climate change and human disturbance.

3.5. *Distribution and Variation Characteristics of* FVC *with Terrain*

The growth and spatial distribution of vegetation are affected by climate, topography, and human activities, among which the topographic factors affect the growth of vegetation by changing the vegetation habitat elements such as water, heat, and soil in local areas. In this study, concerning desertification land and plain forest resources investigation, the altitude, slope and aspect are divided into 7, 6 and 9 grades according to the classification standard of the data dictionary of Xinjiang desertification, and then the distribution and change of *FVC* in the study area in the past 15 years under different terrain conditions are statistically analyzed. The results are shown in Figures 10–12.

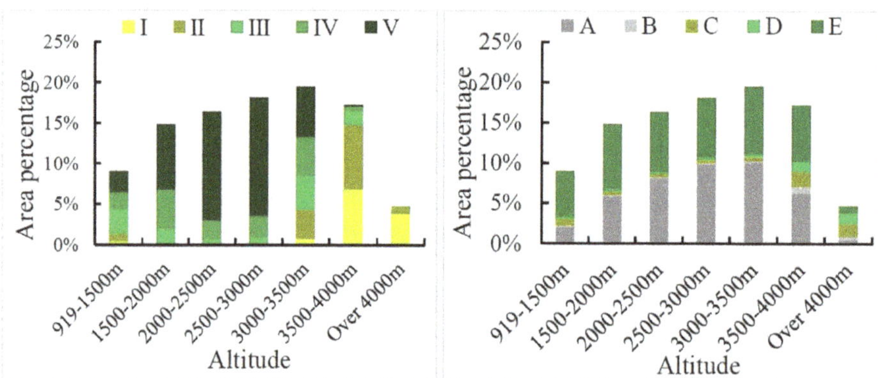

(**a**) FVC distribution (**b**) FVC variation

Figure 10. Distribution (**a**) and variation (**b**) of *FVC* at different altitudes.

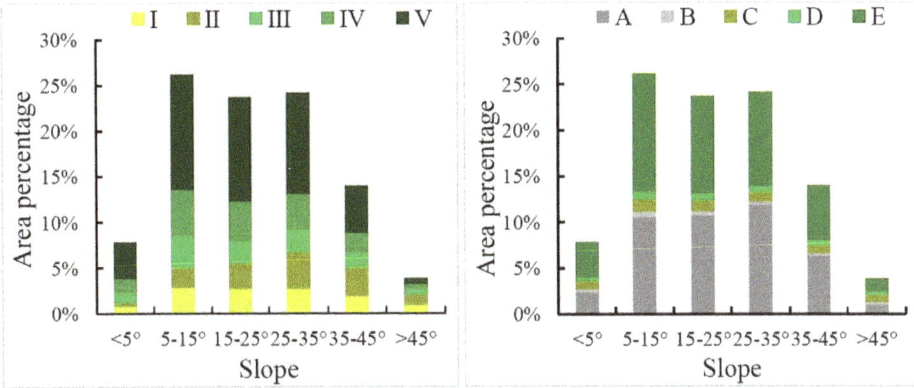

(a) FVC distribution (b) FVC variation

Figure 11. Distribution (**a**) and variation (**b**) of *FVC* for different slopes.

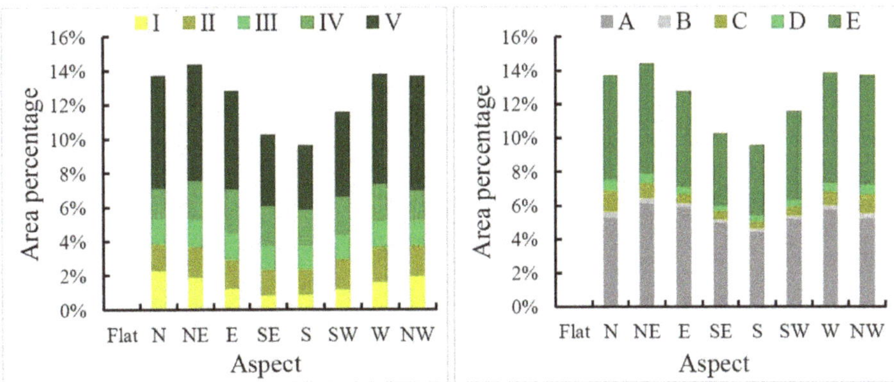

(a) FVC distribution (b) FVC variation

Figure 12. Distribution (**a**) and variation (**b**) of *FVC* for different slope directions.

3.5.1. Distribution and Variation Characteristics of *FVC* with Altitude

As shown in Figure 10a, the vegetation was mainly distributed in the area from 1500 to 4000 m, accounting for 86% of the total vegetation area in the study area. Among them, the area of high *FVC* increases with altitude below 3000 m and is absolutely dominant; in the area from 3000 to 4000 m, the area of very low and low *FVC* shows an increasing trend, and in the area above 4000 m, the vegetation decreases rapidly and is dominated by very low *FVC*, which is mainly influenced by natural conditions. As shown in Figure 10b, the change of vegetation cover is mainly in the area below 4000 m, which is related to the vegetation distribution. In the area from 2000 to 3500 m, the proportion of decrease in vegetation cover is greater than the proportion of increase, indicating a slight degradation trend of vegetation cover in this elevation range; in the remaining elevations, the proportion of increase in vegetation cover is higher than the proportion of decrease, indicating a slight improvement trend of vegetation cover in this elevation range. The area above 4000 m is less affected by human activities, and is mainly an insignificant change.

3.5.2. Distribution and Variation Characteristics of *FVC* with Slope

As shown in Figure 11a, the vegetation was mainly distributed in the area from 5° to 45°, accounting for 88% of the total vegetation coverage in the study area. In each slope class, the proportion of high and medium-high *FVC* area gradually decreases and the proportion of very low and low *FVC* area gradually increases as the slope increases; except for >45° slope class, high *FVC* occupies absolute dominance in all slope classes. As shown in Figure 11b, corresponding to the distribution of vegetation in different slope grades, the changes of *FVC* are also mainly distributed in the area from 5° to 45°. In each slope class, significant decrease, insignificant change and significant increase were basically maintained in low proportions. In the slope range from 25° to 45°, the proportion of decrease in vegetation cover was greater than the proportion of increase, indicating a slight degradation trend of vegetation cover in this slope range; in the rest of the slope classes, the proportion of increase in vegetation cover was greater than the proportion of decrease, indicating a slight improvement trend of vegetation cover in this slope range.

3.5.3. Distribution and Variation Characteristics of *FVC* with Direction

As shown in Figure 12a, the vegetation cover of flat slopes only accounted for 0.05% of the total vegetation cover in the study area, and the difference in the proportion of vegetation distribution in the remaining slope directions was small. Among the slope directions, the vegetation distribution area is higher on the north slope than on the south slope, and higher on the west slope than the east slope. This is because the north and west slopes are windward slopes in the western Tian Shan region, which have more precipitation and a more humid environment suitable for the growth of vegetation. As shown in Figure 12b, the distribution of vegetation change in the flat slope area is the least, and the difference of the distribution ratio of vegetation change in the remaining slope directions is small, which is closely related to the distribution of vegetation. Among the slopes, the decrease proportion of vegetation in the south and southeast slopes is larger than the increase proportion, which indicates that the vegetation cover in the south and southeast shows a slight trend to degradation; the increase proportion of vegetation in the rest of the slopes is larger than the decrease proportion, which indicates that the vegetation cover in these slopes shows a slight trend to improvement.

4. Discussion

In studies related to vegetation cover change in arid and semi-arid areas of Xinjiang, China, low spatial resolution time series data such as MODIS *NDVI* (16 d/250 m/500 m) [21,22] and GIMMS *NDVI* (15 d/8 km) [23–25] are mostly used. Among them, MODIS data are used in vegetation cover change studies after 2000, and longer time scale data are dominated by GIMMS data. Other scholars have used Landsat data for vegetation cover change studies with higher spatial resolution in a 5-year span [18,19]. However, studies based on the above data require a trade-off between temporal continuity and spatial resolution, and cannot achieve long time series to monitor vegetation change trends with high accuracy. In a study by Cai et al. [21] based on MODIS data, vegetation degradation occurred in the Tianshan region of Xinjiang in 2013 and 2014, and vegetation grew rapidly again in 2015, while no abrupt change years were found in a study by Wen et al. [18] with a 5-year span of Landsat data. In this paper, the time-series data obtained based on ESTARFM model can also detect the vegetation degradation in 2013 and 2014 years while having high spatial resolution. On the other hand, in the West Tianshan region, where the topography is complex and the terrain is undulating, low-resolution remote sensing data often produces large errors when analyzing the influence of each topographic factor on vegetation cover [39]. In this paper, we effectively obtained the high-spatial resolution time-series *NDVI* data based on ESTARFM model for the complex terrain area in the western Tianshan Mountains, which provides data support to realize the dynamic monitoring of cover change in the complex terrain area in the western Tianshan Mountains.

In this paper, we conducted a study on vegetation cover change based on ESTARFM *NDVI* data and found that the overall *FVC* in the western Tian Shan showed a weak increasing trend, which was strongly associated with ecological restoration projects implemented by the Chinese government in recent years, but a degradation trend was observed in the years 2013, 2014, and 2020, indicating that the ecological environment is vulnerable to disturbance by natural and human activities [18,19,23]. It was also found that while the overall *FVC* in the study area showed an increase, there was a large variation in the internal changes and a large area of improvement and degradation at the same time, which is basically consistent with previous studies [18,40–42]. Compared with the study by Wang et al. [43], the different criteria for classifying the trend level of change and the difference in the resolution of remote sensing data make the results of the study somewhat different. In the river valley and pre-mountain plain zone, it is mainly a medium vegetation cover area, where water and heat conditions and topography are suitable and human activities are concentrated, presenting a medium vegetation cover with a strong vegetation change trend at the same time. With the increase of altitude, the increase of snow and ice melt water, and the decrease of human activities, the vegetation shows high vegetation cover and the vegetation change trend gradually decreases. The southern part of the study area is the summit part of Tianshan Mountain, where the vegetation cover is low because of the high altitude and the snow and glacier cover above the snow line, which is not suitable for vegetation growth. At the same time, the natural and human influences in this area are less variable, which makes the vegetation change in this area extremely insignificant.

Influenced by Atlantic water vapor, the West Tianshan region receives more precipitation and is one of the regions with rich forest resources in Xinjiang. However, due to the temperate continental semi-arid climate, the ecological environment is fragile, and the degradation of vegetation still occurs in some areas under the influence of climate change and human activities. The elevation range of vegetation degradation is from 2000 to 3500 m, the slope is from 25° to 45°, and the slope direction is south and southeast slope. Human activities are one of the reasons for the degradation of vegetation in this area. Early urban expansion and construction of aqueducts destroyed the original landscape ecology, resulting in a fragile ecological environment with poor self-repair ability, which in turn led to a long-term trend of weak degradation. On the other hand, the influence of each topographic factor on the change of vegetation cover is complex and comprehensive, and the growth and spatial location of vegetation are closely related to the geographical environment. Elevation, slope, and slope direction determine the distribution of vegetation in mountainous areas by influencing the distribution of water and heat. In areas with slopes greater than 25°, precipitation and ice melt water are not easily stored, soil erosion is serious, and the soil is infertile, resulting in a more fragile vegetation ecology, and the ecological management effect is often unsatisfactory, which is the focus of environmental and ecological management projects [44].

5. Conclusions

Based on the ESTARFM model, this paper fused Landsat and MODIS data to obtain the time-series *NDVI* data of the study area for the past 15 years and estimated the *FVC*. Through trend analysis and combining with DEM data, the characteristics of *FVC* changes in the study area and its response to topographic factors were studied, and the main conclusions are as follows:

(1) The time-series *NDVI* data of the study area obtained by using ESTARFM Model can not only greatly improve the spatial resolution, but also better maintain the original spectral information. It has high consistency compared with the real *NDVI* data, indicating that the ESTARFM Model can better simulate the high spatial resolution *NDVI* data in the missing phase, which can monitor and study the dynamic change of vegetation cover in the complex terrain area of Xinjiang in a long time series.

(2) The overall vegetation cover of the study area was at a high level from 2006 to 2020, and the average annual vegetation cover showed a weak increasing trend in fluctuation. Among all *FVC* levels, the area with high *FVC* accounts for the largest, with an average of 45% for many years, and it is generally increasing; the proportion of medium high and medium *FVC* area showed a fluctuating and decreasing trend; the area proportion of low and very low *FVC* changes in waves. In terms of spatial distribution, there are obvious

regional differences, showing the distribution characteristics of high in the north and low in the south.

(3) From the change trend of *FVC*, the area with improved *FVC* in the study area in the past 15 years is larger than the area with degraded *FVC*, accounting for 52.3% and 47.7%, respectively. The results of significance test show that the areas with no significant changes are mainly concentrated in the extremely low *FVC* area, i.e., the southern part of the study area is also the glacier snow coverage area with less interference from human activities; the extremely significant increase mainly occurred in the low and medium *FVC* areas, and the extremely significant decrease mainly occurred in the medium high and high *FVC* areas.

(4) According to different terrain conditions, the main distribution areas of vegetation cover are 1500–4000 m above sea level, 5°–45° slope and all slope directions except flat slope; the areas with a slight degradation trend of *FVC* are mainly 2000–3500 m above sea level, 25°–45° slope and south and southeast slope, while the *FVC* in other areas showed a slight improvement trend.

Author Contributions: Conceptualization, Z.L. and D.C.; Data curation, Z.L., W.F. and N.Z.; Formal analysis, Z.L., S.L. and W.F.; Funding acquisition, D.C.; Methodology, Z.L. and S.L.; Supervision, D.C. and H.L.; Visualization, Z.L. and C.Z.; Writing—original draft, Z.L.; Writing—review & editing, D.C., F.L. and M.Z. All authors have read and agreed to the published version of the manuscript.

Funding: This research was funded by the Major Science and Technology Project of High-Resolution Earth Observation System (grant no. 76-Y50G14-0038-22/23), Key Research and Development Program of Anhui Province (grant no. 2021003; 2022107020028), Major Science and Technology Project of Anhui Province (grant no. 202003a06020002), Collaborative Innovation Project of Universities in Anhui Province (grant no. GXXT-2021-048), Anhui Provincial Special Support Plan (grant no. 2019), Chuzhou Science and Technology Planning Project (grant no. 2021ZD013), Natural Science Foundation of Anhui Province (grant no. 2208085QD107), and National Natural Science Foundation of China (grant no. 42261013).

Data Availability Statement: All data, models, or code generated or used during the study are available from the author by re-quest (liuzhihong@stu.xjnu.edu.cn).

Acknowledgments: We are grateful for the support of Anhui High Resolution Earth Observation System Data Products and Application Software R&D Center, and the basic geographic information data support provided by Tianxi Bureau.

Conflicts of Interest: The authors declare no conflict of interest.

Appendix A

Table A1. Information about Landsat remote sensing data used in this study.

Date Acquired	Sensor	Spatial Resolution	Path/Row	Landsat Scene Identifier
14 August 2006	TM	30 m	146/030, 146/031	LT51460302006226IKR00, LT51460312006226IKR00
17 August 2007	TM	30 m	146/030, 146/031	LT51460302007229IKR00, LT51460312007229IKR00
3 August 2008	TM	30 m	146/030, 146/031	LT51460302008216KHC01, LT51460312008216KHC01
11 July 2011	TM	30 m	146/030, 146/031	LT51460302011192IKR02, LT51460312011192IKR02
22 August 2012	ETM+	30 m	146/030, 146/031	LE71460302012235PFS00, LE71460312012235PFS00
1 August 2013	OLI	30 m	146/030, 146/031	LC81460302013213LGN02, LC81460312013213LGN02
19 July 2014	OLI	30 m	146/030, 146/031	LC81460302014200LGN01, LC81460312014200LGN01
9 August 2016	OLI	30 m	146/030, 146/031	LC81460302016222LGN01, LC81460312016222LGN01
15 August 2018	OLI	30 m	146/030, 146/031	LC81460302018227LGN00, LC81460312018227LGN00
17 July 2019	OLI	30 m	146/030, 146/031	LC81460302019198LGN00, LC81460312019198LGN00
4 August 2020	OLI	30 m	146/030, 146/031	LC81460302020217LGN00, LC81460312020217LGN00

Table A2. Information about the MODIS remote sensing data used in this study.

Date Acquired	Product Type	Spatial Resolution	Horizontal/Vertical Tile Number	Entity Identifier
13 August 2006	MOD09A1	500 m	h23v04	MOD09A1.A2006225.h23v04.006
13 August 2007	MOD09A1	500 m	h23v04	MOD09A1.A2007225.h23v04.006
4 August 2008	MOD09A1	500 m	h23v04	MOD09A1.A2008217.h23v04.006
12 July 2011	MOD09A1	500 m	h23v04	MOD09A1.A2011193.h23v04.006
20 August 2012	MOD09A1	500 m	h23v04	MOD09A1.A2012233.h23v04.006
28 July 2013	MOD09A1	500 m	h23v04	MOD09A1.A2013209.h23v04.006
20 July 2014	MOD09A1	500 m	h23v04	MOD09A1.A2014201.h23v04.006
12 August 2016	MOD09A1	500 m	h23v04	MOD09A1.A2016225.h23v04.006
13 August 2018	MOD09A1	500 m	h23v04	MOD09A1.A2018225.h23v04.006
20 July 2019	MOD09A1	500 m	h23v04	MOD09A1.A2019201.h23v04.006
4 August 2020	MOD09A1	500 m	h23v04	MOD09A1.A2020217.h23v04.006

References

1. Dixon, R.K.; Brown, S.; Houghton, R.A.; Solomon, A.M.; Trexler, M.C.; Wisniewski, J. Carbon pools and flux of global forest ecosystems. *Science* **1994**, *263*, 185–190. [CrossRef] [PubMed]
2. Sitch, S.; Smith, B.; Prentice, I.C.; Arneth, A.; Bondeau, A.; Cramer, W.; Kaplans, J.O.; Levis, S.; Lucht, W.; Sykes, M.T.; et al. Evaluation of ecosystem dynamics, plant geography and terrestrial Carbon cycling in the LPJ dynamic global vegetation mode. *Glob. Change Biol.* **2003**, *9*, 161–185. [CrossRef]
3. Wu, X.R.; Guo, P.; Sun, Y.Q.; Liang, H.; Zhang, X.G.; Bai, W.H. Recent Progress on Vegetation Remote Sensing Using Spaceborne GNSS-Reflectometry. *Remote Sens.* **2021**, *13*, 4244. [CrossRef]
4. Hao, J.W.; Chu, L.M. Vegetation Types Attributed to Deforestation and Secondary Succession Drive the Elevational Changes in Diversity and Distribution of Terrestrial Mosses in a Tropical Mountain Forest in Southern China. *Forests* **2021**, *12*, 961. [CrossRef]
5. Parmesan, C.; Yohe, G. A globally coherent fingerprint of climate change impacts across natural systems. *Nature* **2003**, *421*, 37–42. [CrossRef]
6. Yu, L.X.; Yan, Z.R.; Zhang, S.W. Forest Phenology Shifts in Response to Climate Change over China–Mongolia–Russia International Economic Corridor. *Forests* **2020**, *11*, 757. [CrossRef]
7. Notaro, M.; Mauss, A.; Williams, J.W. Projected vegetation changes for the American Southwest: Combined dynamic modeling and bioclimatic-envelope approach. *Ecol. Appl.* **2012**, *22*, 1365–1388. [CrossRef]
8. Yang, X.C.; Xu, B.; Jin, Y.X.; Qin, Z.H.; Ma, H.L.; Li, J.Y.; Zhao, F.; Chen, S.; Zhu, X.H. Remote sensing monitoring of grassland vegetation growth in the Beijing–Tianjin sandstorm source project area from 2000 to 2010. *Ecol. Indic.* **2015**, *51*, 244–251. [CrossRef]
9. Wang, J.; Xie, Y.W.; Wang, X.Y.; Guo, K.M. Driving factors of recent vegetation changes in Hexi Region, Northwest China based on a new classification framework. *Remote Sens.* **2020**, *12*, 1758. [CrossRef]
10. Hill, M.J.; Guerschman, J.P. Global trends in vegetation fractional cover: Hotspots for change in bare soil and non-photosynthetic vegetation. *Agric. Ecosyst. Environ.* **2022**, *324*, 107719. [CrossRef]
11. Gitelson, A.A.; Kaufman, Y.J.; Stark, R.; Rundquist, D. Novel algorithms for remote estimation of vegetation fraction. *Remote Sens. Environ.* **2002**, *80*, 76–87. [CrossRef]
12. Zhang, J.; Walsh, J.E. Relative impacts of vegetation coverage and leaf area index on climate change in a greener north. *Geophys. Res. Lett.* **2007**, *34*, L15703. [CrossRef]
13. Song, Y.F.; Guo, Z.X.; Lu, Y.J.; Yan, D.H.; Liao, Z.L.; Liu, H.W.; Cui, Y.J. Pixel-Level Spatiotemporal Analyses of Vegetation Fractional Coverage Variation and Its Influential Factors in a Desert Steppe: A Case Study in Inner Mongolia, China. *Water* **2017**, *9*, 478. [CrossRef]
14. Smith, A.M.S.; Kolden, C.A.; Tinkham, W.T.; Talhelm, A.F.; Marshall, J.D.; Hudak, A.T.; Boschetti, L.; Falkowski, M.J.; Greenberg, J.A.; Anderson, J.W.; et al. Remote sensing the vulnerability of vegetation in natural terrestrial ecosystems. *Remote Sens. Environ.* **2014**, *154*, 322–337. [CrossRef]
15. Liu, D.Y.; Jia, K.; Jiang, H.Y.; Xia, M.; Tao, G.F.; Wang, B.; Chen, Z.L.; Yuan, B.; Li, J. Fractional Vegetation Cover Estimation Algorithm for FY-3B Reflectance Data Based on Random Forest Regression Method. *Remote Sens.* **2021**, *13*, 2165. [CrossRef]
16. Phiri, D.; Morgenroth, J. Developments in Landsat Land Cover Classification Methods: A Review. *Remote Sens.* **2017**, *9*, 967. [CrossRef]
17. Wang, Y.L.; Li, M.S. Annually Urban Fractional Vegetation Cover Dynamic Mapping in Hefei, China (1999–2018). *Remote Sens.* **2021**, *13*, 2126. [CrossRef]
18. Wen, G.C.; Zhao, M.J.; Xie, H.B.; Zhang, Y.; Zhang, j. Analysis of land vegetation cover evolution and driving forces in the western part of the Ili River Valley. *Arid Zone Res.* **2021**, *38*, 843–854.
19. Li, H.M.; Bahejiayinaer, T.; Chang, S.L.; Zhang, Y.T. Spatial-temporal change and prediction analysis of vegetation coverage in the northern slope of Tianshan Mountains in 30 years. *Chin. J. Ecol.* **2022**, 1–15. Available online: http://kns.cnki.net/kcms/detail/21.1148.q.20220526.1759.025.html (accessed on 16 November 2022).

20. Chang, M.D.; Wang, X.J.; Yan, L.N.; Ma, K.; Li, Y.K.; Li, J.Y.; Jia, H.T. Temporal and spatial pattern dynamic variations of vegetation cover and management factors in the middle of the northern slope of Tianshan Mountains. *J. Agric. Res. Environ.* **2022**, *39*, 836–846.
21. Cai, C.C.; Huai, Y.J.; Bai, T.; Dong, L. Recent Changes of Grassland Cover in Xinjiang Based on *NDVI* Analysis. *J. Basic Eng.* **2020**, *28*, 1369–1381.
22. Yin, Z.L.; Feng, Q.; Wang, L.G.; Chen, Z.X.; Chang, Y.B.; Zhu, R. Vegetation coverage change and its influencing factors across the northwest region of China during 2000–2019. *J. Desert Res.* **2022**, *42*, 11–21.
23. Wang, H.J.; Li, Z.; Niu, Y.; Li, X.C.; Cao, L.; Feng, R.; He, Q.N.; Pan, Y.P. Evolution and Climate Drivers of *NDVI* of Natural Vegetation during the Growing Season in the Arid Region of Northwest China. *Forests* **2022**, *13*, 1082. [CrossRef]
24. Li, X.W.; Zulkar, H.; Wang, D.Y.; Zhao, T.N.; Xu, W.T. Changes in Vegetation Coverage and Migration Characteristics of Center of Gravity in the Arid Desert Region of Northwest China in 30 Recent Years. *Land* **2022**, *11*, 1688. [CrossRef]
25. Mu, B.; Zhao, X.; Wu, D.H.; Wang, X.Y.; Zhao, J.C.; Wang, H.Y.; Zhou, Q.; Du, X.Z.; Liu, N.J. Vegetation Cover Change and Its Attribution in China from 2001 to 2018. *Remote Sens.* **2021**, *13*, 496. [CrossRef]
26. Gao, F.; Masek, J.; Schwaller, M.; Hall, F. On the blending of the Landsat and MODIS surfaces reflectance: Predicting daily Landsat surface reflectance. *IEEE Tran. Geosci. Remote Sens.* **2006**, *44*, 2207–2218.
27. Zhu, X.L.; Chen, J.; Gao, F.; Chen, X.H.; Masek, J.G. An enhanced spatial and temporal adaptive reflectance fusion model for complex heterogeneous region. *Remote Sens. Environ.* **2010**, *114*, 2610–2623. [CrossRef]
28. Wu, G.J.; Liu, X.H.; Kang, S.C.; Chen, T.; Xu, G.B.; Zeng, X.M.; Kang, H.H. Age-dependent impacts of climate change and intrinsic water-use efficiency on the growth of Schrenk spruce (*Picea schrenkiana*) in the western Tianshan Mountains, China. *For. Ecol. Manag.* **2018**, *414*, 1–14. [CrossRef]
29. Jarihani, A.A.; Mcvicar, T.R.; Vanniel, T.G.; Emelyanova, I.V.; Callow, J.N.; Johansen, K. Blending Landsat and MODIS data to generate multispectral indices: A Comparison of "Index-then-Blend" and "Blend-then-Index" approaches. *Remote Sens.* **2014**, *6*, 9213–9238. [CrossRef]
30. Xiao, J.F.; Moody, A. A comparison of method for estimating fractional green vegetation cover within a desert-to-upland transition zone in central New Mexico, USA. *Remote Sens. Environ.* **2005**, *98*, 237–250. [CrossRef]
31. Qi, J.; Marsett, R.C.; Moran, M.S.; Goodrich, D.C.; Heilman, P.; Kerr, Y.H.; Dedieu, G.; Chehbouni, A.; Zhang, X.X. Spatial and temporal dynamics of vegetation in the San Pedro River basin area. *Agric. For. Meteorol.* **2000**, *105*, 55–68. [CrossRef]
32. Rundquist, B.C. The influence of canopy green vegetation fraction on spectral measurements over native tallgrass prairie. *Remote Sens. Environ.* **2002**, *81*, 129–135. [CrossRef]
33. Li, M.M. The Method of Vegetation Fraction Estimation by Remote Sensing. Master's Thesis, University of Chinese Academy of Sciences (Institute of Remote Sensing Applications), Beijing, China, 2003.
34. Wei, X.D.; Wang, S.N.; Wang, Y.K. Spatial and temporal change of fractional vegetation cover in North-western China from 2000 to 2010. *Geol. J.* **2017**, *53*, 427–434. [CrossRef]
35. Wu, C.S.; Murray, A.T. Estimating impervious surface distribution by spectral mixture analysis. *Remote Sens. Environ.* **2003**, *84*, 493–505. [CrossRef]
36. Suvachananonda, T.; Maruyama, Y. Urban Growth Prediction of Special Economic Development Zone in Mae Sot District, Thailand. *Eng. J.* **2018**, *22*, 269–277.
37. Zhang, H.; Xue, L.Q.; Wei, G.H.; Dong, Z.C.; Meng, X.Y. Assessing Vegetation Dynamics and Landscape Ecological Risk on the Mainstream of Tarim River, China. *Water* **2020**, *12*, 2156. [CrossRef]
38. Sun, X.F.; Yuan, L.G.; Zhou, Y.Z.; Shao, H.Y.; Li, X.F.; Zhong, P. Spatiotemporal change of vegetation coverage recovery and its driving factors in the Wenchuan earthquake-hit areas. *J. Mt. Sci.* **2021**, *18*, 2854–2869. [CrossRef]
39. Qin, J.L.; Xue, L.Q. Spatial and temporal variation characteristics of vegetation in the Manas River Basin in northwest arid region and its spatial relationship with topographical factors. *Ecol. Environ. Sci.* **2020**, *29*, 2179–2188.
40. Shao, S.S.; Shi, Q.D. Spatial and temporal change of vegetation cover in Xinjiang based on *FVC*. *Sci. Silvae Sin.* **2015**, *51*, 35–42.
41. He, B.Z.; Ding, J.L.; Zhang, Z.; Ghulam, A. Experimental analysis of spatial and temporal dynamics of fractional vegetation cover in Xinjiang. *Acta Geogr. Sin.* **2016**, *71*, 1948–1966.
42. Mo, K.L.; Chen, Q.W.; Chen, C.; Zhang, J.Y.; Wang, L.; Bao, Z.X. Spatiotemporal variation of correlation between vegetation cover and precipitation in an arid mountain-oasis river basin in northwest China. *J. Hydrol.* **2019**, *574*, 138–147. [CrossRef]
43. Wang, H.L.; Feng, A.P.; Gao, Y.H.; Wang, X.L. Temporal-spatial dynamic change on maximum vegetation coverage degree of Ili River basin. *Environ. Sci. Technol.* **2018**, *41*, 161–167.
44. He, Z.H.; Zhang, Y.H.; He, Y.; Zhang, X.W.; Cai, J.Z.; Lei, L.P. Trends of Vegetation Change and Driving Factor Analysis in Recent 20 Years over Zhejiang Province. *Ecol. Environ. Sci.* **2020**, *29*, 1530–1539.

forests

MDPI

Article

Study on the Spatial Heterogeneity of the Impact of Forest Land Change on Landscape Ecological Risk: A Case Study of Erhai Rim Region in China

Mengjiao Wang, Yingmei Wu *, Yang Wang, Chen Li, Yan Wu, Binpin Gao and Min Wang

Faculty of Geography, Yunnan Normal University, Kunming 650500, China; 2123130070@ynnu.edu.cn (M.W.); 210058@ynnu.edu.cn (Y.W.); 2023130027@user.ynnu.edu.cn (C.L.); 2023130031@user.ynnu.edu.cn (Y.W.); 1933130005@user.ynnu.edu.cn (B.G.); wangmin1994@user.ynnu.edu.cn (M.W.)

* Correspondence: 2287@ynnu.edu.cn

Abstract: As an important ecological ecotone of water and land ecosystems, the lakeside is characterized by a variety ecosystem services and high vulnerability. Forest land is important in resolving the ecological risks of the lakeside area and building its ecological base. It is important to explore the effect of change in forest land on landscape ecological risk in the lakeside area, alleviate the contradiction between ecological protection and construction and development in the area, and realize sustainable development. The present study attempted to explore the spatial and temporal evolutionary features of forest land in the Erhai rim region from 2000 to 2020 using bivariate spatial autocorrelation and multi-scale geographical weighted regression (MGWR) models. The following are the findings of this investigation of the 2000–2020 period: (1) the forest land area in the region generally decreased, first increasing and then decreasing, and was mainly occupied by cultivated land and artificial surfaces; (2) the total landscape ecological risk in the region presented an upward trend, and medium- and higher-risk areas were the main risk areas, with the latter increasing; (3) the impact of forest land expansion and contraction intensity on landscape ecological risk exhibited spatial and temporal heterogeneity. The main forms of forest land change at different stages differed, and the impacts on landscape ecological risk were also different. Reasonable forest land expansion can effectively alleviate the growth in landscape ecological risk, whereas the shrinkage of forest land would aggravate the landscape ecological risk in the Erhai rim region. Moreover, the findings can offer reference for the exploration of ecological protection and coordinated optimization of economic development in Erhai Lake.

Keywords: forest land change; landscape ecological risk; heterogeneity; multi-scale geographically weighted regression; Erhai rim region

Citation: Wang, M.; Wu, Y.; Wang, Y.; Li, C.; Wu, Y.; Gao, B.; Wang, M. Study on the Spatial Heterogeneity of the Impact of Forest Land Change on Landscape Ecological Risk: A Case Study of Erhai Rim Region in China. *Forests* **2023**, *14*, 1427. https://doi.org/10.3390/f14071427

Academic Editor: Panteleimon Xofis

Received: 12 June 2023
Revised: 28 June 2023
Accepted: 10 July 2023
Published: 12 July 2023

1. Introduction

Forest land is a vital part of the forest ecosystem and a basic part of ecological environment construction [1]. It plays a vital part in regulating the regional climate, regulating the hydrological cycle, maintaining the global carbon balance, protecting biodiversity, improving ecological security barrier functions, and promoting sustainable social development [2–5]. Changes in forest land directly affect the construction of national ecological security patterns and the global ecological environment [6]. In recent years, the protection of forest ecosystems has become a consensus worldwide [7–9]. China has also carried out a series of ecological construction projects around forest land protection, including returning farmland to forests, natural forest protection, and the development of the Three-North Shelter Forests Project. These constructions have effectively restored forest ecosystems, increased the area of forest land, protected the diversity of forest species, and significantly improved the quality of the ecological environment [10–12]. However, with the rapid increase in population and rapid development of industrialization and urbanization, human beings have also increased the development and utilization of forest land, resulting in serious damage to the forest ecosystem, which directly or indirectly brings many risks to the regional ecological environment [13,14].

The organization and structure of the ecosystem are influenced by various land-use patterns and intensities, which have regional and cumulative effects on ecology [15]. Currently, the ecological risks resulting from changes in natural ecosystems and land-use types are important factors affecting regional ecological security [16]. As an important tool for macroecological management, ecological risk assessment has gradually become a prominent topic in research [17]. As a significant component of ecological risk assessment, landscape ecological risk assessment specifically concentrates more on the spatial and temporal heterogeneity of ecological risks and possible adverse results of scale effects. It examines the spatial and temporal differentiation patterns of risks and risk expression of specific spatial patterns for ecological functions and processes [18,19]. Landscape ecological risk assessment is of great significance for research on mountains, rivers, forests, farmlands, lakes, grasslands, and deserts as part of the community of life, and is also an effective means of regional ecological risk prevention and management. Kapustka et al. (2001) used the landscape ecology theory to evaluate ecological risks and proposed corresponding control strategies [20]. Paukert et al. (2011) proposed a landscape-scale ecological risk index from the perspective of landscape structure and land-use change. [21]. Ayre et al. (2012) introduced the Bayesian method for landscape ecological risk assessment to assess the ecological risk in the upper reaches of the Grand River in Oregon [22]. With the landscape ecological risk assessment, which is based on land-use change and is becoming a research hotspot, researchers worldwide have furthered the research on the topic [23,24]. Most research areas are concentrated in key risk control regions, such as urban [25] and administrative [26] areas, watersheds [27], industrial and mining areas [28] and nature reserves [29]; however, landscape ecological risk assessment around lakeside areas needs to be further developed.

In addition, as human understanding of the structure and function of forest land continues to advance, the ecological and environmental issues brought on by changing forest lands have gradually attracted the attention of many scholars. Iroume et al. (2005) studied the relationship between forest coverage and summer floods [30]. Lombaerde et al. (2022) concluded that a forest cover has a buffering effect on the temperature of future climate [31]. Hu et al. (2020) evaluated the effects of poplar ecological retreat initiatives on the water content of Dongting Lake wetlands using the InVEST model [32]. Shi et al. (2007) adopted the soil classification category (SCS) model for exploring the impact of forest land change on watershed runoff in Shenzhen [33]. Shi et al. (2016) investigated how changing forest land affected the value of ecosystem services using a price inversion technique [34]. Yao et al. (2006) used geographic information system (GIS) spatial analysis and traditional statistical analysis to analyze the impact of forest land change on soil erosion [35]. The existing research focuses more on the effect of forest land change on climate regulation, water conservation, soil conservation, and habitat maintenance, and rarely on the spatial heterogeneity of the impact of changing forest land on ecological risk in a landscape.

The lakeside area is an important ecological ecotone between aquatic and terrestrial ecosystems and shows high sensitivity to human activities [36,37]. The total landscape pattern there is relatively fragmented and poorly stabilized. Changes to it are significant and rapid under the influence of natural conditions and human activity [19]. As one of the nine plateau lakes in Yunnan, the resources and environmental conditions of the lake are good, economic development has occurred early, and the degree of urbanization is high. At the same time, the implementation of a series of ecological actions, like closing hillsides to facilitate afforestation and returning farmland to forests, as a part of China's "Natural Forest Resources Protection Project" has greatly improved the forest coverage of Eryuan County in the north of Erhai Lake and Cangshan Mountain in the east coast. Nevertheless, with the acceleration of urbanization, the demand for urban space land, particularly in the central cities in western Yunnan and mountainous cities in Haidong, is increasing daily.

Forest land on the low hills and gentle slopes on the south and east banks of Erhai Lake is occupied, and the land-use types have changed fundamentally, posing a great threat to the ecosystem in the region. Improving the ecological function of lakeside areas has become a focus of increasing attention [38].

Forest land accounts for the largest proportion of the land-use type in the Erhai rim region. Compared with those in other land-use types, changes in forest land can more obviously reflect the effect of human activities on landscape structure and function. However, with the fast development of urbanization and the implementation of ecological projects, the forest land area in the Erhai rim has changed considerably; therefore, what are the changes of forest land and landscape ecological risks in the Erhai rim from 2000 to 2020? How did the woodland changes affect the landscape ecological risk? These ate urgent questions that need to be addressed at present. Therefore, in this study, we considered the Erhai rim region as an example, measured and investigated the intensity of forest land change from 2000 to 2020, constructed a landscape ecological risk model through a landscape pattern index, and discussed the spatial and temporal distribution characteristics of landscape ecological risk in the region. Furthermore, a multi-scale geographically weighted regression (MGWR) model was adopted for revealing the spatial heterogeneity of the impact of forest land change intensity on landscape ecological risk change to enrich empirical studies on the internal mechanisms of forest land change and landscape ecological risk. Our study offers a scientific basis for promoting the exploration of high-quality development paths oriented toward ecological priority and green development in the Erhai rim region.

2. Materials and Methods

2.1. Study Area

The Erhai rim region is located in the western part of Yunnan Province, China, in the central part of Dali Bai and Bai Autonomous Prefecture (100°5′ E–100°17′ E, 25°36′ N–25°58′ N). There is a high altitude in the west and a low altitude in the east of the terrain; the difference in relative altitude is 2554 m, and mountains surround it to the west. From northwest to southeast, it presents an irregular strip, including 18 towns in Dali City and Eryuan County, and the total area is about 2565 km^2 (Figure 1). The Erhai rim region is a typical area with overlapping characteristics, such as the fragile ecological environment of plateau lakes, diversified integration of ethnic cultures, and active rural economic and social development [39]. With the rise in tourism and rapid advancement of urbanization in the region in recent years, this region presents high requirements for ecological and environmental protection. As a vital part of the region's community life, the stability of the function and structure of its mountains, rivers, forests, fields, lakes, and sands is of great significance for the sustainable development of the region. Therefore, it is important to study the spatial heterogeneity of the effect of forest land change on landscape ecological risk in this area to promote the prevention of regional ecological risks.

2.2. Data Source

Considering the time node of China's policy of return of cultivated land to forest and the implementation of China's natural forest resource protection project, land-use data from 2000 to 2020 were selected. The data were derived from the Globeland30 global land cover database (http://www.globallandcover.com/, accessed on 15 December 2022), including three periods of data from 2000, 2010, and 2020, with a spatial resolution of 30 m × 30 m. The dataset covered a long period and had high overall accuracy, and the Kappa coefficient was 0.78 [40]. In accordance with the classification system of Globeland30, there are seven land-use types in the study area, including cultivated land, forest land, grassland, shrubland, wetland, water body, and artificial surface. The World Geodetic System 1984 (WGS84) Universal Transverse Mercator (UTM) Zone 47 Northern Hemisphere projection was used.

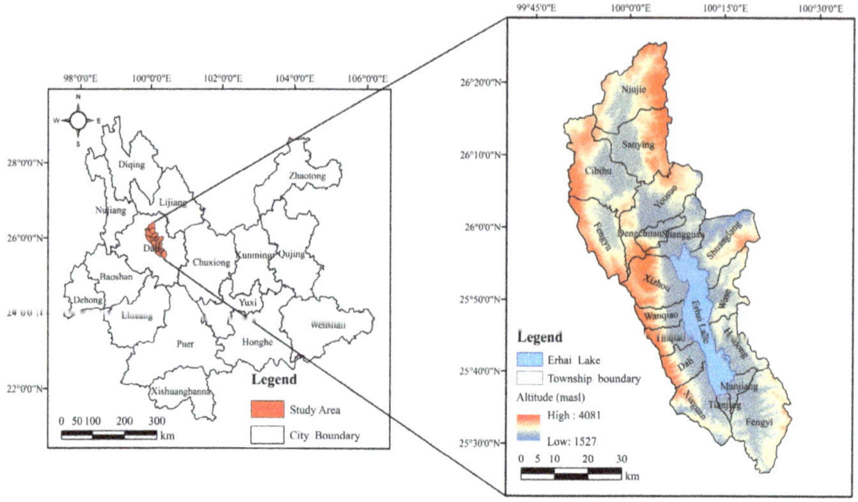

Figure 1. Location of the Erhai rim in China.

2.3. Research Methods

The content of the present study mainly includes the intensity of forest land change, landscape ecological risk and the effect of forest land change intensity on landscape ecological risk, and the relevant research methods are used to conduct the study around the research content as follows (Figure 2):

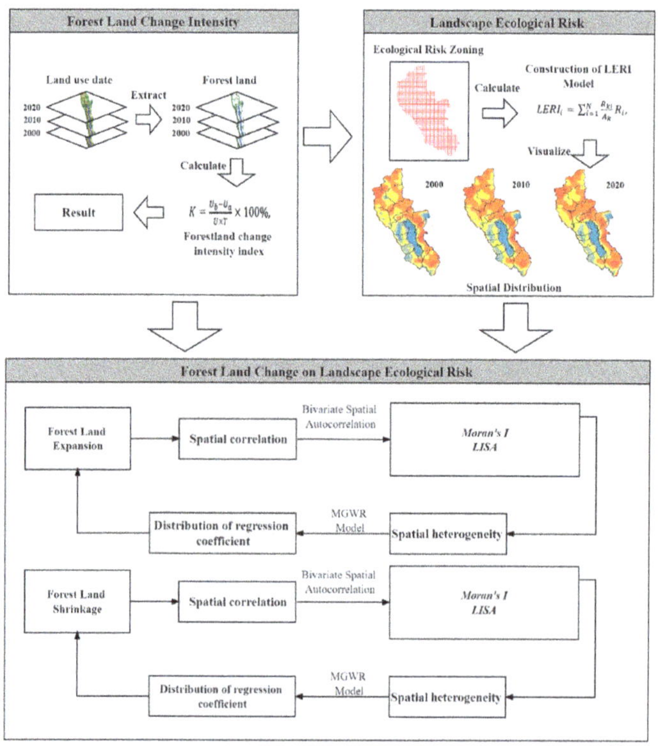

Figure 2. Research processes.

2.3.1. Forest Land Change Intensity Index

The intensity index of forest land change refers to the percentage of the changed area of forest land within a specific period out of the total area of regional land. It is a common quantitative index that reflects the degree and speed of forest land change [41–43]. The formula used to calculate the change intensity index was as follows:

$$K = \frac{U_b - U_a}{U \times T} \times 100\%, \tag{1}$$

where K denotes the intensity index of forest land change during the study period; U_a and U_b are the areas of forest land at the beginning and end of the study period, separately; U refers to the total area of the study zone; and T indicates the length of the study period, generally in years.

2.3.2. Ecological Risk Zoning

With the purpose of reasonably dividing the ecological risk area based on spatial heterogeneity, patch size, and size of the study area, according to the principle of two–five times the average patch [44,45], the study area was classified into 836 2 km × 2 km grids, and the ecological risk value of the center point of each grid was used as the ecological risk index of that ecological risk plot.

2.3.3. Construction of Landscape Ecological Risk Model

As a commonly used quantitative research method, the landscape pattern index method describes landscape patterns and their changes through a combination of multiple indices. In line with previous research results [46,47], the present study used the landscape disturbance index, landscape vulnerability, and landscape loss indices to construct an ecological risk assessment model that could be used to analyze the temporal and spatial evolution features of landscape ecological risk in the Erhai rim region. The formula is written as follows:

$$LERI_i = \sum_{i=1}^{N} \frac{A_{ki}}{A_k} R_i, \tag{2}$$

where $LERI_i$ represents the landscape ecological risk index of the risk area i, N denotes the number of landscape types, A_{ki} represents the area of landscape type i in the kth risk plot, A_k denotes the total area of the kth risk cell, and R_i refers to the loss index of landscape type i.

(1) Landscape Disturbance Index (U_i)

The landscape disturbance index is adopted for describing the degree of interference of different landscape types with external factors. The formula is as follows:

$$U_i = aC_i + bS_i + cF_i, \tag{3}$$

where C_i refers to the landscape fragmentation index; S_i refers to the landscape separation index; F_i is the landscape dominance index; and a, b, and c represent the weights of the corresponding landscape indices, with $a + b + c = 1$. Based on existing research results [48], we assigned the following values: $a = 0.5$, $b = 0.3$, and $c = 0.2$.

(2) Landscape Vulnerability Index (E_i)

The landscape vulnerability index represents the sensitivity of different landscape types to external interference. In line with the existing research results and expert scoring, the vulnerability indices of six landscape types were scored from low to high [49]: forest land = 1; water body = 2; wetland = 3; shrub land = 4; grassland = 5; and cultivated land = 6. The vulnerability indices of each landscape type were obtained after normalization.

(3) Landscape Loss Index (R_i)

The landscape loss index was employed to represent the degree of loss for each landscape type when disturbed. The formula is as follows:

$$R_i = U_i \times E_i. \tag{4}$$

2.3.4. Bivariate Spatial Autocorrelation

Geoda 1.20 is superior in the calculation of spatial autocorrelation; it can be applied to investigate the spatial distribution characteristics of an attribute and the correlations between variables. Generally, the spatial correlation is measured and tested using global and local spatial autocorrelations [50,51]. The formula is as follows:

$$I = \frac{n}{\sum_{i=1}^{n}\sum_{j=1}^{n} W_{ij}} \times \frac{\sum_{i=1}^{n}\sum_{j=1}^{n} W_{ij}(x_i - \bar{x})(x_j - \bar{x})}{\sum_{i=1}^{n}(x_i - \bar{x})^2}, \tag{5}$$

where x_i and x_j are the observed values, $\bar{x}$ represents the average of x_i, W_{ij} is the spatial weight adjacency matrix of spatial cells i and $j(i\ j = 1, 2, 3, \ldots)$, and the global Moran's I value is generally $[-1, 1]$. A Moran's I value greater than 0 suggests that the spatial unit attributes are positively associated. A value of less than 0 suggests that the spatial unit attributes are negatively correlated; a value of 0 suggests that the spatial units are randomly distributed, and there is no correlation. The p-value was used for the significance test—when $p > 0.1$, it indicates that it is not significant, while $p < 0.1$ indicates significance.

$$I_{kl}^{i} = Z_k^i \sum_{j=1}^{n} w_{ij} Z_l^j, \tag{6}$$

where w_{ij} is the spatial connection matrix between spatial elements; $Z_k^i = \frac{x_k^i - \bar{x}_k}{\lambda_k}$; $Z_l^i = \frac{x_l^i - \bar{x}_l}{\lambda_l}$; x_k^i refers to the value of attribute k of spatial cell i; x_l^i is the value of attribute l of spatial cell i; $\bar{x}_k$and $\bar{x}_l$ represent the average of k and l, respectively; and λ_k and λ_l represent the variance of attributes k and l, respectively.

Based on the local Moran's I index, the H-H stands for high–high clusters, the L-L stands for low–low clusters, the H-L stands for high–low clusters, the L-H stands for low–high clusters.

2.3.5. Construction of MGWR Model

MGWR is an improvement over the geographically weighted regression model (GWR). It differs from GWR, which assumes that the local relationship within each model changes on the same spatial scale. MGWR permits the relationship between response variables and different predictors to change at different spatial scales [52,53], which is more in line with spatial heterogeneity [54]. Therefore, the MGWR model is more conducive to an in-depth analysis of the effect of forest land change on the spatial differentiation of landscape ecological risk. The formula for the MGWR model can be expressed as follows:

$$y_i = \sum_{j=1}^{k} \beta_{bwj}(u_i, v_i)x_{ij} + \varepsilon_i, \tag{7}$$

where y_i represents the explained variable; (u_i, v_i) denotes the coordinates of the center point at i; bwj indicates the broadband used by the jth variable regression coefficient; β_{bwj} denotes the regression coefficient of the jth variable at i.

3. Results

3.1. Analysis of Forest land Change Characteristics in the Erhai Rim Region

Based on Figure 3 and Tables 1 and 2, we know that the land-use structure in the Erhai rim region changed obviously, and there were frequent and complex mutual transformation phenomena among various land-use types. The evolutionary features of the landscape pattern were further analyzed in terms of changes in the number of patches, fragmentation, separation and dominance indices of land-use types. Regarding the number of patches, the overall number of patches in the study area decreased, while the number of patches in wetlands, cultivated land and artificial surfaces increased, and the number of patches in forest land, grassland, shrubland and water bodies presented an overall decreasing trend. From the dominance index, forest land, as the dominant land-use type in the study area, has the highest dominance value and has a relatively strong influence on the landscape pattern of the whole area. In 2000, 2010, and 2020, the area of forest land was 112,227.03 hm^2, 113,239.08 hm^2, and 109,728.54 hm^2, accounting for 37.97%, 38.32%, and 37.12% of the overall land area, respectively. It was mainly located in areas with less human activity at the periphery of the study area and crossed with grassland and shrubland. From 2000 to 2020, the change in forest land in the study area presented a development trend of an initial expansion and subsequent contraction, with the total area decreasing by 2498.49 hm^2. From 2000 to 2010, the forest land mainly expanded by 1012.05 hm^2. The number of patches in forest land increased, and its fragmentation, separation, interference, vulnerability, loss, and dominance also increased significantly. From 2010 to 2020, the change in forest land was mainly due to a shrinkage of 3510.54 hm^2. The number of patches in forest land decreased, and the degree of fragmentation, separation, interference, vulnerability, loss, and dominance also decreased significantly. Based on the area of forest land change, the expansion of forest land mainly occurred in Fengyu Town, north of Erhai Lake; the most significant area of forest land shrinkage was on the east and south shores of Erhai Lake, and the area with the smallest change in forest land area was in the Cangshan National Nature Reserve, west of Erhai Lake.

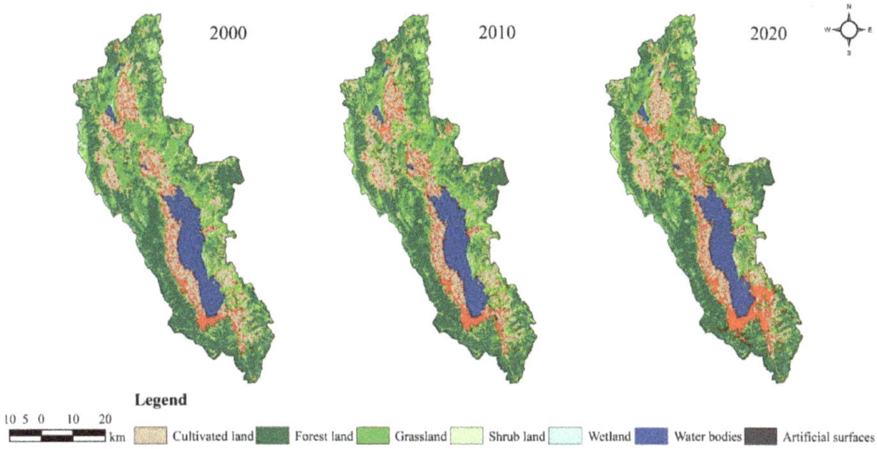

Figure 3. Distribution of land-use types in the Erhai rim at different times.

The conversion matrix of land-use types was used to explore the interconversion of land-use types in the Erhai Rim from 2000 to 2020 (Table 2). Different degrees of mutual transformation were observed between the land-use types in the study area. Moreover, the findings demonstrate that although the implementation of the policy of returning farmland to forests has led to the conversion of cultivated land to forest land in a small part of the Erhai rim region, overall, the transfer to forest land was still mainly from shrubland and grassland. Affected by urbanization, the cultivated land was also converted to artificial surfaces. Among them, the most significant area of forest land shrinkage was in the northeast and south of Erhai Lake, and most of it has been converted to cultivated land and artificial surfaces that are closely related to the economy and society, indicating that the intervention of human activities on forest land in the Erhai rim region is still large.

Table 1. Landscape pattern index 2000–2020 for the Erhai Rim Region.

Landcover Class	Year	Number of Plaques	Area/hm^2	Fragmentation	Separation Degree	Dominance	Disturbance Degree	Vulnerable Degree	Loss Degree
Cultivated land	2000	1968	66,301.02	0.0297	0.0861	0.5884	0.1584	0.2143	0.0339
	2010	1997	67,958.01	0.0294	0.0857	0.5956	0.1595	0.2143	0.0342
	2020	2229	69,641.37	0.0320	0.0895	0.6116	0.1652	0.2143	0.0354
Forest land	2000	17,076	112,227.03	0.1522	0.1950	0.9144	0.3175	0.0357	0.0113
	2010	19,123	113,239.08	0.1689	0.2055	0.9902	0.3441	0.0357	0.0123
	2020	16,962	109,728.54	0.1546	0.1966	0.9642	0.3291	0.0357	0.0118
Grassland	2000	14,998	47,316.78	0.3170	0.2815	0.5164	0.3462	0.1786	0.0618
	2010	15,553	41,723.55	0.3728	0.3053	0.5339	0.3847	0.1786	0.0687
	2020	14,618	39,954.87	0.3659	0.3024	0.5182	0.3773	0.1786	0.0674
Shrub land	2000	26,791	29,889	0.8963	0.4734	0.4519	0.6806	0.1429	0.0972
	2010	28,122	30,961.17	0.9083	0.4765	0.5210	0.7013	0.1429	0.1002
	2020	26,628	28,135.26	0.9464	0.4864	0.5008	0.7193	0.1429	0.1028
Wetland	2000	2	0.81	2.4691	0.7857	0.1766	1.5056	0.1071	0.1613
	2010	13	71.82	0.1810	0.2127	0.0026	0.1548	0.1071	0.0166
	2020	7	23.67	0.2957	0.2719	0.0017	0.2298	0.1071	0.0246
Water bodies	2000	303	26,482.77	0.0114	0.0535	0.3483	0.0914	0.0714	0.0065
	2010	214	26,208.72	0.0082	0.0452	0.2152	0.0607	0.0714	0.0043
	2020	219	26,159.85	0.0084	0.0457	0.2166	0.0612	0.0714	0.0044
Artificial surfaces	2000	810	13,189.95	0.0614	0.1239	0.2676	0.1214	0.2500	0.0303
	2010	793	15,245.01	0.0520	0.1140	0.1744	0.0951	0.2500	0.0238
	2020	1093	21,763.80	0.0502	0.1121	0.2462	0.1080	0.2500	0.0270

Table 2. Land-use type transfer matrix for the Erhai Rim 2000–2020.

Landcover Class	Cultivated Land	Forest Land	Grassland	Shrub Land	Wetland	Water Bodies	Artificial Surfaces	hm^2 The 2000
Cultivated land	53,218.80	1824.75	2640.69	752.31	11.70	216.09	6970.77	65,635.11
Forest land	2983.50	96,770.96	3960.18	6348.51	0	44.28	1136.16	111,243.59
Grassland	7513.83	3834.45	29,710.35	3658.41	2.97	20.25	2329.65	47,069.91
Shrub land	2426.31	6112.26	3282.12	17,057.07	0.45	35.01	832.50	29,745.72
Wetland	0.09	0	0	0	0	0.72	0	0.81
Water bodies	170.73	162.63	90	141.57	5.67	25,559.54	88.47	26,218.61
Artificial surfaces	2593.98	75.87	117.45	58.86	2.88	24.21	10,174.59	13,047.84
2020	68,907.24	108,780.92	39,800.79	28,016.73	23.67	25,900.10	21,532.14	292,961.59

3.2. Analysis of Spatial and Temporal Variation Characteristics of Landscape Ecological Risk

In 2000, 2010, and 2020, the mean landscape ecological risk index values in the Erhai rim region were 0.0448, 0.0453, and 0.0457, respectively, indicating a total upward trend. This shows that with the advancement of urbanization and tourism, the ecological risk in the Erhai rim has increased. Meanwhile, to compare the grade changes of landscape ecological risk in the Erhai rim, the spatial distribution of landscape ecological risk in 2000, 2010, and 2020 was acquired via Kriging interpolation of the landscape risk index of 836 landscape ecological risk communities. To show and compare the changes in ecological risk intuitively, we combined the natural breakpoint method and divided it into five risk grades based on the values of 2010 (Figure 4): the lowest risk area ($LERI < 0.024$), lower risk area ($0.024 \leq LERI \leq 0.034$), medium risk area ($0.034 \leq LERI \leq 0.043$), higher risk area ($0.043 \leq LERI \leq 0.051$), and the highest risk area ($LERI > 0.051$).

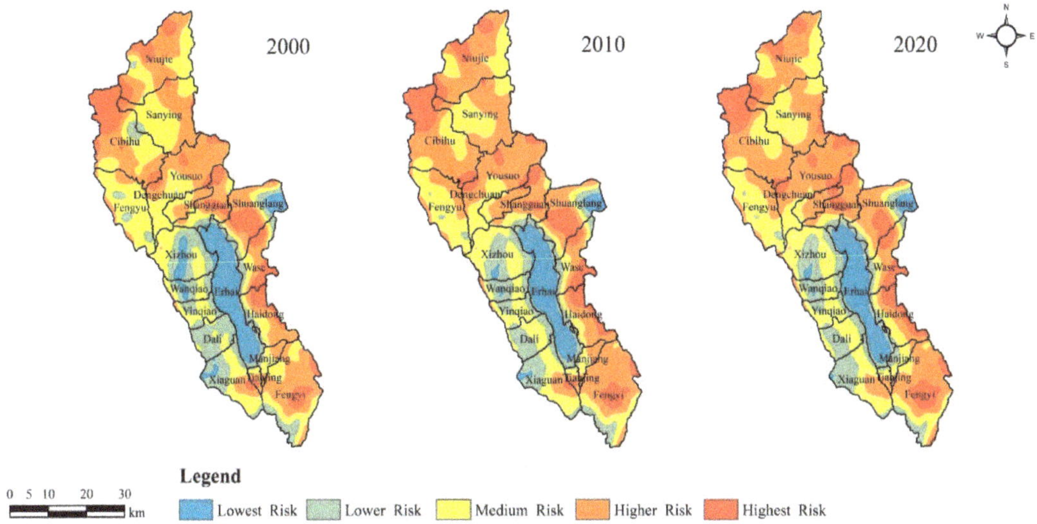

Figure 4. Distribution of ecological risks in Erhai rim region at different periods.

From the area share of ecological risk in each landscape (Table 3), from 2000 to 2020, the overall ecological risk in the Erhai rim was high, with the area share of medium-risk and higher-risk areas accounting for more than 60%, whereas the area share of low ecological risk area was less than 10%. From the perspective of the area change at each landscape's ecological risk level, the higher-risk and highest-risk areas presented a gradual upward trend. Among them, the higher-risk area increased the most, from 92,044.67 hm^2 in 2000 to 105,638.47 hm^2 in 2020, an increase of 13,593.8 hm^2, and the proportional area ratio also increased from 31.50% in 2000 to 36.11% in 2020, becoming the largest ecological risk area in this region. However, the lowest-, lower-, and medium-risk areas tended to be lower. Among them, the medium-risk areas experience the most significant decrease, by 5969.81 hm^2. The area change in the landscape ecological risk level shows that the landscape ecology in the Erhai rim region generally tends to deteriorate. In the future, in order to prevent an increase in risk levels, more attention should be focused on the ecological protection of areas with medium risk and above. From 2000 to 2020, the spatial distribution of landscape ecological risk level in the Erhai rim region was comparatively stable and formed a "multi-polar" distribution pattern in space. The distribution of ecological risks in the landscape showed significant spatial heterogeneity. The landscape ecological risk in the Haixi area was significantly lower than that in the Haidong area, and it was mainly in the lowest- and lower-risk areas. From 2000 to 2020, the spatial distribution of the lowest-risk areas in the Erhai rim region remained essentially unchanged, being mainly concentrated in the waters of the Erhai Sea, the Cang Mountains on the western side of the Erhai Lake, and the Jizu Mountain area on the northeastern side, and remained highly consistent with the scope of the nature reserve. Within the study period, the west coast of Erhai Lake was dominated by low- and medium-risk areas, and the lowest-risk areas were distributed in a ring. However, with the development of tourism in Xizhou Town, land-use types have changed greatly, the landscape pattern has become increasingly complex, and the ecological risk areas have deteriorated, showing the transformation of the lowest- and lower-risk areas to higher-level ecological risk areas. The townships in Eryuan County north of Erhai Lake were mainly highest-risk areas, relatively highest-risk areas, and medium-risk areas, and there existed little change between the ecological risk levels. The ecological risks on the eastern and southern sides of Lake Erhai changed significantly. The highest-risk areas on the eastern side of Erhai Lake are mainly were distributed in mountainous forest areas with relatively fragmented landscapes. Although Dali Bai Autonomous Prefecture has increased its ecological protection and restoration efforts in recent years, they are affected by karst landforms. The soil is barren, the survival rate of tree species is low, ecological restoration is difficult, the landscape vulnerability index is high, and the ecosystem has poor resistance to external interference. Therefore, the risk of ecosystem deterioration increased, which is manifested by the transformation of the landscape ecological risk level to a higher level, with the increase in ecological risk south of Erhai Lake mainly due to the construction of innovative industrial parks. There have been significant conversions of forest and agricultural land into artificial surfaces. High-intensity human activities make landscape patches on the edge of artificial surface expansions more fragmented, leading to an increase in high-risk areas. In contrast, in the courtyard office area of the innovation industrial park on the south bank of Erhai Lake, due to stable land development, the space is connected to a piece, the landscape patches are gradually aggregated and regular, and the landscape fragmentation is reduced. This leads to a reduction in landscape loss; thus, the landscape ecological risk value is reduced from the highest to a lower risk area. Despite a number of ecological conservation initiatives having been conducted, high-intensity economic activities and special geological features in the region have led to a continuous reduction in forest land, shrubland, and grassland, a significant reduction in lake ecosystem functions, and an increased risk of ecosystem deterioration.

Table 3. Ecological risk level of Erhai rim region in each period.

Types		2000	2010	2020
Lowest Risk	Area/hm^2	27,613.96	27,613.95	22,165.54
	Proportion/%	9.45	9.45	7.58
Lower Risk	Area/hm^2	43,794.88	43,794.90	39,047.11
	Proportion/%	14.99	14.99	13.35
Medium Risk	Area/hm^2	93,532.15	93,532.13	87,562.34
	Proportion/%	32.01	32.01	29.93
Higher Risk	Area/hm^2	92,044.67	92,044.69	105,638.47
	Proportion/%	31.50	31.50	36.11
Highest Risk	Area/hm^2	35,255.70	35,255.70	38,155.50
	Proportion/%	12.06	12.06	13.04

3.3. *Impact Analysis of Forest Land Change on Landscape Ecological Risk Based on MGWR Model*

Forest land, the most obvious type of land use in Erhai rim region, has the largest area and is crucial as an ecological barrier. Its change may threaten regional ecological security. From 2000 to 2020, the main forms of forest land change in the Erhai Lake area were forest land expansion and shrinkage, and the intensity index can better characterize the degree and speed of forest land change; therefore, this study will explore the impact of forest land expansion intensity and shrinkage contraction intensity on landscape ecological risk and offer a scientific foundation for the future development of differentiated forest land protection strategies in the Erhai Lake area of China.

3.3.1. Effects of Forest Land Expansion Intensity on Landscape Ecological Risk

(1) Spatial correlation between forest land expansion intensity and landscape ecological risk

This study used the Geoda spatial analysis tool and the k-nearest neighbor spatial weight matrix. Taking the land expansion intensity index as the first variable and the landscape ecological risk value as the second variable, the bivariate global spatial autocorrelation Moran's I index of forest land expansion intensity and landscape ecological risk in the Erhai rim region from 2000 to 2010 and 2010 to 2020 were 0.053 and -0.053, respectively, and significant at the 0.005 level. There existed an obvious positive spatial correlation between forest land expansion intensity and landscape ecological risk in 2000–2010, and the increase in forest land expansion intensity would aggravate the increase in landscape ecological risk; nevertheless, there existed a significant negative spatial relationship between forest land expansion intensity and landscape ecological risk in the study area in 2010–2020. The spatially significant negative correlation between the intensity of forest land expansion and landscape ecological risk in the study area from 2010 to 2020 was significant, and the increase in forest land expansion intensity did not aggravate the increase in landscape ecological risk in this period. The bivariate global spatial autocorrelation test suggests that there is a significant spatial dependence effect between forest land expansion intensity and landscape ecological risk. Consistent with the results, there was an obvious space-dependent effect between forest land expansion intensity and landscape ecological risk.

We used the Geoda spatial analysis tool to draw a bivariate local index of a spatial autocorrelation (LISA) cluster diagram with the purpose of further exploring the local spatial correlation features of the impact of forest land changes on landscape ecological risk in the Erhai rim region (Figure 5). From 2000 to 2010, the effect of forest land expansion intensity on landscape ecological risk in the Erhai rim region was mainly low forest land expansion intensity–low landscape ecological risk area in local spatial agglomeration mainly distributed on the west and east coasts of Erhai Lake, around the towns of Cuose and Haidong and southeast of Fengyi Town. From 2010 to 2020, the impact of forest land expansion intensity on landscape ecological risk was mostly based on low forest land expansion intensity–low landscape ecological risk areas and high forest land expansion intensity–low landscape ecological risk areas. The low forest land expansion intensity–low landscape ecological risk area is mostly concentrated in Xizhou Town, Dali Town, Xiaguan Town, and Shuanglang Town on the east coast of Erhai Lake. The high forest land expansion intensity–low landscape ecological risk areas are distributed in Shuanglang Town, Yinqiao Town, Dali Town, and Xiaguan Town.

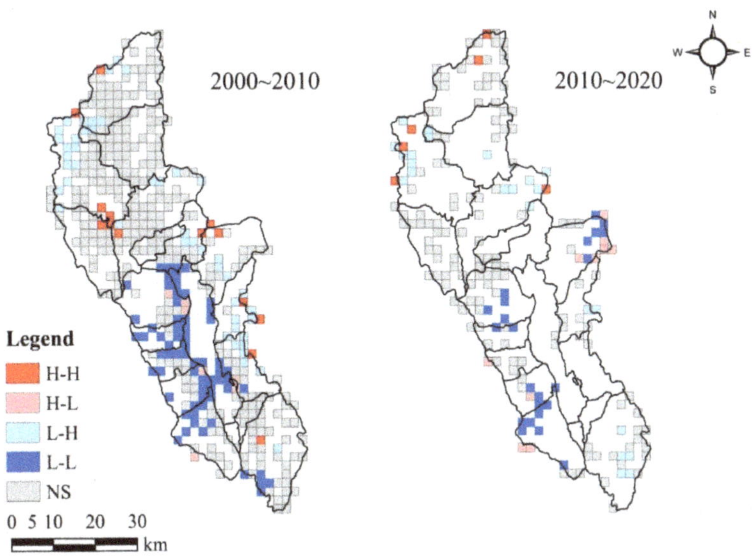

Figure 5. Spatial LISA cluster diagram of forest land expansion intensity and landscape ecological risk in the Erhai rim region.

(2) Spatial heterogeneity of the effect of forest land expansion intensity on landscape ecological risk

In the current work, the forest land expansion intensity index was used as the explanatory variable, and the landscape ecological risk value was the dependent variable (Figure 6). The MGWR model was adopted for analyzing the two variables. From the perspective of time scale, from 2000 to 2010, the regression coefficient of the impact of forest land expansion intensity on landscape ecological risk in the Erhai rim region was mainly positive, and the positive area accounted for 75.40%; that is, the forest land expansion intensity and the number of forest land landscape patches increased, the landscape fragmentation was large, and the landscape ecological risk value continued to increase. During the period from 2010 to 2020, all the regression coefficients of the influence of forest land expansion intensity on landscape ecological risk in the Erhai rim region were negative; that is, the increase in forest land expansion intensity causes the forest land to present a concentrated and contiguous distribution, which will not aggravate the increase in landscape ecological risk but will promote the reduction in landscape ecological risk. In line with the aspect of spatial distribution, the spatial differences in the impact of forest land expansion on landscape ecological risk were large. During the period from 2000 to 2010, the wider area of the Erhai Rim was positive and the smaller area was negative. The high-value areas are mostly concentrated in Xizhou, Wanqiao, and Yinqiao Towns on the west bank of Erhai Lake, and in Xiaguan and Fengyi Towns on the south bank. The expansion of forest land in these areas was mainly due to cultivated land and artificial surfaces. The intensity of forest land expansion was large, the distribution of forest land changed from concentration to dispersion, and the degree of landscape fragmentation and separation and the landscape ecological risk increased. The low value area is distributed in Shangguan, Shuanglang, and Wase Towns northeast of Erhai Lake. Forest land expansion was mostly in areas where the distribution of forest land and grassland was relatively concentrated. It is mainly due to the optimization and adjustment of land-use types within the forest land. The intensity of forest land expansion hs increased, the area of forest land increased significantly, fragmentation and separation within forest land decreased, and the landscape ecological risk decreased accordingly. From 2010 to 2020, the impact of forest land expansion intensity on landscape ecological risk in the Erhai rim region was negative, and the difference between regression coefficients was small, indicating that forest land expansion stabilized during this period and had little impact on landscape ecological risk. Meanwhile, the regression coefficient of the impact of forest land expansion on landscape ecological risk was shown to be high in the southwest and low in the northeast and was distributed in the northwest–southeast band.

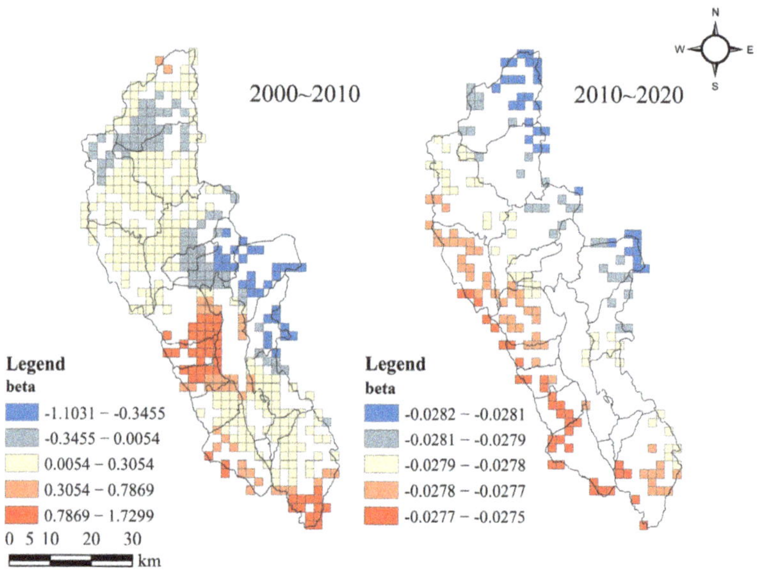

Figure 6. Distribution of regression coefficient between forest land expansion intensity and landscape ecological risk in the Erhai rim region.

3.3.2. Effects of Forest Land Shrinkage on Landscape Ecological Risk

(1) Spatial correlation between forest land shrinkage intensity and landscape ecological risk

Similarly, the forest land shrinkage intensity index was used as the first variable, and the landscape ecological risk value was the second variable. The Geoda spatial analysis tool was employed to analyze the bivariate global spatial autocorrelation Moran's I index of the effect of forest land shrinkage intensity on landscape ecological risk, −0.048 and 0.059, respectively, and was of significance at the 0.005 level. There existed an obvious negative spatial correlation between forest land shrinkage intensity and landscape ecological risk in the study area from 2000 to 2010, indicating that forest land shrinkage intensity increased and landscape ecological risk decreased. From 2010 to 2020, there was a significant positive spatial correlation between forest land shrinkage intensity and landscape ecological risk in the Erhai rim region, and an elevation in the forest land shrinkage intensity aggravated the increase in landscape ecological risk. Based on the obtained results, there was a significant spatial relationship between the intensity of forest land shrinkage and landscape ecological risk regarding spatial dependence.

From the bivariate local spatial autocorrelation LISA clustering map of the effect of forest land shrinkage intensity on landscape ecological risk (Figure 7), from 2000 to 2010 and 2010 to 2020, the impact was mainly in low forest land shrinkage intensity–low landscape ecological risk areas in the local spatial agglomeration. From 2000 to 2010, the low forest land shrinkage intensity–low landscape ecological risk areas were mostly distributed in nature reserves with higher altitude and better forest coverage, namely the Cangshan Mountain on the west coast of Erhai Lake and the Jizu Mountain in Shuanglang Town on the northeast shore of Erhai Lake. In addition, from 2010 to 2020, the area of low forest land shrinkage intensity–low landscape ecological risk increased obviously and was mainly distributed in the Cangshan Mountains and flat urban areas on the west shore of Erhai Lake, with a small-scale distribution on the east coast of Erhai Lake and south of Fengyi Town.

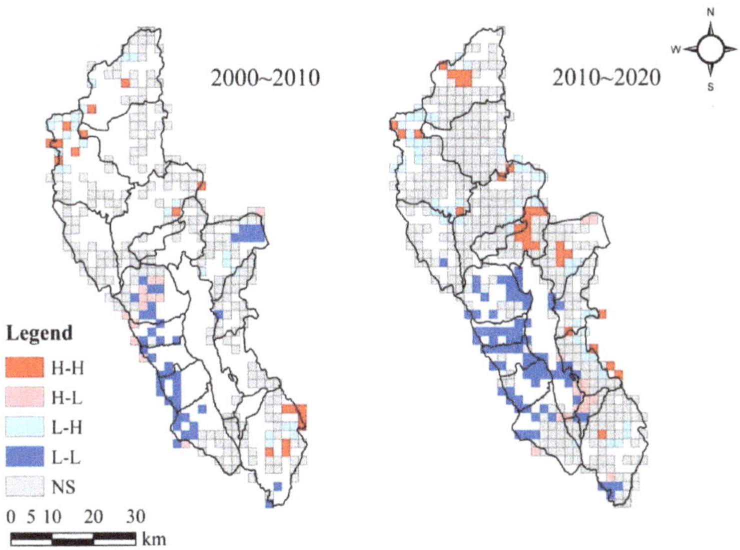

Figure 7. Spatial LISA clustering map of forest land shrinkage intensity and landscape ecological risk in the Erhai rim region.

(2) Spatial heterogeneity of the impact of forest land shrinkage intensity on landscape ecological risk

The forest land shrinkage intensity index was used as the explanatory variable to further investigate the spatial heterogeneity of the impact of forest land shrinkage intensity on landscape ecological risk, and the landscape ecological risk value was the dependent variable (Figure 8). The MGWR model was used for the regression analysis. From the aspect of time scale, from 2000 to 2020, the regression coefficient of the impact of forest land shrinkage intensity on landscape ecological risk in the Erhai rim region was mainly positive, and the regression coefficient rose from 0.20 to 0.42. The positive impact of forest land shrinkage intensity on landscape ecological risk was even more significant. However, from the perspective of spatial distribution, the effect of forest land shrinkage intensity on landscape ecological risk showed significant spatial differences. From 2000 to 2010, except for some negative grids in Xizhou, Fengyu, Cibihu, and Xiaguan Towns and Niujie Township, all regions were positive. The land-use types in the low-value areas were mainly forest land and grassland, and most of the forest land was converted to grassland. Grasslands increased over a large area, the fragmentation of landscape patches decreased, and the landscape ecological risk was low. The high-value area is primarily distributed northeast of Shuanglang Town. Forest land shrinkage in this area was mainly concentrated at the edge of the junction between cultivated and forest land. Forest land is occupied by cultivated land; the landscape fragmentation and degree of interference are large, and the landscape ecological risk is high. From 2010 to 2020, the range and depth of the impact of forest land shrinkage intensity on landscape ecological risk in the Erhai rim region increased, and the spatial differences were more complex. The high-value areas were mainly distributed in the towns of Xizhou, Wanqiao, Yinqiao, and Xiaguan and southwest of Fengyi and Cibihu on the western shore of Erhai Lake. The high-value distribution areas are mostly at the edge of the junction of forest land and other land-use types. The degree of landscape fragmentation and separation is high, and the shrinkage of forest land can lead to an increase in the landscape ecological risk in this area. The low-value areas were mostly located in the marginal areas of the eastern part of the study area. These areas had smaller forest land shrinkage and higher landscape ecological risk, indicating that the higher landscape ecological risk in these areas may have been caused by other factors.

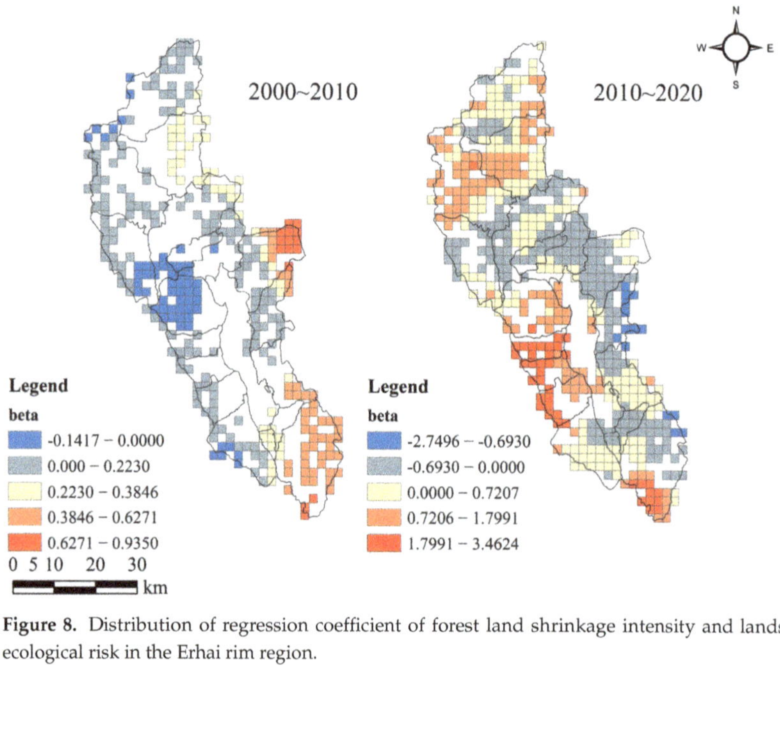

Figure 8. Distribution of regression coefficient of forest land shrinkage intensity and landscape ecological risk in the Erhai rim region.

4. Discussion

As one of the three major forest regions in China and one of the key areas of the ecological security barrier in Southwest China, the Erhai rim region is abundant in ecological resources. However, with the development of urbanization and tourism, cultivated land and artificial surfaces occupy a large amount of ecological space. The area of ecological lands, such as forest land, grassland, shrub land, and water bodies, is decreasing, posing a threat to the security of the regional ecological environment. As the most important land-use type in the Erhai rim region, forest land plays a critical part in maintaining regional ecological functions, including water and soil conservation, water quality improvement, habitat maintenance, and other ecological functions. However, its changes were influenced by natural factors and human activities, which has a great effect on the landscape pattern and landscape ecological risk of Erhai rim region.

(1) The construction of an ecological civilization in the Erhai rim has achieved remarkable results, and government policies and economic development have had a profound impact on forest land changes. The years 2000, 2010, and 2020 are important time points for the implementation of China's natural forest resource protection project, and forest land change in the Erhai rim region also shows important characteristics at these stages. From 2000 to 2010, due to the effective implementation of the first phase of the National Natural Forest Resources Protection Project and the policy of returning cultivated land to forest land, the area of forest land in the Erhai rim increased from 112,227.03 hm^2 in 2000 to 113,239.08 hm^2 in 2010, showing an expansion in forest land change. From 2010 to 2020, despite the implementation of the second phase of the Natural Forest Resource Protection Project, construction projects such as the central city of western Yunnan, mountainous city of Haidong, Dali Expressway, railway, and innovative industrial park were launched because of the acceleration of urbanization. The forest land on the low mountainous gentle slope and the suburban area was required for urban and industrial construction land, and the area occupied ecological land such as forest land, grassland, shrubland, and water bodies was significantly reduced. Most of it was converted to artificial surfaces and cultivated land closely related to human activities. During the period from 2000 to 2020, the spatial distribution of forest land changes in the Erhai rim region varied. The expansion of forest land mainly occurred in Fengyu Town, north of Erhai Lake. Large-scale planting of walnut trees had a vital impact on the stability and increase in forest land. The shrinkage of forest land was mainly in the construction area of Haidong Mountain City and an innovative industrial park, in line with the economic development of the Erhai rim region. The change in forest land area in the Cangshan National Nature Reserve on the western shore of Erhai Lake was not obvious due to the protection of the ecological environment in the Erhai rim region that guaranteed the protection of forest land and improvement of ecological environment quality.

(2) The landscape ecological risk in the Erhai rim region deteriorated overall but partially improved. The landscape ecological risk value in the Erhai rim region presented a gradual upward trend, and the ecological risk level in the region was mainly due to medium- and higher-risk areas. Clearly, the areas of higher and highest risk increased, whereas the lowest- areas, lower- areas, and medium-risk areas decreased. There existed significant spatial heterogeneity in the distribution of ecological risk levels. The ecological risk levels in the Haixi area were mainly the lowest- and lower-risk areas and the total ecological risk was significantly lower than that in the Haidong area. The lowest-risk areas were mostly concentrated in nature reserves such as Erhai Lake, Cangshan Mountain, and Jizu Mountain, where the landscape pattern was relatively smooth, whereas the highest-risk and higher-risk areas were mostly distributed in the northern, eastern, and southern parts of Erhai Lake. The landscape patterns of these areas were complex; in most of them, the landscape patches of forest land, grassland, and shrubland were staggered, and land-use changes were the most frequent and significant. Landscape patches exhibited high fragmentation, high separation, and

poor connectivity. The interaction between different landscapes was blocked, resulting in high landscape ecological risk. In some areas, the ecological environment has exhibited a positive development trend. The Patio Office of the Innovative Industrial Park on the southern shore of Erhai Lake has a stable artificial surface development; landscape patches were spatially connected, landscape fragmentation decreased, and the ecological risk level changed from high to low.

(3) Forest land changes in the Erhai rim region profoundly affect the ecological risk in the region. The bivariate global spatial autocorrelation Moran's *I* index of forest land change and landscape ecological risk showed a significant positive correlation between forest land expansion and landscape ecological risk, which gradually became negative over time. The correlation between forest land shrinkage and landscape ecological risk was the opposite and gradually changed from negative to positive. High forest land expansion intensity–low landscape ecological risk areas and low forest contraction intensity–low landscape ecological risk areas indicate that ecological protection measures in the Erhai rim region are effective. Areas with low forest land expansion intensity and high landscape ecological risk, and high forest contraction intensity and high landscape ecological risk need to focus on forest land change dynamics to strengthen ecological restoration in the future. The driving mechanisms of landscape ecological risk changes caused by high forest land expansion intensity, high landscape ecological risk areas, high forest contraction intensity, low landscape ecological risk areas, and low forest contraction intensity and high landscape ecological risk areas deserve further exploration. From 2000 to 2020, the impact of forest land change on landscape ecological risk in the Erhai rim region was spatially heterogeneous. Reasonable forest land expansion can effectively alleviate the increase in landscape ecological risk and is closely related to the expansion of forest land. Forest land expansion has gradually shifted from fragmentation to concentration, and its impact on landscape ecological risk has gradually changed from positive to negative. The negative area spread from the northeast and north of Erhai Lake to the entire study area. The shrinkage of forest land aggravates the increase in landscape ecological risk, the positive impact of forest land shrinkage on landscape ecological risk is becoming increasingly significant, and the positive area is expanding. Therefore, in order to increase the area of forest land and lower the regional eco-logical risk, the Erhai rim region should highlight the development strategy of forest land with ecological priority, pay attention to the formulation of forest land protection and utilization planning, strive to deal with the conflicts between forest land resources, urban construction and agricultural production, etc., and promote intensive forest land expansion on the basis of keeping the "red line" of forest land. In addition, researchers should also pay attention to the selection of tree species in the karst landscape of the eastern coast of the Erhai Sea to improve the survival rate of trees and forest coverage in the region, thus promoting the high-quality development path of the Erhai rim region with ecological priority and green development as the guides.

The present study investigated the spatial heterogeneity of the impact of forest land change on landscape ecological risk in the Erhai rim region, and these results are highly reliable and provide significant guidance for forest land protection in the Erhai rim region. However, deficiencies remain in the research process and are subject to improvement.

(1) With regard to the setting of the research scale, a variety of scales can be used for comparative research in the future to explore the correlation and influence between different variables at different scales, to enhance the accuracy of ecological protection planning, and scientifically coordinate the correlation between ecological protection and economic development.
(2) Regarding the construction of the landscape ecological risk model, the present study only considered the area weights of each landscape from the aspect of land use without considering the influence of other ecological factors, thus reducing the ecological

meaning of representation of landscape ecological risk; future research needs to be further supplemented and improved.

(3) This study only discussed the influence of forest land change intensity on landscape ecological risk and did not consider the influence of other forest land change indices on landscape ecological risk. At the same time, research on impact models needs to be further explored, and a better impact model needs to be developed to investigate the impact of forest land change on landscape ecological risk.

5. Conclusions

In accordance with the land-use data of the Erhai rim region from 2000 to 2020, the present study analyzed the change in forest land in the region from 2000 to 2020, evaluated and analyzed the landscape ecological risk with a landscape ecological risk model, bivariate spatial autocorrelation, and an MGWR regression model. The degree of influence of forest land change intensity on landscape ecological risk in the Erhai rim region was discussed. The key conclusions are as follows:

(1) Forest land is the main land-use type in the Erhai rim region and is mostly distributed in the periphery of the study area with less human activity and cross-distribution with grassland and shrubland. From 2000 to 2020, the area of forest land decreased by 2498.49 hm^2, with significant spatial and temporal heterogeneity. From 2000 to 2010, the change in forest land in the study area was dominated by expansion, which mostly occurred in Fengyu Town in the northern Erhai Lake. From 2010 to 2020, the demand for urban construction land increased, forest land shrank significantly, and most lands were converted into cultivated and construction land. The most evident shrinkage occurred in the relatively flat areas on the east and south shores of Erhai Lake.

(2) During the period from 2000 to 2020, the overall landscape ecological risk in the Erhai rim region presented an upward trend. The landscape ecological risk levels in the region were mostly medium- and high-risk areas. The landscape ecological risk levels showed a "multi-polar" distribution pattern, and the changes in higher-risk and highest-risk areas increased, while the lowest-, lower-, and medium-risk areas decreased.

(3) From 2000 to 2020, forest land changes made an obvious impact on landscape ecological risk. The influence of forest land expansion intensity on landscape ecological risk gradually changed from positive to negative. The high-value areas were mostly distributed on the west and south shores of Erhai Lake, whereas the low-value areas were mainly concentrated on the east and north shores of Erhai Lake. The effect of forest land shrinkage intensity on landscape ecological risk is positive and significant. Local small-scale areas have negative effects. The high-value areas changed from the northeast of Shuanglang Town and Fengyi Town on the south shore of Erhai Lake to the north and south of the study area and the west coast of Erhai Lake.

Author Contributions: Conceptualization, M.W. (Mengjiao Wang), Y.W. (Yingmei Wu) and Y.W. (Yang Wang); methodology, software, and data curation, M.W. (Mengjiao Wang), C.L. and Y.W. (Yan Wu); writing—original draft preparation, M.W. (Mengjiao Wang), Y.W. (Yingmei Wu) and Y.W. (Yang Wang), B.G. and M.W. (Min Wang); writing—review and editing, M.W. (Mengjiao Wang), Y.W. (Yingmei Wu) and Y.W. (Yang Wang); visualization, M.W. (Mengjiao Wang) and C.L. All authors have read and agreed to the published version of the manuscript.

Funding: The current research was financially supported by the National Natural Science Foundation of China (42071381; 41761037; 41961019). Ten Thousand Talent Plans for Young Top-notch Talents of Yunnan Province (YNWR-QNBJ-2019-200).

Data Availability Statement: The data supporting the findings of the present study are available from the corresponding author upon reasonable request.

Conflicts of Interest: The authors declare no conflict of interest.

Abbreviation

LERI: landscape ecological risk; MGWR, multi-scale geographically weighted regression; GWR, geographically weighted regression.

References

1. Bonan, G. Forests, climate, and public policy: A 500-year interdisciplinary odyssey. In *Annual Review of Ecology, Evolution, and Systematics*; Futuyma, D.J., Ed.; Annual Review of Ecology Evolution and Systematics; Annual Reviews: Palo Alto, CA, USA, 2016; Volume 47, pp. 97–121.
2. Nesha, K.; Herold, M.; De Sy, V.; Duchelle, A.E.; Martius, C.; Branthomme, A.; Garzuglia, M.; Jonsson, O.; Pekkarinen, A. An Assessment of Data Sources, Data Quality and Changes in National Forest Monitoring Capacities in the Global Forest Resources Assessment 2005–2020. *Environ. Res. Lett.* **2021**, *16*, 054029. [CrossRef]
3. Zhen, S.; Zhao, Q.; Liu, S.; Wu, Z.; Lin, S.; Li, J.; Hu, Y. Detecting Spatiotemporal Dynamics and Driving Patterns in Forest Fragmentation with a Forest Fragmentation Comprehensive Index (FFCI): Taking an Area with Active Forest Cover Change as a Case Study. *Forests* **2023**, *14*, 1135. [CrossRef]
4. Yonaba, R.; Tazen, F.; Cissé, M.; Mounirou, L.A.; Belemtougri, A.; Ouedraogo, V.A.; Koïta, M.; Niang, D.; Karambiri, H.; Yacouba, H. Trends, sensitivity and estimation of daily reference evapotranspiration ET0 using limited climate data: Regional focus on Burkina Faso in the West African Sahel. *Theor. Appl. Climatol.* **2023**, *153*, 947–974. [CrossRef]
5. Lèye, B.; Zouré, C.O.; Yonaba, R.; Karambiri, H. Water resources in the sahel and adaptation of agriculture to climate change: Burkina faso. In *Climate Change and Water Resources in Africa: Perspectives and Solutions Towards an Imminent Water Crisis*; Diop, S., Scheren, P., Niang, A., Eds.; Springer International Publishing: Cham, Switzerland, 2021; pp. 309–331.
6. Xu, X.; Liu, J.; Zhuang, D.; Zhang, S. Analysis on Spatial-Temporal Characteristics and Driving Factors of Woodland Change in the Northeastern China Based on 3S Technology. *Sci. Geogr. Sin.* **2004**, *24*, 55–60. [CrossRef]
7. Jackson, B.; Sparks, J.L.D.; Brown, C.; Boyd, D.S. Understanding the Co-occurrence of Tree Loss and Modern Slavery to Improve Efficacy of Conservation Actions and Policies. *Conserv. Sci. Pract.* **2020**, *2*, 13. [CrossRef]
8. Moffette, F.; Alix-Garcia, J.; Shea, K.; Pickens, A.H. The Impact of Near-real-time Deforestation Alerts Across the Tropics. *Nat. Clim. Chang.* **2021**, *11*, 172–178. [CrossRef]
9. Xiong, B.; Chen, R.S.; Xia, Z.L.; Ye, C.; Anker, Y. Large-scale Deforestation of Mountainous Areas during the 21(st) Century in Zhejiang Province. *Land Degrad. Dev.* **2020**, *31*, 1761–1774. [CrossRef]
10. Li, X.; Hai, Q.; Zhu, Z.; Zhang, D.; Shao, Y.; Zhao, Y.; Li, H.; Vandansambuu, B.; Ning, X.; Chen, D.; et al. Spatial and Temporal Changes in Vegetation Cover in the Three North Protection Forest Project Area Supported by GEE Cloud Platform. *Forests* **2023**, *14*, 295. [CrossRef]
11. Li, W.; Zinda, J.A.; Zhang, Z. Does the "Returning Farmland to Forest Program" Drive Community-Level Changes in Landscape Patterns in China? *Forests* **2019**, *10*, 933. [CrossRef]
12. Pang, Y.; Meng, S.; Shi, K.; Yu, T.; Wang, X.; Niu, X.; Zhao, D.; Liu, L.; Feng, M.; Qin, X.; et al. Forest Coverage Monitoring in the Natural Forest Protection Project Area of China. *Acta Ecol. Sin.* **2021**, *41*, 5080–5092.
13. Ma, S.; Qiao, Y.P.; Wang, L.J.; Zhang, J.C. Terrain Gradient Variations in Ecosystem Services of Different Vegetation Types in Mountainous Regions: Vegetation Resource Conservation and Sustainable development. *For. Ecol. Manag.* **2021**, *482*, 118856. [CrossRef]
14. Ma, X.; Wu, H.; Qin, B.; Wang, L. Spatiotemporal Change of Landscape Pattern and Its Eco-environmental Response in the Yangtze River Economic Belt. *Sci. Geogr. Sin.* **2022**, *42*, 1706–1716. [CrossRef]
15. Li, C.; Wu, Y.; Gao, B.; Zheng, K.; Wu, Y.; Li, C. Multi-scenario Simulation of Ecosystem Service Value for Optimization of Land Use in the Sichuan-Yunnan Ecological Barrier, China. *Ecol. Indic.* **2021**, *132*, 108328. [CrossRef]
16. Chen, X.; Ding, Z.; Yang, J.; Chen, X.; Chen, M. Ecological Risk Assessment and Driving Force Analysis of Landscape in the Compound Mine-urban Area of the Northern Peixian County. *Chin. J. Ecol.* **2022**, *41*, 1796–1803. [CrossRef]
17. Zhou, P.; Meng, J. Progress of Ecological Risk Management Research: A Review. *Acta Ecol. Sin.* **2009**, *29*, 2097–2106. [CrossRef]
18. Cao, Q.; Zhang, X.; Ma, H.; Wu, J. Review of Landscape Ecological Risk and An Assessment Framework Based on Ecological Services: ESRISK. *Acta Geogr. Sin.* **2018**, *73*, 843–855. [CrossRef]
19. Peng, J.; Dang, W.; Liu, Y.; Zong, M.; Hu, X. Review on Landscape Ecological Risk Assessment. *Acta Geogr. Sin.* **2015**, *70*, 664–677. [CrossRef]
20. Kapustka, L.; Galbraith, H.; Luxon, M.; Yocum, J. Using Landscape Ecology to Focus Ecological Risk Assessment and Guide Risk Management Decision-making. *Toxicol. Ind. Health* **2001**, *17*, 236–246. [CrossRef]
21. Paukert, C.; Pitts, K.; Whittier, J.; Olden, J. Development and Assessment of A Landscape-scale Ecological Threat Index for the Lower Colorado River Basin. *Ecol. Indic.* **2011**, *11*, 304–310. [CrossRef]
22. Ayre, K.K.; Landis, W.G. A Bayesian Approach to Landscape Ecological Risk Assessment Applied to the Upper Grande Ronde Watershed, Oregon. *Hum. Ecol. Risk Assess.* **2012**, *18*, 946–970. [CrossRef]
23. Liu, Y.; Xu, W.H.; Hong, Z.H.; Wang, L.G.; Ou, G.L.; Lu, N. Assessment of Spatial-Temporal Changes of Landscape Ecological Risk in Xishuangbanna, China from 1990 to 2019. *Sustainability* **2022**, *14*, 10645. [CrossRef]

24. Zhu, K.W.; He, J.; Zhang, L.X.; Song, D.; Wu, L.J.; Liu, Y.Q.; Zhang, S. Impact of Future Development Scenario Selection on Landscape Ecological Risk in the Chengdu-Chongqing Economic Zone. *Land* **2022**, *11*, 964. [CrossRef]
25. Li, R.; Wei, W. A Study on Landscape Ecological Risk Assessment in New Urban Districts: A Case Study of Chenggong District, Kunming City. *J. Kunming Univ. Sci. Technol. Nat. Sci.* **2020**, *45*, 124–132. [CrossRef]
26. Meng, X.; Ren, Z.; Zhang, C. Study on Land Use Change and Ecological Risk in Xianyang City. *Arid. Zone Res.* **2012**, *29*, 137–142. [CrossRef]
27. Li, X.; Li, J. Analysis on Regional Landscape Ecological Risk Based on GIS --A Case Study along the Lower Reaches of the Weihe River. *Arid. Zone Res.* **2008**, *25*, 899–903. [CrossRef]
28. Wu, J.; Qiao, N.; Peng, J.; Huang, X.; Liu, J.; Pan, Y. Spatial Variation of Landscape Eco-risk in Open Mine Area. *Acta Ecol. Sin.* **2013**, *33*, 3816–3824. [CrossRef]
29. Gaines, K.F.; Porter, D.E.; Dyer, S.A.; Wein, G.R.; Pinder, J.E.; Brisbin, I.L. Using Wildlife as Receptor Species: A Landscape Approach to Ecological Risk Assessment. *Environ. Manag.* **2004**, *34*, 528–545. [CrossRef]
30. Iroume, A.; Huber, A.; Schulz, K. Summer Lows in Experimental Catchments with Different Forest Covers, Chile. *J. Hydrol.* **2005**, *300*, 300–313. [CrossRef]
31. De Lombaerde, E.; Vangansbeke, P.; Lenoir, J.; Van Meerbeek, K.; Lembrechts, J.; Rodríguez-Sánchez, F.; Luoto, M.; Scheffers, B.; Haesen, S.; Aalto, J.; et al. Maintaining forest cover to enhance temperature buffering under future climate change. *Sci. Total Environ.* **2022**, *810*, 151338. [CrossRef] [PubMed]
32. Hu, W.; Li, G.; Gao, Z.; Jia, G.; Wang, Z.; Li, Y. Assessment of the impact of the Poplar Ecological Retreat Project on water conservation in the Dongting Lake wetland region using the InVEST model. *Sci. Total Environ.* **2020**, *733*, 139423. [CrossRef] [PubMed]
33. Shi, P.; Yuan, Y.; Zheng, J.; Wang, J.; Ge, Y.; Qiu, G. The Effect of Land Use/Cover Change on Surface Runoff in Shenzhen Region, China. *Catena* **2007**, *69*, 31–35. [CrossRef]
34. Shi, X.; Chen, K.; Jie, C.; Long, T. Evaluation on Service Value of Forest Ecosystem in Jilin Province. *Bull. Soil Water Conserv.* **2016**, *36*, 312. [CrossRef]
35. Yao, H.; Cui, B. The Effect of Land Use and Its Change on Soil Erosion of the Lancang River Watershed in Yunnan Province. *Acta Sci. Circumstantiae* **2006**, *26*, 1362–1371. [CrossRef]
36. Wu, Y.; Li, C.; Gao, B.; Wang, M.; Wu, Y.; Zheng, K. Construction of Urban Ecological Security Pattern in Highland Lakes Cites: The Case of Dali City. *Acta Ecol. Sin.* **2023**, *2023*, 1–14. [CrossRef]
37. Shao, J.; Zhou, J. Practice and Explore of Lakeside Urbanization in the Context of "Two-oriented Society" in Wuhan Urban Agglomeration. *Urban Dev. Stud.* **2012**, *19*, 39–45. [CrossRef]
38. Ye, C.; Li, C.; Deng, T. Structures and Ecological Functions of Lake Littoral Zones. *Res. Environ. Sci.* **2015**, *28*, 171–181. [CrossRef]
39. Ding, W. A Study on the Characteristics of Climate Change around the Erhai Area, China. *Resour. Environ. Yangtze Basin* **2016**, *25*, 599–605. [CrossRef]
40. Jun, C.; Ban, Y.F.; Li, S.N. Open access to Earth land-cover map. *Nature* **2014**, *514*, 434. [CrossRef]
41. Liu, S.; Shen, H. A GIS based Model of Urban Land Use Growth in Beijing. *J. Geogr.* **2000**, *55*, 407. [CrossRef]
42. Wang, Y.; Sun, R. Impact of Land Use Change on Coupling Coordination Degree of Regional Water-energy-food System: A Case Study of Beijing-Tianjin-Hebei Urban Agglomeration. *J. Nat. Resour.* **2022**, *37*, 582–599. [CrossRef]
43. Jia, H.; Wang, R.; Li, H.; Diao, B.; Zheng, H.; Guo, L.; Liu, L.; Liu, J. The Changes of Desertification and Its Driving Factors in the Gonghe Basin of North China over the Past 10 Years. *Land* **2023**, *12*, 998. [CrossRef]
44. Wei, H.; Xu, L.; Li, X.; Li, J. Landscape Ecological Risk Assessment and Its Spatiotemporal Changes of the Boston Lake Basin. *Environ. Sci. Technol.* **2018**, *41*, 345–351. [CrossRef]
45. Lu, L.; Zhang, J.; Sun, C.; Wang, X.; Zheng, D. Landscapeecological Risk Assessment of Xi river Basin Based on Land-use Change. *Acta Ecol. Sin.* **2018**, *38*, 5952–5960. [CrossRef]
46. Xie, H. Regional Eco-risk Analysis of Based on Landscape Structure and Spatial Statistics. *Acta Ecol. Sin.* **2008**, *28*, 5020–5026. [CrossRef]
47. Gao, B.; Wu, Y.; Li, C.; Zheng, K.; Wu, Y.; Wang, M.; Fan, X.; Ou, S. Multi-Scenario Prediction of Landscape Ecological Risk in the Sichuan-Yunnan Ecological Barrier Based on Terrain Gradients. *Land* **2022**, *11*, 2079. [CrossRef]
48. Zhang, Y.; Xie, X. Regional Ecological Risk Assessment in Nansi Lake Based on RS and GIS. *Acta Ecol. Sin.* **2015**, *35*, 1371–1377. [CrossRef]
49. Gao, B.; Li, C.; Wu, Y.; Zheng, K.; Wu, Y. Landscape Ecological Risk Assessment and Influencing Factors in Ecological Conservation Area in Sichuan-Yunnan Provinces, China. *J. Appl. Ecol.* **2021**, *32*, 1603–1613. [CrossRef]
50. Wu, Y.; Wu, Y.M.; Li, C.; Gao, B.P.; Zheng, K.J.; Wang, M.J.; Deng, Y.H.; Fan, X. Spatial Relationships and Impact Effects between Urbanization and Ecosystem Health in Urban Agglomerations along the Belt and Road: A Case Study of the Guangdong-Hong Kong-Macao Greater Bay Area. *Int. J. Environ. Res. Public Health* **2022**, *19*, 16053. [CrossRef]
51. Chang, J.; Sun, P.J.; Wei, G.E. Spatial Driven Effects of Multi-Dimensional Urbanization on Carbon Emissions: A Case Study in Chengdu-Chongqing Urban Agglomeration. *Land* **2022**, *11*, 1858. [CrossRef]
52. Fotheringham, A.S.; Yang, W.B.; Kang, W. Multiscale Geographically Weighted Regression (MGWR). *Ann. Am. Assoc. Geogr.* **2017**, *107*, 1247–1265. [CrossRef]

53. Shen, T.; Yu, H.; Zhou, L.; Gu, H.; He, H. On Hedonic Price of Second-Hand Houses in Beijing Based on Multi-Scale Geographically Weighted Regression:Scale Law of Spatial Heterogeneity. *Econ. Geogr.* **2020**, *40*, 75–83. [CrossRef]
54. Yu, H.; Fotheringham, A.; Li, Z.; Oshan, T.; Kang, W.; Wolf, L. Inference in Multiscale Geographically Weighted Regression. *Geogr. Anal.* **2020**, *52*, 87–106. [CrossRef]

forests

MDPI

Article

Characteristics of Vegetation Change and Its Climatic and Anthropogenic Driven Pattern in the Qilian Mountains

Yanmin Teng [1], Chao Wang [2], Xiaoqing Wei [3], Meirong Su [4,*], Jinyan Zhan [5,*] and Lixiang Wen [5]

1 Research Center for Eco-Environmental Engineering, Dongguan University of Technology, Songshan Lake, Dongguan 523808, China; tengym@dgut.edu.cn
2 School of Labor Economics, Capital University of Economics and Business, Beijing 100089, China; wangc@cueb.edu.cn
3 School of Energy and Water Resources, Shenyang Institute of Technology, Shenyang 113122, China
4 Key Laboratory for City Cluster Environmental Safety and Green Development of the Ministry of Education, School of Ecology, Environment and Resources, Guangdong University of Technology, Guangzhou 510006, China
5 State Key Laboratory of Water Environment Simulation, School of Environment, Beijing Normal University, Beijing 100875, China
* Correspondence: sumr@gdut.edu.cn (M.S.); zhanjy@bnu.edu.cn (J.Z.)

Abstract: The Qilian Mountains (QLM) are an essential ecological security barrier in northwest China. Identifying the driven pattern of vegetation change is crucial for ecological protection and restoration in the QLM. Based on high-resolution vegetation coverage (VC) data in the QLM from 1990 to 2018, linear trend analysis was employed to examine the spatiotemporal dynamics of VC in the QLM, while correlation analysis was utilized to establish relationships between VC change and environmental factors. Multiple correlation analysis and residual analysis were adopted to recognize the climatically and anthropogenically driven pattern of VC change. The results showed that VC in the QLM presented a remarkable upward trend in volatility from 1990 to 2018. The significant increase areas accounted for 59.32% of the total, mainly distributed in the central and western QLM, and the significant decrease areas accounted for 9.18%, mostly located in the middle and eastern QLM. VC change showed a significant positive correlation with precipitation change and annual average temperature, while it exhibited a significant negative correlation with annual average precipitation, current VC status, livestock density, and slope. Climate change played a leading role in the increase of VC, and the impact of precipitation was significantly higher than that of temperature. Affected by climate change, the VC of alpine steppes and temperate steppes increased the most. Under the human interference, VC decreased significantly in 9.2% of the region, of which shrubs fell the most, followed by alpine meadows and forests. This study can provide certain guidance for local ecological protection and restoration efforts.

Keywords: vegetation coverage; climate change; human activities; vegetation degradation; Qinghai–Tibetan Plateau

Citation: Teng, Y.; Wang, C.; Wei, X.; Su, M.; Zhan, J.; Wen, L. Characteristics of Vegetation Change and Its Climatic and Anthropogenic Driven Pattern in the Qilian Mountains. *Forests* **2023**, *14*, 1951. https://doi.org/10.3390/f14101951

Academic Editor: Maciej Pach

Received: 11 August 2023
Revised: 12 September 2023
Accepted: 21 September 2023
Published: 26 September 2023

1. Introduction

As a natural link between soil, atmosphere, and water, vegetation plays an important role in terrestrial ecosystems by regulating carbon cycling and energy exchange [1,2]. Vegetation coverage (VC) can reflect the regional comprehensive characteristics of climate, topography and human activities, and is easily affected by climatic and anthropogenic factors [3,4]. Therefore, it is an important indicator for monitoring the impact of climate change and human activities on ecosystems [5,6], and it has been widely used in large-scale ecological environment change monitoring and assessment [7–9].

In terms of identifying VC change, researchers mostly adopt linear regression trend analysis [10–14], and a few adopt the ordinary least squares method [15]. At present the,

GeoDetector model is one of the most commonly used methods to detect the drivers of vegetation change and has received extensive attention from researchers [16–18]. GeoDetector has many advantages, including the ability to analyze both numerical and qualitative data, as well as to identify the interaction between two driving factors [10,19,20]. However, it can only analyze the influence intensity of driving factors statistically and is unable to display the spatial heterogeneity of driving factors' influence on a map [12,21]. Hence, studies have usually adopted correlation analysis (or partial correlation analysis) to spatially clarify the driving effect of climate variables on vegetation change [11,13]. In addition, the geographic weighted regression model is also frequently used to spatially identify the comprehensive driving effects of multiple factors [15,22]. As for human factors, residual analysis is often adopted to spatially identify the influences of human intervention on vegetation change [14,23].

Among the climatic factors, temperature and precipitation are the most important [15,24,25]. The effects of climatic factors on VC change differ spatially, and precipitation dominates in arid and semi-arid regions, while temperature dominates in humid southern regions [26]. For the Qinghai–Tibet Plateau (QTP), precipitation plays a decisive role in VC change in the Three River Source Region, Heihe River Basin, and Yarlung Zangbo River Basin [10,21,27]. However, studies also show that annual mean temperature is the dominant factor driving VC change in the northeastern Tibetan Plateau [12,28]. In addition to precipitation and temperature, solar radiation and wind speed are also important driving factors affecting vegetation change [29]. Moreover, elevation and soil type were found to be the main factors affecting VC in the Qilian Mountains (QLM) [17,30]. For elevation gradient, VC in the QLM increased first and then decreased with the increase of altitude [30]. From the long-term change, VC showed an increasing trend at low altitudes (below 3200 m) and gentle slopes (below 15°), and a decreasing trend at high altitudes (above 3700 m) and steep slopes (above 25°) [31]. For soil type, black felt soil has abundant humus, which can provide rich nutrients for vegetation [17]. Human activities also have a significant impact on VC change, especially in the semi-humid region of China [19]. Land use change is a dominant factor, especially in semi-humid, semi-arid, and arid areas [21,32]. In addition, GDP and population density also have a certain impact [10].

Within the QTP, climate change in the QLM is more significant, and the vegetation is more susceptible to climate change as well [12,33]. Since the 1980s, VC in the QLM has shown an overall improvement trend, with significant increases in the central and western regions and partial degradations in the central and eastern regions [34–37]. However, due to different analysis periods, there are significant differences in the degraded areas identified by several studies. In addition, studies further analyzed the correlation between vegetation change and climate change in the QLM and found that precipitation was the main factor leading to vegetation change, followed by temperature [34–37]. Moreover, the correlation between VC and temperature in winter and spring was the highest, and the correlation between VC and precipitation in summer was the highest in the previous period [35].

Climate change has promoted VC increase in the QLM on a large scale. However, due to the impacts of human activities such as overgrazing, mining, and hydropower development, vegetation degradation and land desertification in local areas of the QLM were prominent in recent decades [38,39], thus affecting the overall role of the ecological security barrier [40]. To control the degraded lands and restore regional ecological functions, the local government has implemented a series of ecological protection and restoration projects in the QLM, such as a grazing-forbidding project, a grassland–livestock balance project, and a grassland ecological award and compensation project [41,42]. These projects reduced livestock numbers to a certain extent, increased the construction of ecological engineering projects including fences and artificial grasslands, and promoted vegetation restoration [43]. However, the climatically and anthropogenically driven patterns of VC change in this region are not yet clear. Therefore, we aim to clarify the temporal and spatial characteristics of VC change in the QLM and recognize its main natural and anthropogenic influencing factors. Further, we will identify the VC change zones driven by climate change and human activities, as well as the VC change types under different vegetation types. This study has a certain guiding significance for targeted implementation of ecological protection and restoration measures in the QLM.

2. Materials and Methods

2.1. Study Area

The QLM (93°30′–103°00′ E, 35°43′–39°36′ N) is located at the northeast edge of the Qinghai–Tibet Plateau, with a total area of about 193,300 km^2. It is the source of Heihe River, Datong River, Huangshui River, Shule River, and Shiyang River (Figure 1a). The QLM is high in the central and northwest and low in the southeast, and most of the areas are about 3500–5000 m above sea level (Figure 1a). The QLM has a typical continental climate: the eastern part is warm and humid, having a continental semi-arid alpine grassland climate, and the western part is cold and dry, with a continental arid desert climate [44]. The main vegetation types included alpine meadow, alpine steppe, temperate steppe, temperate desert, alpine desert, and shrub, in addition to small areas of coniferous forest, broad-leaved forest, and cropland (Figure 1b). Based on previous studies [45,46], we obtained the characteristics of different vegetation types and their dominant plant species: the main tree species for forests are Simon poplar (*Populus simonii* Carr.), white birch (*Betula platyphylla* Suk), aspen (*Populus davidiana* Dode), Qinghai spruce (*Picea crassifolia*), Chinese pine (*Pinus tabuliformis*), and Qilian juniper (*Juniperus przewalskii* Kom.); shrubs are led by cold-tolerant shrubs, including *salix gilashanica*, *caragana jubata*, bush cinqefoil (*Potentilla fruticosa* L.), and sea buckthorn (*Hippophae rhamnoides* L.); as a good summer pasture, alpine meadow develops under moderate moisture conditions and is dominated by perennial mesophytes (such as *Kobresia*); alpine steppe is mainly composed of cold- and drought-resistant perennial herbs (such as *Carex*) and cushion plants; temperate steppe is dominated by herbs suitable for warm and arid climates (such as *Stipa*); alpine desert is largely composed of cold and xerophytic shrubs, accompanied by a certain number of cushion herbaceous plants; and temperate desert is dominated by xerophytic tufted grasses and shrubs. The hydrothermal regime in the QLM varies greatly, with an annual mean precipitation of 67–758 mm and annual mean temperature of −16.3–6 °C (Figure 1c,d). Affected by climatic conditions, VC in the QLM shows a significant downward trend from east to west (Figure 1e). Similarly, the population is concentrated in the lower elevations of the eastern QLM (Figure 1f).

2.2. Data Sources

Monthly and annual meteorological data from 1990 to 2018 were obtained from the National Tibetan Plateau Data Center (TPDC, https://data.tpdc.ac.cn (accessed on 8 May 2022)), including temperature and precipitation, with a spatial resolution of 1 km. These data were generated from the global high-resolution climate dataset published by WorldClim through the Delta spatial downscaling scheme [47]. VC data of seven periods (1990, 1995, 2000, 2005, 2010, 2015, and 2018) at 30 m spatial resolution were provided by the TPDC (accessed on 8 March 2020). Annual VC data at 1 km spatial resolution from 1990 to 2018 were derived from the Resource and Environmental Science and Data Center (RESDC, https://www.resdc.cn (accessed on 19 April 2022)). Vegetation type data with a 1 km spatial resolution and rural settlement shape files were obtained from the RESDC (accessed on 24 October 2020). A digital elevation model (DEM) product with a spatial resolution of 30 m was provided by Qi et al. [48] (accessed on 4 July 2021). Road networks data were obtained from the TPDC (accessed on 15 September 2021) and derived from a 1:100,000 ADC_WorldMap (2014), including major highways, roads, and railways. Grazing intensity data in 2010 with a 1 km spatial resolution were acquired from the Food and Agriculture Organization of the United Nations (http://www.fao.org/geonetwork/srv/en/main.home (accessed on 19 May 2020)). Population density data in 2015 at 1 km resolution were obtained from the Chinese population spatial distribution dataset provided by the RESDC (accessed on 4 July 2021).

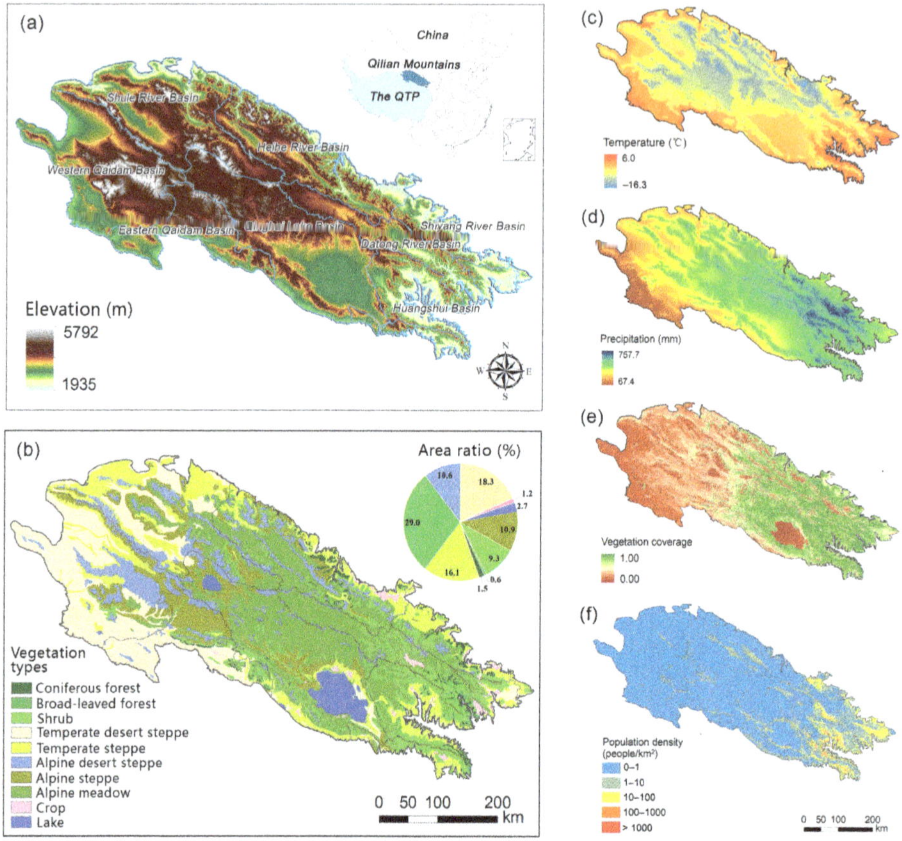

Figure 1. Basic information of the study area. (**a**) Location of the QLM and a DEM map with basin boundaries, (**b**) vegetation types, (**c**) temperature, (**d**) precipitation, (**e**) vegetation coverage, and (**f**) population density.

2.3. Research Methods

2.3.1. Change Trends of VC, Temperature and Precipitation

Based on the seven-period, 30 m resolution VC data, excluding cultivated land and non-vegetated areas such as desert and water bodies, we identified the high-resolution spatial pattern of vegetation change in the QLM. The change trends of VC, temperature, and precipitation at a grid scale over the QLM were analyzed by using linear regression trend analysis. After the data were preprocessed, trend analysis and significance testing of each factor were conducted via MATLAB (R2019b). The main calculation formula is as follows [49]:

$$S = \frac{n \times \sum_{i=1}^{n} i \times X_i - \left(\sum_{i=1}^{n} i\right) \times \left(\sum_{i=1}^{n} X_i\right)}{n \times \sum_{i=1}^{n} i^2 - \left(\sum_{i=1}^{n} i\right)^2} \quad (1)$$

where S is the change rate of VC, temperature, and precipitation; n is the number of years. A positive value indicates an overall upward trend, while a negative value indicates an overall downward trend.

Further, based on the boundary data of the QLM, we extracted the average values of 1 km resolution annual VC, temperature, and precipitation year by year in ArcGIS. Then, we used the extracted data to create line plots and obtained the change trends of VC, temperature, and precipitation in time series from 1990 to 2018.

2.3.2. Correlation Analysis between VC Change and Climate Change

Pearson correlation analysis was used to determine the correlations between VC change and temperature and precipitation change. The main calculation methods are as follows:

$$R = \frac{\sum_{i=1}^{n}(X_i - \overline{X})(Y_i - \overline{Y})}{\sqrt{\sum_{i=1}^{n}(X_i - \overline{X})^2 * \sum_{i=1}^{n}(Y_i - \overline{Y})^2}} \tag{2}$$

where R is the correlation coefficient; n is total research years; X_i is VC of the year i; Y_i is temperature and precipitation of the year i; and $\overline{X}$ is the mean VC from 1990 to 2018. $\overline{Y}$ is the average temperature or precipitation from 1990 to 2018. When $R > 0$, each factor is positively correlated, when $R < 0$, each factor is negatively correlated, and the correlation increases as the R value approaches 1 or -1.

Partial correlation analysis can reveal the impact intensity of one factor on another while other factors remain unchanged [11,13]. Multiple correlation analysis can analyze the correlation between two or more factors and a certain factor, so as to identify the comprehensive influences of multiple factors on this factor [50]. In this study, partial correlation and multiple correlation analysis of VC, temperature, and precipitation were carried out to identify the dominant areas of VC change affected by different climatic factors in the QLM. The calculation formulas of the partial correlation analysis and t-test are as follows:

$$R_{xy,z} = \frac{R_{xy} - R_{xz}R_{yz}}{\sqrt{(1 - R_{xz}^2) * (1 - R_{yz}^2)}} \tag{3}$$

$$t = \frac{R_{xy,z} * \sqrt{n - m - 1}}{\sqrt{1 - R_{xy,z}^2}} \tag{4}$$

where $R_{xy,z}$ is the partial correlation coefficient between the dependent variable x and the independent variable y when the independent variable z remains constant; R_{xy}, R_{xz}, and R_{yz} are correlation coefficients between factors x, y, and z; n is sample size; and m is the degree of freedom.

The calculation formulas of the multiple correlation analysis and F-test are as follows:

$$R_{x,yz} = \sqrt{1 - (1 - R_{xy}^2) * (1 - R_{xz,y}^2)} \tag{5}$$

$$F = \frac{R_{x,yz}^2}{1 - R_{x,yz}^2} \frac{n - m - 1}{m} \tag{6}$$

where $R_{x,yz}$ is the multiple correlation coefficient between the dependent variable x and the independent variables y and z; n is sample size; and m is the degree of freedom.

2.3.3. Identification of Climate-Driven Zones

With reference to previous research [51], we formulated the zoning criteria for driving factors of VC change in the QLM (Table 1). The climate-driven types include the temperature-driven zone, precipitation-driven zone, temperature and precipitation co-driven zone, and non-climate-driven zone.

Table 1. Zoning criteria for driving factors of VC change.

VC Change Types	Zoning Criteria		
	$R_{xy,z}$	$R_{xz,y}$	$R_{x,yz}$
Temperature-driven type	$t \geq t_{0.05}$		$F \geq F_{0.05}$
Precipitation-driven type		$t \geq t_{0.05}$	$F \geq F_{0.05}$
Co-driven type	$t \leq t_{0.05}$	$t \leq t_{0.05}$	$F \geq F_{0.05}$
Non-climate driven type			$F \leq F_{0.05}$

Note: $R_{xy,z}$ and $R_{xz,y}$ represent partial correlation coefficients between VC and temperature and precipitation, respectively; $R_{x,yz}$ represents the multiple correlation coefficient between VC and climatic factors; t and F represent the statistical values of the t-test and F-test, respectively; 0.05 is the significance level.

2.3.4. Residual Analysis of VC Change

Residual analysis is a quantitative analysis method to analyze the impact of human activities on VC change [52] and is widely used in large-scale vegetation change analysis [53,54]. Based on temperature and precipitation data, multivariate linear regression was performed to fit the predicted VC, that is, the expected VC only under the influence of climatic factors. The effect of anthropogenic factors on vegetation can be obtained by subtracting the real VC value from the predicted value. The calculation formula is as follows:

$$VC_{\text{predicted}} = a \times T + b \times P + c \tag{7}$$

$$\varepsilon = VC_{\text{real}} - VC_{\text{predicted}} \tag{8}$$

where $VC_{\text{predicted}}$ and VC_{real} are the predicted VC value based on regression models and the observed VC value based on VC data with a resolution of 30 m, respectively; T and P are temperature and precipitation, respectively; a, b, and c are model parameters; and ε is the residual, that is, the impacts of human activities on VC. When $\varepsilon > 0$, human activities promote the increase of VC, and when $\varepsilon < 0$, human activities cause the decrease of VC. The greater the absolute value, the stronger the influences of human activities on vegetation.

3. Results

3.1. *Change Features of VC*

From 1990 to 2018, VC in most areas of the QLM showed an increasing trend, with 47.05% and 12.27% of the areas significantly and extremely significantly increasing in VC, respectively (Figure 2a). At the same time, VC decreased in local areas, with 7.54% and 1.64% of the areas significantly and extremely significantly decreasing in VC, respectively. For the spatial distribution, areas with significant increase in VC were mainly distributed in the western QLM with weak human disturbances, mainly involving the Shule River Basin, Qinghai Lake Basin and the eastern and western sections of Qaidam Basin. Areas where VC decreased remarkably were mostly located in the central and eastern QLM with high human activity intensity, mainly involving the Heihe River Basin, Datong River Basin, Huangshui River Basin, Shiyang River Basin, and Qinghai Lake Basin. In terms of time dynamics, VC in the QLM showed a significant fluctuating upward trend from 1990 to 2018, with an average annual growth rate of about 0.55% (Figure 2b).

Furthermore, we extracted the mean VC change rate under different vegetation types (Figure 2c). The results showed that the change rates of alpine and temperate grasslands were positive, with values of 0.97% and 0.47%, respectively. On the contrary, the VC of forests, shrubs, and alpine meadows showed a decreasing trend, with shrubs showing the largest change (−0.18%), followed by alpine meadows (−0.15%), and forests (−0.06%). Hence, VC increase in the QLM was mainly dominated by alpine grassland and temperate grassland, and forests, shrubs and alpine meadows degraded to some extent under human interference.

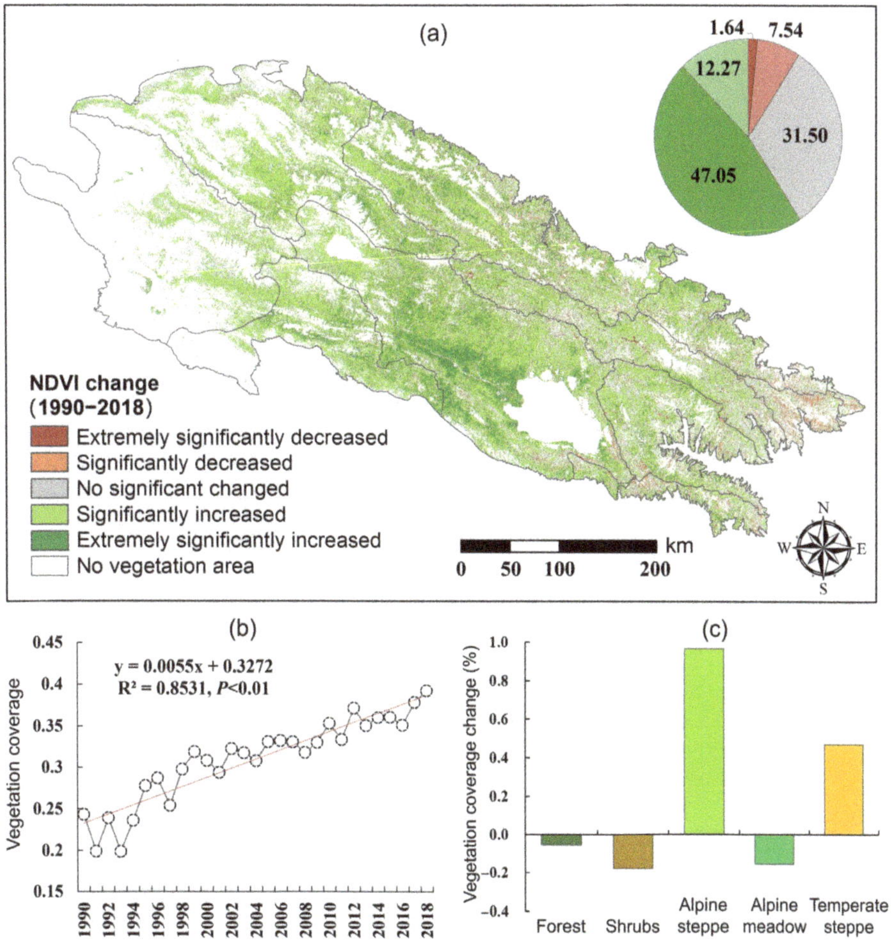

Figure 2. Vegetation coverage (VC) change in the QLM from 1990 to 2018. (**a**) Spatial variation, (**b**) temporal change, and (**c**) changes under different vegetation types.

We extracted the VC change rate and various environmental variables and then conducted Pearson correlation analysis to recognize the main influencing factors of VC change. The results showed that VC change was significantly negatively correlated with elevation, slope, livestock density, current VC status, annual mean precipitation, and precipitation change rate, while it was significantly positively correlated with annual mean temperature, distance from roads, and distance from rural settlements (Table 2). Among them, the correlation between VC change rate and annual mean precipitation was the highest (−0.583), followed by VC, livestock density, precipitation change rate, annual mean temperature, slope, distance from roads, distance from rural settlements, elevation, and temperature change rate. Therefore, areas with significant vegetation degradation tend to be distributed in areas with good ecological environments (high precipitation and VC, and obvious precipitation increase) and high human activity intensity (high livestock density and close to rural settlements and roads).

Table 2. Correlation between VC change and environmental factors.

Environmental Factors	Elevation	Slope	Distance from Rural Settlements	Distance from Roads	Livestock Density
Correlation coefficient	−0.022 *	−0.167 **	0.064 *	0.079 *	−0.265 **
Environmental Factors	**Current Vegetation Coverage Status**	**Annual Mean Temperature**	**Annual Mean Precipitation**	**Temperature Change Rate**	**Precipitation Change Rate**
Correlation coefficient	−0.327 **	0.175 **	−0.583 **	−0.013	−0.257 **

Note: * represents $p < 0.05$, ** represents $p < 0.01$.

3.2. The Impact of Climate Change on VC Change

From 1990 to 2018, temperature and precipitation showed a significant increase trend, with average annual increases of about 0.027 °C (Figure 3a) and 2.38 mm (Figure 3b), respectively. Regarding the spatial pattern, the regions with obvious temperature and precipitation increases were mainly located in the northwest QLM (Figure 3a) and the high-altitude area in the central QLM (Figure 3b), respectively. The impact of temperature change on VC change is relatively small and exhibits a scattered distribution feature (Figure 3c). Approximately 8.6% and 5.4% of the areas had significant positive and negative correlations between temperature change and VC change, respectively. In contrast, the impact of precipitation change on VC change is more significant, and it is concentrated in the central QLM (Figure 3d). About 45.2% and 1.2% of the areas showed significant positive and negative correlations between precipitation change and VC change, respectively.

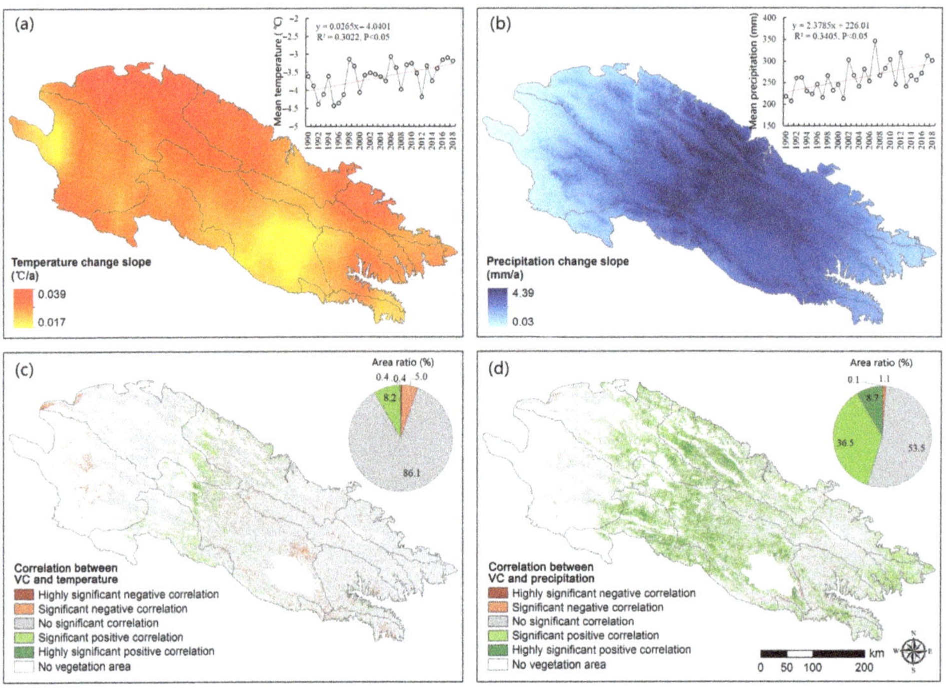

Figure 3. Temperature change (**a**), precipitation change, (**b**) and their correlation (**c**,**d**) with vegetation coverage (VC) change in the QLM. The significant and highly significant correlations are the situations under $p < 0.05$ and $p < 0.01$, respectively.

The partial correlation analysis results were consistent with those of the correlation analysis, but the effect of solely temperature or precipitation was slightly stronger (Figure 4a,b). The multiple correlation analysis shows that 54.2% and 5.4% of the regional VC changes in the QLM were significantly and extremely significantly positively correlated with temperature and precipitation changes, respectively (Figure 4c). Therefore, the increases of temperature and precipitation were the main factors leading to the VC increase in the QLM.

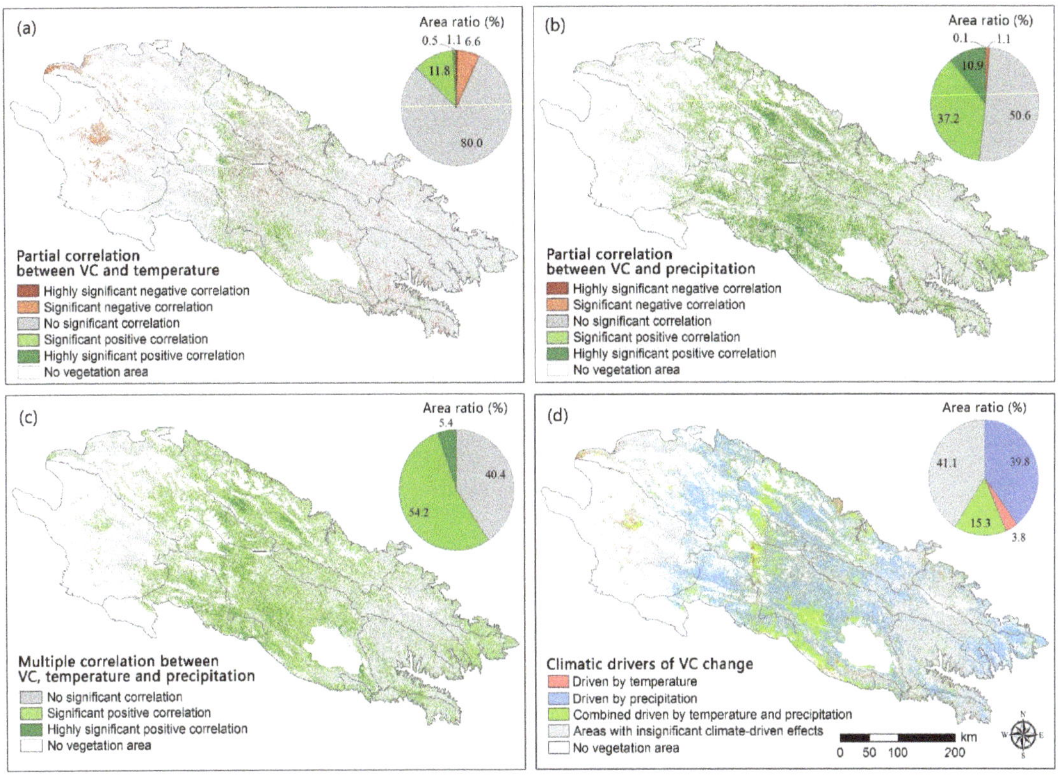

Figure 4. Partial correlation (**a**,**b**) and multiple correlation (**c**) between vegetation coverage (VC) change and climate change, as well as climate-driven patterns (**d**) of VC change in the QLM. The significant and highly significant correlations are the situations under $p < 0.05$ and $p < 0.01$, respectively.

Based on the results of partial and multiple correlation analysis, the climate-driven types of VC change in the QLM were identified, including the temperature-driven zone, precipitation-driven zone, temperature and precipitation co-driven zone, and non-climate-driven zone (Figure 4d). The climate-driven zone accounted for 58.9% of the total QLM. Among the zones, the precipitation driven zone accounted for 39.8% of the total, and was largely distributed in the central high-altitude region and the eastern edge of the QLM. The co-driven zone occupied 15.3%, mostly being distributed in the central QLM. The temperature-driven zone only accounted for 3.8%, and the distribution was scattered.

3.3. *Effects of Human Activities on VC Change*

The areas where human activities had an obvious negative effect on VC change accounted for 17.3% of the study area, mainly distributed in the southeastern QLM, especially in the lower reaches of Huangshui Basin, Qinghai Lake Basin, and Datong River Basin (Figure 5a,b). The areas where human activities played a significant positive role accounted for 17.7%, mainly distributed in the midwestern and northeastern QLM, especially in the western part of the Heihe River Basin and Qaidam Basin, as well as the upper reaches of Qinghai Lake Basin and Shule River Basin. Spatially, the regions where the residuals decreased remarkably were largely distributed in the central and northern QLM, involving the Heihe River Basin, Datong River Basin and Shiyang River Basin (Figure 5c). In terms of dynamic changes, the proportions with significant reductions and increases in residuals were 8.6% and 14.8%, respectively (Figure 5d). Hence, the vegetation degradation area caused by human disturbances was smaller than the ecological restoration area in the QLM.

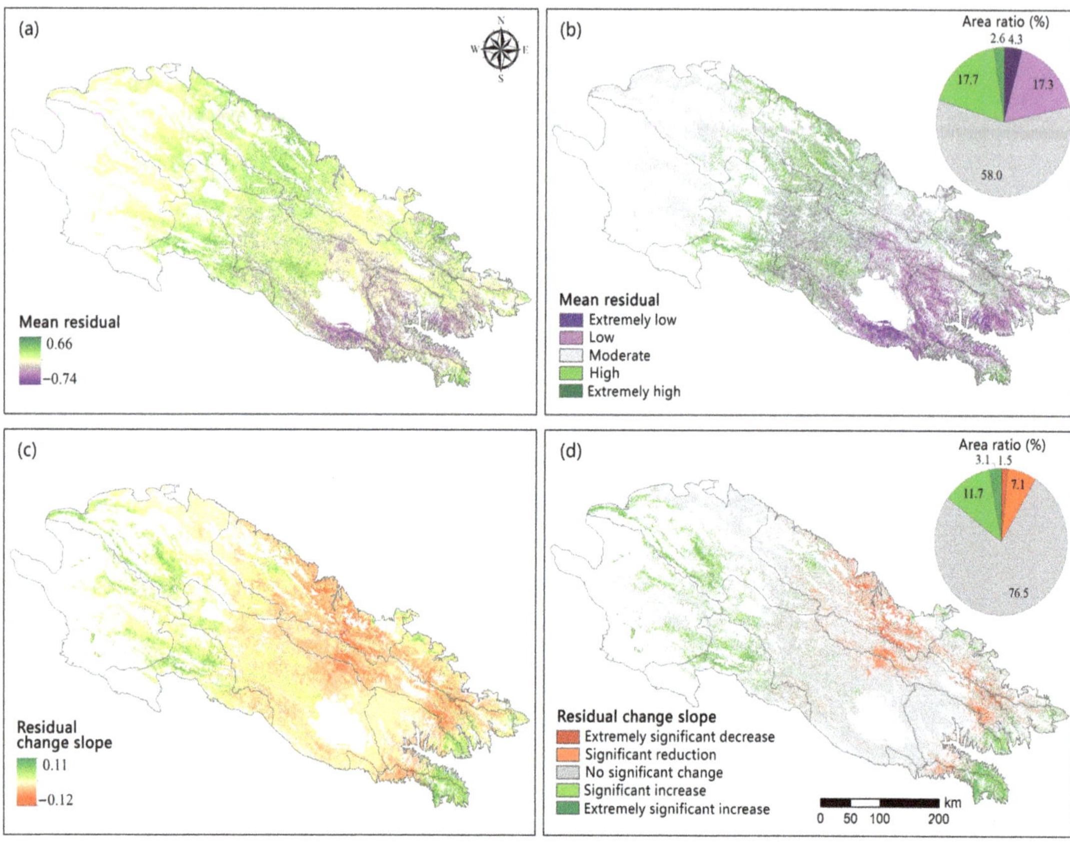

Figure 5. Spatial distribution of mean residual (**a**,**b**) and residual slope (**c**,**d**) of vegetation cover change in the QLM.

3.4. Climatically and Anthropogenically Driven Pattern of VC Change

The climatically and anthropogenically driven pattern (Figure 6) of VC change in the QLM during 1990–2018 was obtained by superimposing the VC change map (Figure 2), the complex correlation map (Figure 4c), and the residual map (Figure 5a). Climate change played a significant role in the improvement of VC, and the area driven by climate change is 37.9% of the total. In addition, climate change and human activities together driving the increase of the VC area accounted for 10.7% of the total. Within these regions, human activities further promoted vegetation restoration with the contribution of climate change. Therefore, VC increased significantly in nearly half of the QLM under the influence of climate change. However, 9.2% of the areas showed a significant VC decrease under the interference of human activities. In addition, 10.6% of the areas showed a significant improvement in VC, and were largely distributed in the western part of the QLM where human activities were weak.

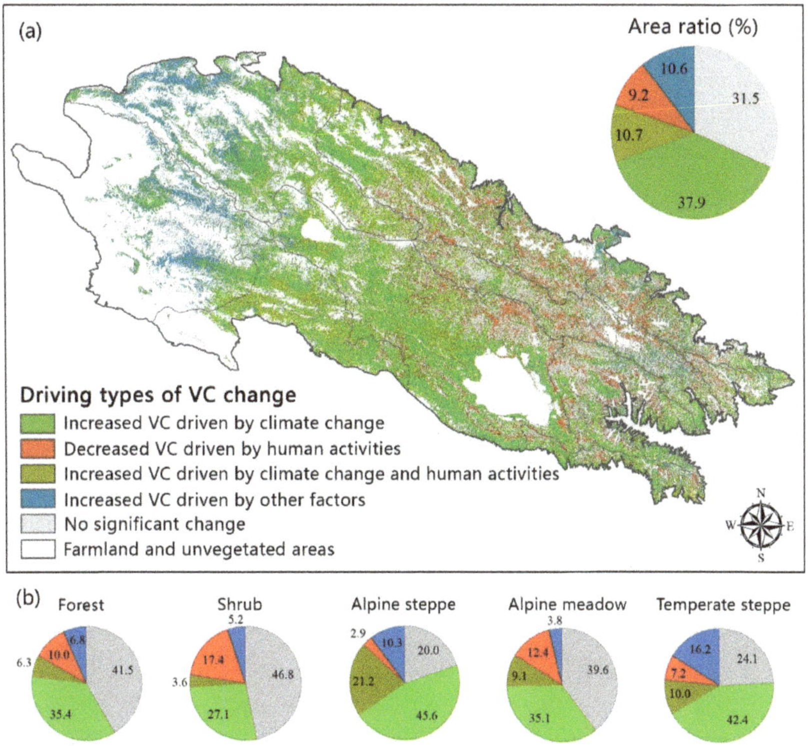

Figure 6. Climatically and anthropogenically driven pattern of vegetation coverage (VC) change (**a**) and percentages of VC change types under different vegetation types (**b**) in the QLM.

Based on the vegetation type (Figure 1a) and climatically and anthropogenically driven pattern of VC change (Figure 5a) in the QLM, we obtained the area proportion of VC change types for different vegetation types (Figure 6). Forests and shrubs are relatively stable ecosystems, with 41.5% and 46.8% of the regions having no significant change in VC, respectively. Affected by climate change, 35.4% of forests had significant improvement in VC, while another 6.3% of forests had significantly increased VC under the combined effects of climatic and anthropogenic factors. In addition, about 10% of forests had significantly reduced VC due to human activities. Shrubs were most obviously affected by human activities, with 17.4% of the area showing significant degradation due to human activities. Meanwhile, shrubs were least affected by climate change, with 30.7% of the areas exhibiting significant increases in VC due to the comprehensive impacts of climate change and human activities. The impact of climate change on alpine steppes was most significant, with 66.8% of the VC increase driven by climate change. Human activities had extremely low interference, with only 2.9% of the areas degraded by human activities. Affected by human activities, 12.4% of alpine meadows significantly degraded. In addition, caused by the comprehensive impacts of climatic and anthropogenic factors, approximately 48% of alpine meadows significantly increased in VC. The influence of climate change on temperate steppes was second only to that on alpine steppes, with 52.4% of the regions significantly increased in VC due to climate change and human activities. However, 7.2% of the areas also showed a significant decrease in VC caused by human activities.

4. Discussion

4.1. Impacts of Natural and Human Factors on Vegetation Coverage Change

The overall VC in the QLM improved remarkably from 1990 to 2018, which was largely attributed to climate change and ecological restoration. Our research showed that climate change played a leading role in the VC increase in the QLM, and the influence of precipitation was obviously greater than that of temperature. One possible reason is that the climate in the QLM is relatively dry, and vegetation growth is more sensitive to precipitation change [29]. In addition, active artificial restoration, such as afforestation and mine restoration, also played a significant role in improving VC in local areas [38,42]. The climate warming and humidifying has significantly increased VC in the QLM, but extensive vegetation degradation still exists. The results of the correlation analysis and residual analysis showed that human activities were the main factors causing vegetation degradation in the QLM (Table 2, Figure 6). Of all human activities, mining development and hydropower construction were the leading factors in the rapid deterioration of vegetation conditions [55,56]. Meanwhile, we found a very strong positive correlation between vegetation degradation and grazing intensity, which indicates that grazing activities may be one of the most important factors [57]. In addition, VC change was significantly negatively correlated with elevation and slope. That is, vegetation degradation tends to occur in areas with relatively low elevations and gentle slopes, which may be related to the strong human activities in these areas [58].

For vegetation types, forests, shrubs, and alpine meadows distributed in areas with superior natural conditions and intense human activities experienced a certain amount of degradation (Figures 2 and 6). A relevant study also showed that the alpine meadows, shrubs, and forests in the QLM were most affected by human interference due to the increase of tourism, overgrazing, and other disturbances [59]. In some areas, forests had a single-stand structure and underwent degradation due to drought, forest fires, and infection with pests and diseases [60]. In addition, research has found that overgrazing is the main factor causing a decrease in VC in shrubs, sparse forests, and young forests, especially in shrubs with VC ranging from 30% to 40% [61]. Therefore, it is still necessary to reduce grazing intensity in the future, especially for alpine meadows and shrubs with significant degradation. Meanwhile, it is particularly important to conduct ecological engineering in areas with good natural conditions and relatively dense populations to promote local vegetation restoration.

4.2. Limitations and Prospects

Consistent with previous studies [36,37], the areas of significant increase and decrease in VC were mainly distributed in the western and eastern QLM, respectively. However, there have been significant spatial differences in the VC change among diverse studies. On the one hand, this may be due to the different research periods and the significant discrepancies in VC change between different years. On the other hand, differences in data sources and resolution can also lead to differences in analysis results. The 30 m resolution VC data in this study come from reflectance data of red and near-infrared channels from Landsat5, Landsat8, and Sentinel 2 [62]. Although various datasets have been widely used in vegetation change analysis, there are large deviations between different sensors. Therefore, comparative analysis of multi-source data and multiple time scales may make the results more robust and accurate. Affected by human activities, land use types in the QLM have changed significantly from 1990 to 2018 [63], and vegetation types will also change accordingly. In this research, vegetation type data were extracted from the 1:1,000,000 Chinese Vegetation Atlas, which reflects the distribution of 11 vegetation type groups in China around the year 2000. Therefore, these data can only reflect the status of vegetation types in general, and it is difficult to accurately show the current distribution of various vegetation types. In addition, this study only analyzed the impacts of the two essential climatic factors, precipitation and temperature, on vegetation change. The direct and indirect effects of various climatic variables on vegetation dynamics are complicated.

Thus, further studies should be carried out by combining various climatic factors, and the interaction relationships of climatic factors should be identified.

5. Conclusions

This study analyzed the temporal and spatial characteristics of VC change in the QLM from 1990 to 2018 and recognized its main natural and anthropogenic influencing factors. Furthermore, we identified a climate-driven pattern of VC change and obtained the VC change zones driven by climate change and human activities. In addition, we also analyzed VC change intensity and types under different vegetation types. The main conclusions are as follows:

(1) VC in the QLM showed an overall fluctuating increase trend from 1990 to 2018. Areas with significant increases were mainly distributed in the central and western QLM, and regions with significant decreases were mostly located in the central and eastern QLM. The decrease in VC is more pronounced in areas with high annual precipitation, high VC, and high livestock density. For vegetation types, the VC of alpine steppes and temperate steppes increased significantly, while forests, shrubs, and alpine meadows deteriorated.

(2) The increase of VC in the QLM was primarily driven by climate change, and the effect of precipitation increasing was more obvious than that of temperature increasing. Among them, the precipitation-driven type was mostly distributed in the central high-altitude region and the eastern margin, the co-driven zone of temperature and precipitation was largely located in the central QLM, and the temperature-driven zone was relatively scattered. For vegetation types, alpine grasslands were the most affected by climate change, followed by temperate grasslands, forests, and alpine meadows.

(3) Vegetation degradation in the QLM was primarily attributed to anthropogenic factors. Areas where human activities had an obvious negative effect were largely distributed in the eastern part of the QLM, especially the Huangshui River Basin and the areas around Qinghai Lake. The regions with enhanced human disturbances were mainly distributed in the north–central QLM, mainly involving the Heihe River Basin, Datong River Basin and Shiyang River Basin. In terms of vegetation type, shrubs were most disturbed by human activities, followed by alpine meadows and forests.

Author Contributions: Conceptualization, Y.T.; data curation, X.W.; formal analysis, Y.T.; funding acquisition, M.S.; methodology, L.W.; project administration, M.S. and J.Z.; resources, C.W.; supervision, M.S. and J.Z.; validation, X.W.; visualization, L.W.; writing—original draft, Y.T.; writing—review and editing, C.W. and J.Z. All authors have read and agreed to the published version of the manuscript.

Funding: The research was funded by the Second Scientific Expedition to the Qinghai–Tibet Plateau (grant number 2019QZKK0405-05), the Guangdong Basic and Applied Basic Research Foundation (grant number 2022A1515110665), and the China Postdoctoral Science Foundation (grant number 2022M722522).

Data Availability Statement: All data used in this study are available from the author by request (tengym_simlab@163.com).

Acknowledgments: We are grateful to the editors and reviewers for their constructive work.

Conflicts of Interest: The authors declare no conflict of interest.

References

1. Liu, H.; Zheng, L.; Yin, S. Multi-perspective analysis of vegetation cover change and driving factors of long time series based on climate and terrain data in Hanjiang River Basin, China. *Arab. J. Geosci.* **2018**, *11*, 509. [CrossRef]
2. Zhou, Z.; Ding, Y.; Shi, H.; Cai, H.; Fu, Q.; Liu, S.; Li, T. Analysis and prediction of vegetation dynamic change in China: Past, present and future. *Ecol. Indic.* **2020**, *117*, 11. [CrossRef]
3. Pettorelli, N.; Vik, J.O.; Mysterud, A.; Gaillard, J.M.; Tucker, C.J.; Stenseth, N.C. Using the satellite-derived NDVI to assess ecological responses to environmental change. *Trends Ecol. Evol.* **2005**, *20*, 503–510. [CrossRef] [PubMed]
4. Huang, S.; Zheng, X.; Ma, L.; Wang, H.; Huang, Q.; Leng, G.; Meng, E.; Guo, Y. Quantitative contribution of climate change and human activities to vegetation cover variations based on GA-SVM model. *J. Hydrol.* **2020**, *584*, 124687. [CrossRef]

5. Ma, Z.; Yu, H.; Cao, C.; Zhang, Q.; Hou, L. Spatiotemporal characteristics of fractional vegetation coverage and its influencing factors in China. *Resour. Environ. Yangtze Basin* **2020**, *29*, 1310–1321. (In Chinese)
6. You, G.; Liu, B.; Zou, C.; Li, H.; Xu, W. Sensitivity of vegetation dynamics to climate variability in a forest-steppe transition ecozone, north-eastern Inner Mongolia, China. *Ecol. Indic.* **2021**, *120*, 106833. [CrossRef]
7. Tian, H.; Cao, C.; Chen, W.; Bao, S.; Yang, B.; Ranga, B.M. Response of vegetation activity dynamic to climatic change and ecological restoration programs in Inner Mongolia from 2000 to 2012. *Ecol. Eng.* **2015**, *82*, 276–289. [CrossRef]
8. Wang, S.; Zhang, B.; Xie, G.D.; Zhai, X.; Sun, H.L. Vegetation cover changes and sand-fixing service responses in the Beijing–Tianjin sandstorm source control project area. *Environ. Dev.* **2020**, *34*, 100455. [CrossRef]
9. Jin, K.; Wang, F.; Han, J.; Shi, S.; Ding, W. Effects of climate change and human activities on vegetation NDVI change in China from 1982 to 2015. *J. Geogr. Sci.* **2020**, *75*, 961–974. (In Chinese)
10. Gao, S.; Dong, G.; Jiang, X.; Nie, T.; Yin, H.; Guo, X. Quantification of natural and anthropogenic driving forces of vegetation changes in the Three-River Headwater Region during 1982–2015 based on Geographical Detector model. *Remote Sens.* **2021**, *13*, 4175. [CrossRef]
11. Han, Y.; Lin, Y.; Zhou, P.; Duan, J.; Cao, Z. Dynamic change, driving mechanism and spatiotemporal prediction of the normalized vegetation index: A case study from Yunnan Province, China. *Front. Ecol. Evol.* **2023**, *11*, 1177849. [CrossRef]
12. Liu, C.; Li, W.; Wang, W.; Zhou, H.; Liang, T.; Hou, F.; Xu, J.; Xue, P. Quantitative spatial analysis of vegetation dynamics and potential driving factors in a typical alpine region on the northeastern Tibetan Plateau using the Google Earth Engine. *Catena* **2021**, *206*, 105500. [CrossRef]
13. Ma, Y.; Guan, Q.; Sun, Y.; Zhang, J.; Yang, L.; Yang, E.; Li, H.; Du, Q. Three-dimensional dynamic characteristics of vegetation and its response to climatic factors in the Qilian Mountains. *Catena* **2022**, *208*, 105694. [CrossRef]
14. Wu, H.; Zhang, J.; Bao, Z.; Wang, G.; Wang, W.; Yang, Y.; Wang, J.; Kan, G. The impacts of natural and anthropogenic factors on vegetation change in the Yellow-Huai-Hai River Basin. *Front. Earth Sci.* **2022**, *10*, 959403. [CrossRef]
15. Zhao, Y.; Sun, R.; Ni, Z. Identification of natural and anthropogenic drivers of vegetation change in the Beijing-Tianjin-Hebei megacity region. *Remote Sens.* **2019**, *11*, 1224. [CrossRef]
16. Chen, X.; Zhao, X.; Zhang, J.; Wang, R.; Lu, J. Variation of NDVI spatio-temporal characteristics and its driving factors based on geodetector model in Horqin Sandy Land, China. *J. Plant Ecol.* **2023**, *47*, 1–12. (In Chinese)
17. Tong, S.; Cao, G.; Yan, X.; Diao, E.; Zhang, Z. Spatial-temporal evolution of vegetation cover and its driving factors on the south slope of the Qilian Mountains, China from 2000 to 2020. *Mt. Res.* **2022**, *40*, 491–503. (In Chinese)
18. Yu, S.; Du, W.; Zhang, X.; Hong, Y.; Liu, Y.; Hong, M.; Chen, S. Spatio-temporal changes in NDVI and its driving factors in the Kherlen River Basin. *Chin. Geogr. Sci.* **2023**, *33*, 377–392. [CrossRef]
19. Dong, Y.; Yin, D.; Li, X.; Huang, J.; Su, W.; Li, X.; Wang, H. Spatial–temporal evolution of vegetation NDVI in association with climatic, environmental and anthropogenic factors in the Loess Plateau, China during 2000–2015: Quantitative analysis based on Geographical Detector model. *Remote Sens.* **2021**, *13*, 4380. [CrossRef]
20. Nie, T.; Dong, G.; Jiang, X.; Lei, Y. Spatio-temporal changes and driving forces of vegetation coverage on the Loess Plateau of Northern Shaanxi. *Remote Sens.* **2021**, *13*, 613. [CrossRef]
21. Zhu, L.; Meng, J.; Zhu, L. Applying Geodetector to disentangle the contributions of natural and anthropogenic factors to NDVI variations in the middle reaches of the Heihe River Basin. *Ecol. Indic.* **2020**, *117*, 106545. [CrossRef]
22. Teng, Y.; Zhan, J.; Su, M.; Xu, C. Effects of climate and afforestation on carbon sequestration change in northern China. *Land Degrad. Dev.* **2023**, *34*, 4109–4122. [CrossRef]
23. Gao, W.; Zheng, C.; Liu, X.; Lu, Y.; Chen, Y.; Wei, Y.; Ma, Y. NDVI-based vegetation dynamics and their responses to climate change and human activities from 1982 to 2020: A case study in the Mu Us Sandy Land, China. *Ecol. Indic.* **2022**, *137*, 108745. [CrossRef]
24. Li, H.; Zhang, H.; Li, Q.; Zhao, J.; Guo, X.; Ying, H.; Wang, S. Vegetation productivity dynamics in response to climate change and human activities under different topography and land cover in Northeast China. *Remote Sens.* **2021**, *13*, 975. [CrossRef]
25. Tao, Z.; Wang, H.; Liu, Y.; Xu, Y.; Dai, J. Phenological response of different vegetation types to temperature and precipitation variations in northern China during 1982–2012. *Remote Sens.* **2017**, *38*, 3236–3252. [CrossRef]
26. Zhang, M.; Liu, X.; Nazieh, S.; Wang, X.; Nkrumah, T.; Hong, S. Spatiotemporal distribution of grassland NPP in Gansu province, China from 1982 to 2011 and its impact factors. *PLoS ONE* **2020**, *15*, e0242609. [CrossRef]
27. Sun, W.; Wang, Y.; Fu, Y.H.; Xue, B.; Wang, G.; Yu, J. Spatial heterogeneity of change in vegetation growth and their driving forces based on satellite observations of the Yarlung Zangbo River basin in the Tibetan plateau. *J. Hydrol.* **2019**, *574*, 324–332. [CrossRef]
28. Ren, W.; Wang, X.; Liu, M.; Wang, D. Temporal and spatial changes and the driving factors of vegetation NPP in Shiyang River Basin. *Arid. Zone Res.* **2023**, *40*, 818–828. (In Chinese)
29. Zhang, L.; Yan, H.; Qiu, L.; Cao, S.; He, Y.; Pang, G. Spatial and temporal analyses of vegetation changes at multiple time scales in the Qilian Mountains. *Remote Sens.* **2021**, *13*, 5046. [CrossRef]
30. Zhang, H.; Li, M.; Song, J. Analysis of driving factors of vegetation NDVI change in Qilian Mountain National Park based on geographic detector. *Chin. J. Ecol.* **2021**, *40*, 2530–2540.
31. Li, J.; Gong, C. Effects of terrain factors on vegetation cover change in National Park of Qilian Mountain. *Bull. Soil Water Conserv.* **2021**, *41*, 228–237. (In Chinese)

32. Zhang, X.; Wang, J.; Gao, Y.; Wang, L. Variations and controlling factors of vegetation dynamics on the Qingzang Plateau of China over the recent 20 years. *Geogr. Sustain.* **2021**, *2*, 74–85. [CrossRef]
33. Wang, H.; Liu, D.; Lin, H.; Montenegro, A.; Zhu, X. NDVI and vegetation phenology dynamics under the influence of sunshine duration on the Tibetan plateau. *Int. J. Climatol.* **2015**, *35*, 687–698. [CrossRef]
34. Chen, J.; Jia, W.; Zhao, Z.; Zhang, Y.; Liu, Y. Research on temporal and spatial variation characteristics of vegetation cover of Qilian Mountains from 1982 to 2006. Adv. *Earth Sci.* **2015**, *30*, 834–845. (In Chinese)
35. Wu, Z.; Jia, W.; Zhao, Z.; Zhang, S.; Liu, Y.; Chen, J. Spatial-temporal variations of vegetation and its correlation with climatic factors in Qilian Mountains from 2000 to 2012. *Arid Land Geo.* **2015**, *38*, 1241–1252. (In Chinese)
36. Jiang, Y.; Du, W.; Huang, J.; Zhao, H.; Wang, C. Analysis of vegetation changes in the Qilian Mountains during 2000–2015. *J. Glaciol. Geocryol.* **2017**, *39*, 1130–1136. (In Chinese)
37. Qiu, L.; Zhang, L.; He, Y.; Diao, Z.; Chen, Y. Remote sensing monitoring on vegetation dynamic change in Qilian Mountain from 2000 to 2017. *Remote Sens. Inf.* **2019**, *34*, 97–107. (In Chinese)
38. Li, X.; Gou, X.; Wang, N.; Sheng, Y.; Jin, H.; Qi, Y.; Song, X.; Hou, F.; Li, Y. Tightening ecological management facilitates green development in the Qilian Mountain. *Chin. Sci. Bull.* **2019**, *64*, 2928–2937. (In Chinese)
39. Li, Y.; Li, J.; Are, K.S.; Huang, Z.; Yu, H.; Zhang, Q. Livestock grazing significantly accelerates soil erosion more than climate change in Qinghai-Tibet Plateau: Evidenced from 137Cs and 210Pbex measurements. *Agric. Ecosyst. Environ.* **2019**, *285*, 106643. [CrossRef]
40. Li, Y.; Li, Z.; Feng, Q.; Wei, W.; Yang, J.; Lv, Y.; Gui, J.; Yuan, R.; Zhang, B. Research on the development of the ecological protection of the Qilian Mountains based on ecological redline. *Acta Ecol. Sin.* **2019**, *39*, 2343–2352. (In Chinese)
41. Juan, L.; Thomas, M.M.; Hao, W.; Byron, V.W.; George, B.S.; Charudutt, M.; Zhi, L.; Steven, R.B. Climate refugia of snow leopards in High Asia. *Biol. Conserv.* **2016**, *203*, 188–196.
42. Gui, J.; Wang, X.; Li, Z.; Zou, H.; Li, A. Research on the response of vegetation change to human activities in typical cryosphere areas: Taking the Qilian Mountains as an example. *J. Glaciol. Geocryol.* **2019**, *41*, 1235–1243. (In Chinese)
43. Wang, T.; Gao, F.; Wang, B.; Wang, P.; Wang, Q.; Song, H.; Yin, C. Status and suggestions on ecological protection and restoration of Qilian Mountains. *J. Glaciol. Geocryol.* **2017**, *39*, 229–234. (In Chinese)
44. Zhang, W.; Cheng, W.; Li, B.; Gong, C.; Zhao, M.; Wang, N. Simulation of the permafrost distribution on Qilian Mountains over past 40 years under the influence of climate change. *Geogr. Res.-Aust.* **2014**, *33*, 1275–1284. (In Chinese)
45. An, J.; Niu, Y.; Che, Z.; Hao, H. Grey correlation analysis of water conservation function of typical vegetation types in the alpine region of the Qilian Mountains. *J. Central S. Univ. Forest. Tech.* **2023**, *43*, 93–101. (In Chinese)
46. Hu, Z.; Song, X.; Tan, L.; Liu, H.; Wen, W. Spatio-temporal variation characteristics and its driving factors of NDVI at county scale for an inland arid grassland during 2001—2020. *Bull. Soil Water Conserv.* **2022**, *42*, 213–221. (In Chinese)
47. Peng, S. High-spatial-resolution monthly temperatures dataset over China during 1901–2017. *Earth Syst. Sci. Data* **2019**, *11*, 1931–1946. (In Chinese) [CrossRef]
48. Qi, Y.; Zhang, J.; Zhou, S. The DEM data of 30m in Qilian Mountain (2018). *Natl. Tibet. Plateau Data Cent.* **2019**. (In Chinese) [CrossRef]
49. Pan, T.; Zou, X.; Liu, Y.; Wu, S.; He, G. Contributions of climatic and non-climatic drivers to grassland variations on the Tibetan Plateau. *Ecol. Eng.* **2017**, *108*, 307–317. [CrossRef]
50. Wang, Y.; Sun, Y.; Wang, Z. Spatial-temporal change in vegetation cover and climate factor drivers of variation in the Haihe River Basin 1998–2011. *Resour. Sci.* **2014**, *36*, 594–602. (In Chinese)
51. Jia, J.; Niu, J.; Lin, X.; Zhu, Z.; Wu, S. Temporal and spatial variations of NDVI and its driving factors in the Yanghe Watershed of Northern China. *J. Beijing For. Univ.* **2019**, *41*, 106–115. (In Chinese)
52. Evans, J.; Geerken, R. Discrimination between climate and human-induced dryland degradation. *J. Arid Environ.* **2004**, *57*, 535–554. [CrossRef]
53. Liu, B.; Sun, Y.; Wang, Z.; Zhao, T. Analysis of the vegetation cover change and the relative role of its influencing factors in North China. *J. Nat. Resour.* **2015**, *30*, 12–23. (In Chinese)
54. Luo, M.; Guli, J.; Guo, H.; Guo, H.; Zhang, P.; Meng, F.; Liu, T. Spatial-temporal variation of growing-season NDVI and its responses to hydrothermal condition in the Tarim River Basin from 2000 to 2013. *J. Nat. Resour.* **2017**, *32*, 50–63. (In Chinese)
55. Qian, D.; Yan, C.; Xiu, L.; Feng, K. The impact of mining changes on surrounding lands and ecosystem service value in the southern slope of Qilian Mountains. *Ecol. Complex.* **2018**, *36*, 138–148. [CrossRef]
56. Li, Y.; Li, Z.; Zhang, X.; Yang, A.; Gui, j.; Xue, J. Spatial and temporal changes in vegetation cover and response to human activities in Qilian Mountain National Park. *Acta Ecol. Sin.* **2023**, *43*, 219–233. (In Chinese)
57. Wang, X.; Yang, D.; Zhang, L.; Zhang, L.; Xie, F.; Guo, P. The causes of grassland degradation and restoration strategies in Qilian Mountain National Park. *J. Grass Forage Sci.* **2020**, *6*, 81–86. (In Chinese)
58. Li, X.; Gao, J.; Zhang, J. A topographic perspective on the distribution of degraded meadows and their changes on the Qinghai-Tibet Plateau, West China. *Land Degrad. Dev.* **2018**, *29*, 1574–1582. [CrossRef]
59. Duan, Q.; Luo, L.; Zhao, W.; Zhuang, Y.; Liu, F. Mapping and evaluating human pressure changes in the Qilian Mountains. *Remote Sens.* **2021**, *13*, 2400. [CrossRef]
60. Quan, J. The dilemma and countermeasures of natural forest protection in Qilian Mountain Nature Reserve. *Mod. Horticult.* **2023**, *46*, 174–176. (In Chinese)

61. Yuan, H.; Wang, L.; Guo, S.; Wang, S.; Jin, M.; Wang, Y. Investigation and analysis of forest area change in Qilian Mountain National Nature Reserve in Gansu Province. *For. Sci. Technol.* **2022**, *6*, 65–70. (In Chinese)
62. Zhong, B.; Wu, J. Landsat-based continuous monthly 30 m × 30 m land surface FVC dataset in Qilian Mountain area (1986–2017). *Natl. Tibet. Plateau Data Cent.* **2019**. [CrossRef]
63. Fu, J.; Cao, G.; Guo, W. Land use change and its driving force on the southern slope of Qilian Mountains from 1980 to 2018. *Chin. J. Appl. Ecol.* **2020**, *31*, 2699–2709. (In Chinese)

forests

MDPI

Article

Determination Factors for the Spatial Distribution of Forest Cover: A Case Study of China's Fujian Province

Jiayun Dong [1,2], Congyi Zhou [1], Wenyuan Liang [3] and Xu Lu [4,*]

1 College of Economics and Management, Nanjing Forestry University, Nanjing 210037, China
2 Institute of Geographic Sciences and Natural Resources Research, Chinese Academy of Sciences, Beijing 100101, China
3 Wageningen School of Social Sciences, Wageningen University and Research, 6708 PB Wageningen, The Netherlands
4 Jiangsu Institute of Science and Technology Information, Nanjing 210037, China
* Correspondence: luxu126@126.com

Abstract: Understanding the determination factors of the spatial distribution of forest cover is crucial for global forest governance. This study contributed a nuanced case, focusing on the determination factors for the spatial distribution of forest cover in Fujian Province, China, in 2020. In order to achieve this, a high-resolution GIS-based data set was used, and spatial auto-correlation and geographic detector approaches were adopted. Three findings are presented in the results. First, the spatial distribution of forest cover is affected by natural conditions. In regions with more precipitation, higher altitude, or cooler temperatures, forest cover is higher. The relationship between the spatial distribution of forest cover and slope is an inverted-U shape. Second, socioeconomic factors have a greater explanatory capacity. In particular, regions with dense populations or roads have less forest cover. Third, there is an inverted-U-shaped relationship between the spatial distribution of forest cover and GDP per capita. With the growth of GDP per capita, forest cover is first positive, but subsequently negative. The results indicate that natural factors could shape the spatial distribution of forest cover, while socioeconomic factors could play a more significant role in the spatial distribution of forest cover.

Keywords: forest spatial distribution; geographic detector; driving factors

Citation: Dong, J.; Zhou, C.; Liang, W.; Lu, X. Determination Factors for the Spatial Distribution of Forest Cover: A Case Study of China's Fujian Province. *Forests* **2022**, *13*, 2070. https://doi.org/10.3390/f13122070

Academic Editor: Sandra Oliveira

Received: 17 October 2022
Accepted: 2 December 2022
Published: 5 December 2022

1. Introduction

The spatial distribution of forest cover has been uneven and dynamic across the globe. While tropical regions still experience deforestation, forest cover has expanded in Europe after a period of decline [1–4]. In recent decades, China has reversed the decline of forest cover and experienced net forest cover expansion [5], while its spatial distribution of forest cover is still uneven. For example, the forest cover of traditional forest regions in northeast China is shrinking, and forest cover in central and western China is still low, while forest cover in the south of China is gradually increasing [6].

The dynamic and uneven spatial distribution of forest cover prompts a need to identify the determination factors. With the advancement of GIS-based techniques, recent studies have shown an urgent need to unravel potential determination factors in various regions [7,8].

A great number of scholars have investigated the close relationship between topographical and climatic factors and the spatial distribution of forest cover [9–12]. For example, the forest cover of northeastern China is clustered in mountainous and hilly regions, and is affected by terrain and landform [13]. In addition, precipitation and altitude have a correlation, and they determine the spatial distribution of forest cover in the Qaidam Basin. Regions with higher precipitation in the Qaidam Basin have greater forest cover, and the largest forest cover is found between 3000 and 3500 m above sea level [14]. Additionally, forest cover decreases with an increase in temperature and a decrease in precipitation [9].

Against the background of climate change, the global spatial distribution of forest cover will continue to evolve. Specifically, in regions with high altitudes and heavy precipitation in the Boreal Zone, Atlantic Zone, and Continental Zone, rising temperatures will increase forest cover, while in regions with low elevations and poor precipitation in the Mediterranean, forest degradation will be severe [15]. Therefore, the natural environment can exert control over forest cover [16].

Recent studies have paid increasing attention to humans' capacity to act upon the forest cover [17,18]. The transformation between a variety of land types brought on by the change in land use is the primary determination factor for the spatial distribution of forest cover [19]. In the context of population expansion, the spread of arable land and construction land has condensed the geographical extent of forests [20]. Subsequently, population shift, which is mainly manifested by the large-scale migration of rural populations to cities, has become prevalent [21], reducing the pressure on forests. Additionally, agricultural land productivity has increased, which has encouraged the conversion of agricultural land in forest margin regions to forests [22,23], or vice versa. With the growth of per capita income, the core demand of the public for forests has begun to convert from wood supply toward ecological services [24]. A great number of countries are gradually reducing deforestation and promoting afforestation. Electricity consumption is becoming increasingly popular, and a transition from traditional wood fuel to electricity and fossil energy has decreased the demand for fuel wood [25]. In particular, road infrastructure construction significantly reduces transportation costs, resulting in an expansion of the scope of forests available for lucrative cutting activities [26,27]. The influence of road infrastructure construction on the spatial distribution of forest cover cannot be overlooked [28,29].

China's forest cover has steadily increased since the 1980s. However, from a spatial perspective, forest cover varies between regions [5]. Specifically, some provinces in southern China have taken the lead in increasing national forest cover. Fujian Province, located on China's southeastern coast, is a typical case, with the percentage of forest cover increasing from 37 percent to 66.8 percent over the past four decades [30]. The spatial distribution of forest cover in Fujian Province reveals substantial variation between diverse regions. The issue of what factors determine the spatial distribution for forest cover has not yet been resolved.

Before 1949, socioeconomic retrogression and war in Fujian Province led to the destruction of original forests [5]. Over the next three decades, the percentage of forest cover in Fujian declined by ten percent [30]. With the rapid economic growth and remarkable social progress starting in the eastern coastal cities since the 1980s, forest cover of Fujian province has been steadily increasing. Additionally, transportation infrastructure, especially rural roads, has made great progress in the last twenty years, which lays the foundation for the flow of labor between regions and the substitution of fuel wood [31]. As a result, population emigration creates beneficial conditions for forest restoration in rural regions [19]. While observing the change in forest cover, the variation of socioeconomic factors is an important consideration [4]. The spatial distribution of forest cover reflects forest cover change intuitively. Will these socioeconomic factors influence the spatial distribution of forest cover? In addition, the eastern regions of Fujian Province are flat, while the central and western regions are mainly hills and mountains. With the variation in topography, the climate conditions also change [32]. Therefore, Fujian is a perfect case to explore the determination factors for the spatial distribution of forest cover. This paper used a spatial auto-correlation approach to analyze spatial distribution characteristics of forest cover at the county level. In addition, the geographic detector approach was adopted to investigate the determination factors for the spatial distribution of forest cover. Clarifying the determinants for the spatial distribution of forest cover can help achieve forest growth in developing countries experiencing population growth and economic development.

The rest of the paper is organized as follows. Section 2 describes the data sources and research methods. Section 3 describes our research results. Section 4 discusses the determination factors for the spatial distribution of forest cover. Section 5, finally, presents our conclusions.

2. Data Sources and Research Methods

2.1. Data Sources

Forest cover data for the Fujian province in 2020 was obtained from 30 m- resolution land use data. The natural environmental condition data, road density data, and land-use data were obtained from the Chinese Academy of Sciences' resources and environment science data center. The land use data encompassed six categories of land: agricultural land, forestland, grassland, water area, construction land, and undeveloped land [33]. In order to measure the spatial distribution of forest cover in Fujian Province, forest refers to four sub-categories: closed forest, open forest, shrub, and other forest [16]. Forest cover in this study refers to the ratio of closed forest area to total land area. Due to the availability of climatic data, the average annual temperature and precipitation for each county-level jurisdiction were instead retrieved from the data from 2019. The social and economic data (population density, industry output, grain output, GDP, forestry production, and electricity consumption) were acquired from the statistical yearbook of Fujian Province and from the statistical yearbook of prefecture-level cities of Fujian Province.

2.2. Research Methods

2.2.1. Global Spatial Auto-Correlation

The spatial distribution of forest cover at the county level in Fujian Province was described using global spatial auto-correlation. The range $[-1, 1]$ was used for the Global Moran's I statistic. A significant and positive Moran's I indicates that the region with a larger (smaller) amount of forest cover is spatially clustered. In contrast, if Moran's I is strongly negative, there is a considerable geographical difference between the region and the surrounding area in terms of forest cover. If Moran's I is near zero, forest cover is distributed randomly, or there is no geographical link. The formula for calculating Moran's I is as follows:

$$I = \left(\frac{n}{\sum_i \sum_j W_{ij}}\right)\left[\frac{\sum_i \sum_j W_{ij}(x_i - \overline{x})(x_j - \overline{x})}{\sum_i (x_i - \overline{x})^2}\right] \quad (1)$$

where W_{ij} is the spatial weight matrix, n is the number of regional elements, x_i is the observed value of the ith regional element, and $\overline{x}$ is the average observed value.

2.2.2. Local Spatial Auto-Correlation

The local Moran's I index reveals the similarity or connection of forest cover between a county-level jurisdiction and the neighboring county-level jurisdiction in Fujian Province. In addition, the global Moran's I index is decomposed to the county level of Fujian province. Local spatial auto-correlation results include four types: H-H clustering type, L-L clustering type, H-L clustering type, and L-H clustering type. H-H clustering type refers to the high value of forest cover in the reference county-level jurisdiction, which is surrounded by county-level jurisdictions with a similarly high value of forest cover. L-L clustering type indicates that the reference county-level jurisdiction has a low value and is surrounded by other county-level jurisdictions with low values. H-L clustering type indicates that the reference unit has a high value and is surrounded by low-value units, whereas L-H clustering type indicates that the reference unit has a low value and is surrounded by high-value units. The formula of the local Moran's I index is expressed as follows:

$$I_i = \frac{y_i - \overline{y}}{s^2} \sum_{j=1}^{n} W_{ij}(y_i - \overline{y}) \quad (2)$$

where s^2 is the discrete variance of y_i, $\overline{y}$ is the mean, and W_{ij} is the spatial weight matrix.

2.2.3. Geographical Detector

The geographical detector serves as a statistical method suitable for testing associations between a geographical phenomenon and its potential determination factors [34]. The geographical detector has proven its advantage in making no strict assumptions for explaining determination factors of geographical phenomena, which is suitable for exploratory studies. Based on spatial differentiation, this fundamental theory evaluates the consistency between potential determination factors and geographical phenomena. The potential determination factor should have a similar spatial distribution as the geographical phenomenon if potential determination factors have critical effects on the geographical phenomenon [35]. We selected 12 indicators from the natural environmental factors and socioeconomic factors (Table 1). Then, we applied the geographical detector to analyze potential determination factors in the forest cover in Fujian Province.

Table 1. Determination factors of the spatial distribution of forest cover.

	Determination Factors	Units	Max	Min
Natural environmental factors	Precipitation	Mm	1948.4	1199.7
	Average annual temperature	°C	21.0	16.4
	Elevation	M	2138	−6
	Slope	°	82.3	0
Socioeconomic factors	Population density	person/km^2	18,301	68
	Output value of tertiary industry	billion	17.58	0.34
	Grain output per unit area	ton/ha	7.03	4.58
	GDP per capita	RMB (yuan)	307,557	50,909
	Per capita disposable income of rural households	RMB (yuan)	29,323	14,449
	Forestry output value	RMB (yuan)	251,356	0
	Household electricity consumption	million kw·h	2372.4	147.3
	Road density	km/km^2	113.41	1.21

The mechanism of the geographical detector is as follows. First, Fujian was divided into n units, and the forest cover of every unit was recorded: $y_1, y_2 \ldots \ldots y_n$. In fact, the potential determination factors must be categorical variables in the geographical detector. However, the potential determination factors selected in this paper are all continuous variables. The natural discontinuity grading technique was taken to transform continuous variables into categorical variables. Based on the highest value, mean value, and standard deviation of continuous variables, the natural discontinuity grading technique was used to divide a continuous value into nine geographical strata (X_i, i = 1, 2 ... 9). In particular, stratum 1 (category 1) suggests that the actual value of the determination factor was the smallest, whereas stratum 9 (category 9) suggests that the actual value was the largest. Then, the distribution of the forest cover was overlaid with the geographical stratum X. Every unit recorded the forest cover and each potential determination factor's category. Then, two approaches for the geographical detector were adopted: factor detector and risk detector.

Factor detector calculates the explanatory power of each potential determination factor for the forest cover's spatial distribution. In this study, q is defined as the difference between one and the ratio of accumulated dispersion variance of the forest cover of each sub-region to that of the entire study region. If factor X affects the forest cover, the dispersion variance of the forest cover in each sub-region will be small, whereas the variance between sub-regions will be large. In other words, when the factor completely explains the spatial distribution of forest cover, and the variance $\sigma^2_{x,i} = 0$, then the variance $\sigma^2 \neq 0$ and $q = 1$; in contrast, when a factor is completely unrelated to forest cover, then $q = 0$. The q value ranges between [0, 1]. The effect of a determination factor on the spatial distribution of forest cover increases with q. Equation (3) shows the expression for the q value for the factor detector:

$$q = 1 - \frac{1}{N\sigma^2}\sum_{i=1}^{9} n_{x,i}\sigma_{x,i}^2 \tag{3}$$

where N is the total number of units over the entire region, σ^2 is the variance between sub-regions, $n_{x,i}$ is the number of units in sub-region X_i, and $\sigma_{x,i}^2$ is the dispersion variance of the forest cover in each sub-region. The q value indicates the explanatory power of a determination factor on the distribution of forest cover.

The risk detector is usually used to examine the difference in average values between sub-regions of factor X. However, we focus on calculating the average values of forest cover in each sub-region X_i. Equation (4) shows the expression for the average values of forest cover:

$$Y_{x,i} = \frac{\sum_{k=1}^{m} y_m}{m} \tag{4}$$

where $Y_{x,i}$ is the mean of forest cover in sub-region X_i, m is the number of units in sub-region X_i, and y_m is the forest cover of each unit.

3. Research Results

3.1. Spatial Distribution of Forest Cover

We categorized forest cover into five levels based on the natural discontinuity grading method: lower, low, middle, high, and higher. The forest cover in the majority of county-level jurisdictions is middle or above, with 51 county-level jurisdictions accounting for 65.4 percent. Only a small number of county-level jurisdictions (27, representing 34.6%) did not meet the middle level of forest cover. In particular, high forest covers are concentrated in the central and western county-level jurisdictions of Fujian Province. Forest cover showed a declining trend eastward (Figure 1).

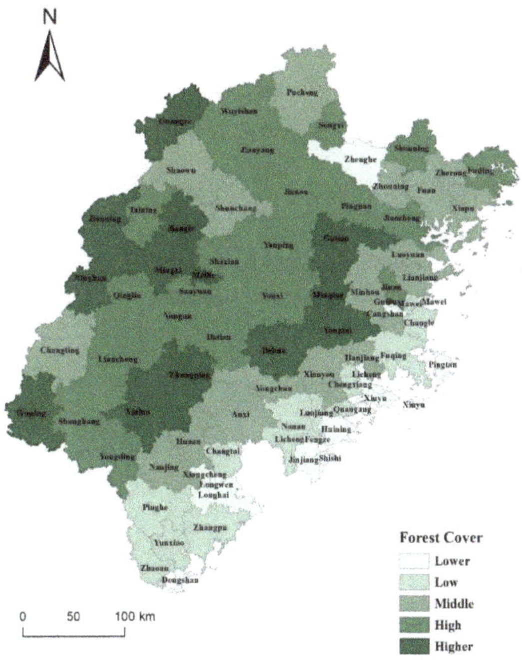

Figure 1. Spatial distribution of forest cover in Fujian Province in 2020.

3.2. Global Spatial Auto-Correlation

ArcGIS software was used to examine global spatial auto-correlation. Moran's I value for forest cover at the county level in Fujian province was 0.4246. Thus, the spatial distribution of forest cover in Fujian Province showed a positive correlation and had certain spatial agglomeration distribution characteristics (Table 2).

Table 2. Results of global spatial auto-correlation of forest cover in Fujian Province.

	Moran's I	Z Score	p Value
Forest cover	0.424560	6.120910	0.000000

3.3. Local Spatial Auto-Correlation

Based on the local spatial auto-correlation analysis of forest cover at the county level in Fujian Province, the results showed that most county-level jurisdictions demonstrated non-significant spatial patterns (see Table 3). The corresponding spatial agglomeration distribution map showed how forest cover was distributed (see Figure 2). In particular, the positive correlation type includes H-H clustering type and L-L clustering type, while the negative correlation type includes H-L clustering type and L-H clustering type.

Table 3. Summary of local spatial auto-correlation categories and the number of districts and counties with forest cover in Fujian Province.

Auto-Correlation Type	Forest Cover	
	Number	Ratio
H-H cluster	14	17.95%
H-L cluster	0	0
L-H cluster	1	1.28%
L-L cluster	10	12.82%
Not significant	53	67.95%
Total	78	100%

In conjunction with Table 3 and Figure 2, the local spatial auto-correlation analysis of forest cover indicated that H-H and L-L accounted for 14 and 10 county-level jurisdictions, or 17.95% and 12.82% of the total number of county-level jurisdictions, respectively. H-L and L-H clustering types accounted for 0 and 1 county-level jurisdictions, or 1.28% of the total number of county-level jurisdictions, among those with negative correlation values. Fifty-three county-level jurisdictions were insignificant, constituting 67.95% of the total. The spatial distribution of positive and negative correlation types varied tremendously. The H-H cluster type mostly created a loop-shaped block distribution in the center, and was located in the western portions of Sanming City, Fuzhou City, the southwestern portion of Ningde City, and the confluence of the southwestern portions of Nanping City and Sanming City. The majority of L-L clusters were located in the south of Quanzhou City, and in the northeast and south of Zhangzhou City. The majority of L-H clusters were found in the northwest portion of Longyan City.

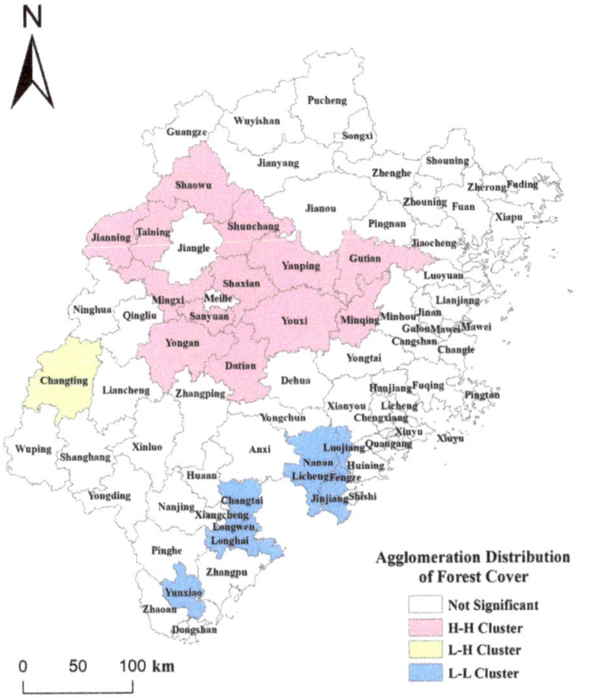

Figure 2. Spatial agglomeration distribution result of forest cover in Fujian Province in 2020.

3.4. Geographic Detector Results

3.4.1. Factor Detector Results

Several variables affected the spatial distribution of forest cover in Fujian Province. The factor detector results from the geographical detector revealed the q and p values of each variable that might impact the spatial distribution of forest cover, which were then ranked according to the explanatory power of each determination factor (Table 4). Overall, the q value of socioeconomic factors was greater than that of natural environmental factors. In other words, natural environmental factors affected the spatial distribution of forest cover, but their explanatory power was less powerful than that of socioeconomic factors.

Table 4. Results of factor detector for the spatial distribution of forest cover in Fujian Province.

Determination Factors	Forest Cover		
	q Value	p Value	Rank
Precipitation (X_1)	0.266	0.000	10
Average annual temperature (X_2)	0.309	0.000	4
Elevation (X_3)	0.277	0.000	7
Slope (X_4)	0.073	0.000	12
Population density (X_5)	0.566	0.000	1
Output value of tertiary industry (X_6)	0.306	0.000	5
Grain output per unit area (X_7)	0.273	0.000	8
GDP per capita (X_8)	0.267	0.000	9
Per capita disposable income of rural households (X_9)	0.316	0.000	3
Forestry output value (X_{10})	0.242	0.000	11
Household electricity consumption (X_{11})	0.304	0.000	6
Road density (X_{12})	0.418	0.000	2

Among the natural environmental factors, climate factors had the greatest influence on the spatial distribution of forest cover. The slope had comparably low explanatory power for forest cover, and the q value was significantly less than 0.1 (q = 0.045). The explanatory power of slope was the lowest among all selective factors, diminishing the explanatory power of topography for the spatial distribution of forest cover. However, temperature had a greater explanatory power for the spatial distribution of forest cover (q = 0.309) compared to other natural environmental conditions. The explanatory power of elevation and precipitation was, respectively, 0.277 and 0.266. Therefore, the impacts of elevation and precipitation on the spatial distribution of forest cover were nearly identical. In general, natural environmental factors, such as annual average temperature, elevation, and precipitation, had significant impacts on the spatial distribution of forest covers. Dire environmental circumstances were the limiting factors of forest cover.

Among socioeconomic factors, population density had the strongest explanatory power for the spatial distribution of forest cover (q = 0.566), indicating that population density played a key role in the spatial distribution of forest cover. Moreover, the effect of road density on the spatial distribution of forest cover should not be overlooked. The explanatory power of road density ranked second, only lower than population density (q = 0.418). The spatial distribution of forest cover was closely associated with economic development in various locations. The explanatory power of rural household per capita disposable income and per capita GDP for the spatial distribution of forest cover was 0.316 and 0.267, respectively. Moreover, household electricity consumption and tertiary industry production value had nearly identical impacts on the spatial distribution of forest cover. The explanatory power of household electricity consumption and tertiary industry production value was 0.306 and 0.304. Similarly, the explanatory power of grain output per unit area and forestry output value was 0.273 and 0.242. Overall, socioeconomic factors played an essential role in the spatial distribution of forest cover.

3.4.2. Risk Detector Results

According to the results of the risk detector (see Figure 3), the forest cover was higher in regions with more precipitation or higher altitude. Within a specific range, the forest cover constantly grew with the increase in regional slope, but when the slope surpassed a particular threshold, forest cover began to decrease. There was no evident change between forest cover and annual average temperature; however, the high value of forest cover was relatively concentrated in county-level jurisdictions with a lower annual average temperature.

Among the social and economic factors, population density, tertiary industry output value, and road density may be detrimental to regional forest covers. According to the results in Figure 3, county-level jurisdictions with denser populations or roads were prone to lower forest covers. The output value of the tertiary industry and forest cover also showed a similar relationship. Lifestyle usually interacted with population size and economic structure. Regions with a high output value for the tertiary industry possibly had better development, which tended to attract a large number of migrants. Therefore, the forest cover in regions with higher household electricity consumption also showed a downward trend. The spatial distribution of forest cover had an inverted U-shaped relationship with regional per capita GDP, as well as with household per capita disposable income level. Specifically, the higher the forestry output value was, the higher the forest cover was, indicating that the development of the forestry industry could promote regional forest cover. Moreover, the spatial distribution of forest cover and grain yield per unit area exhibited M-shaped fluctuations.

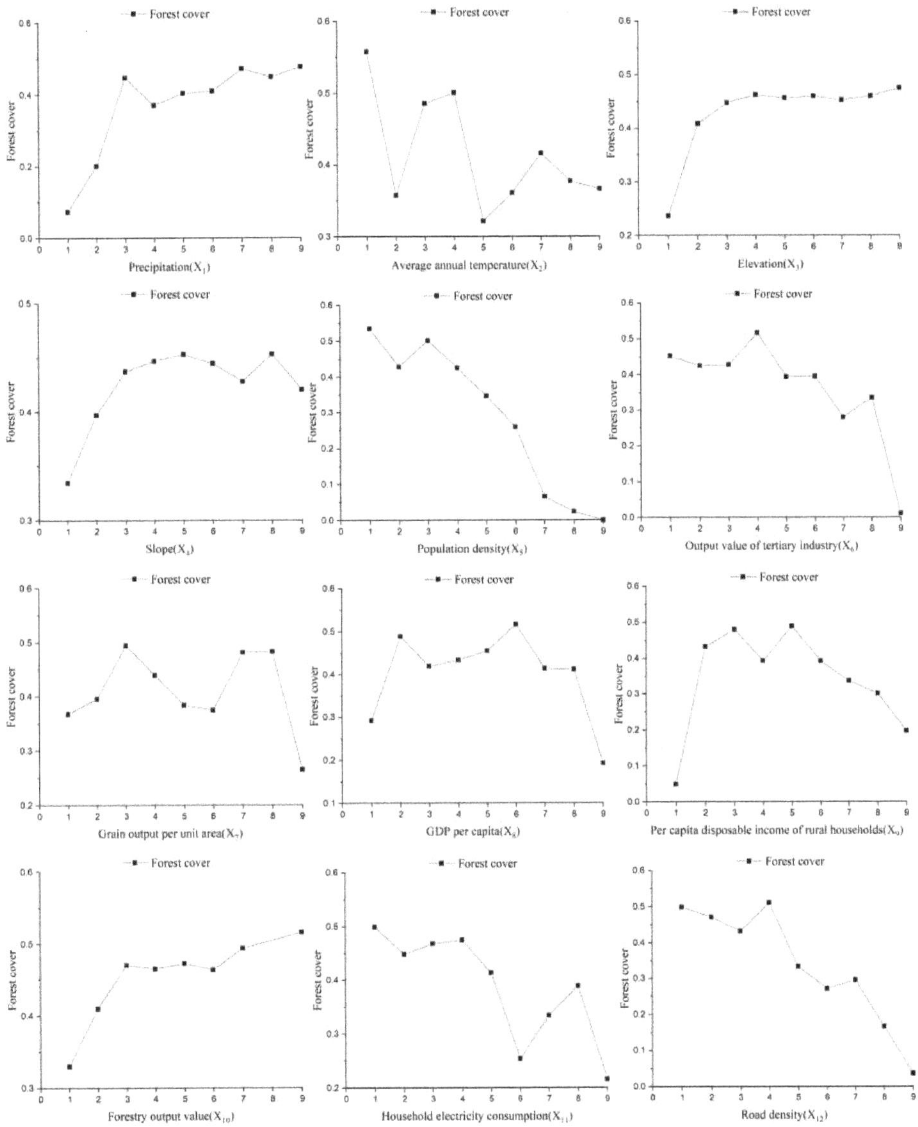

Figure 3. Results of risk detector for the spatial distribution of forest cover in Fujian Province.

4. Discussion

Understanding the determinators for the spatial distribution of forest cover is crucial for forest governance. Through the case study on Fujian Province of China, twelve indicators, regarding the natural environmental conditions and socioeconomic factors, were chosen, and the determination factors for the spatial distribution of forest cover were identified. Compared with previous results, our findings have both similarities and differences.

To a certain extent, the natural environmental conditions had the explanatory power for the spatial distribution of forest cover. Regions with adequate precipitation, suitable temperature, steep terrain, and high elevation are prone to having a higher forest cover.

The eastern regions of Fujian Province consist primarily of plains with flat topography and limited precipitation. The middle and western regions of Fujian Province are dominated by mountains and hills, with a complex topography and high elevation. Influenced by the elevation, precipitation increases gradually from east to west in Fujian Province. Consequently, the middle and western regions of Fujian Province develop a favorable natural environment for forest growth. Forest covers have formed agglomeration distribution in the central and western regions of Fujian Province. Therefore, diversities in natural environmental conditions result in the uneven distribution of forest cover.

Among social and economic factors, population density is the most powerful in explaining the spatial distribution of forest cover. The higher the population density is, the lower the forest cover will be. The finding confirms Wright and Sandel's assertion that humans have the largest impact on existing forest cover [36,37]. On the other hand, Sloan stressed the synergistic influence of population density and other factors on forest cover [38]. This study discovered a particular association between population density, road density, the production value of the tertiary industry, and the spatial distribution of forest cover. Similarly, regions with higher road density or higher tertiary industrial output value have lower forest cover. The development of roads in the east of Fujian Province is useful for those who wish to move from the central and western regions to the eastern regions of Fujian Province. The road weakens the centrifugal force caused by the transportation cost of spatial distance. The migration of a large number of people has provided labor for the development of the tertiary industry in the eastern regions of Fujian Province [39]. Population aggregation and industry growth give rise to the expansion of built-up land and industrial land; as a result, deforestation speeds up [40]. Furthermore, the extension of human production and living space in the east has led to a rise in household electricity consumption. Consequently, regions with high household electricity consumption have a comparatively low forest cover [21,41]. In the central and western regions of Fujian Province, the high altitude and complicated terrain naturally protects forests from destruction. Due to the poor road construction in the mid-west of Fujian Province, a growing number of residents are choosing to emigrate, and the pressure of population on forest cover has gradually decreased. This result is also consistent with the literature, which suggests that roads are the primary cause of deforestation [42–45].

The relationship between forest cover and per capita GDP, as well as per capita income of rural households, has an inverted U shape. Economic development increases the opportunity cost of farming, and rural residents gradually prefer to engage in industrial activities and live in urban areas [21,39]. Consequently, agricultural land becomes vacant and is, naturally, turned into forest. Woodland, on the other hand, is changed into industrial land and urban land to meet the demands of industrial expansion and urban construction. The eastern regions of Fujian Province possess the endowment advantages of economic development. In the past several decades, the eastern regions of Fujian Province have realized the circular accumulation of resource elements and industrial growth. Meanwhile, various factor endowments in the central and western regions have gradually become attracted to the eastern regions of Fujian Province. The central and western regions of Fujian Province have formed conditions for the recovery and growth of forest resources, which further explains the agglomeration distribution of forest cover in the mid-west. This result confirms Caravaggio's conclusion that the potential for economic growth to improve the growth of forest cover is limited [46]. In other words, the forest cover will reach a turning point of growth and cease increasing when the economic level reaches a certain point.

The spatial distribution of forest cover can be explained by the forestry production value. In particular, a region with a higher forestry output value has a higher forest cover. While the satisfaction of local timber demand depends on the availability of local timber, the boom in timber demand can play a key role in promoting the increase in forest cover [47]. Similarly, forest cover is currently high in the mid-west and low in the eastern regions of Fujian Province, according to the output of timber production. With the timber demand increasing, the coverage of forests will continue growing in the central and western regions of Fujian Province.

Despite the valuable findings of this study, several limitations do exist and deserve future advancement. First, considering that the results and the determination factors of the distribution of forest cover fluctuate depending on the historical context, cross-sectional analysis is insufficiently persuasive. Second, more determination factors could be incorporated into the analysis with the increasing availability of socioeconomic data.

5. Conclusions

This paper focused on the determination factors for the spatial distribution of forest cover. We used a 30 m resolution GIS-based data set and socioeconomic data set. Spatial auto-correlation and geographic detector approaches were adopted. This paper revealed the influence of natural environmental and socioeconomic factors on the spatial distribution of forest cover at a county-level jurisdiction. The results indicate that natural environmental factors could shape the spatial distribution of forest cover, while socioeconomic variables could play a more significant role in the spatial distribution of forest cover. This is primarily due to the limited control that natural environmental factors have over forest growth, whereas the progress of human beings' capacity to interfere with forests can further change the spatial distribution of forest cover. Concerning the effect of time on the spatial distribution of forest cover, future studies could investigate the factors influencing the spatial distribution of forest cover in various temporal scenarios.

Author Contributions: Conceptualization, J.D., W.L. and X.L.; methodology, J.D., W.L. and X.L.; validation, J.D. and C.Z.; formal analysis, J.D. and C.Z.; data curation, J.D. and C.Z.; writing—original draft preparation, J.D. and C.Z.; writing—review and editing, W.L. and X.L.; visualization, C.Z.; funding acquisition, J.D. All authors have read and agreed to the published version of the manuscript.

Funding: This research was funded by China Postdoctoral Science Foundation, grant number 2018-1563.

Data Availability Statement: Not applicable.

Acknowledgments: We are grateful to the anonymous reviewers for their constructive reviews.

Conflicts of Interest: The authors declare no conflict of interest.

References

1. Mather, A.S. Forest transition theory and the reforesting of Scotland. *Scott. Geogr. J.* **2004**, *120*, 83–98. [CrossRef]
2. Barbier, E.B.; Burgess, J.C. The economics of tropical deforestation and land use: An introduction to the special issue. *Land Econ.* **2001**, *77*, 155–171. [CrossRef]
3. Laso Bayas, J.C.; See, L.; Georgieva, I.; Schepaschenko, D.; Danylo, O.; Dürauer, M.; Bartl, H.; Hofhansl, F.; Zadorozhniuk, R.; Burianchuk, M. Drivers of tropical forest loss between 2008 and 2019. *Sci. Data* **2022**, *9*, 146. [CrossRef] [PubMed]
4. Geist, H.J.; Lambin, E.F. Proximate Causes and Underlying Driving Forces of Tropical DeforestationTropical forests are disappearing as the result of many pressures, both local and regional, acting in various combinations in different geographical locations. *BioScience* **2002**, *52*, 143–150. [CrossRef]
5. Li, L.; Chhater, A.; Liu, J. Multiple drivers and pathways to China's forest transition. *For. Policy Econ.* **2019**, *106*, 101962. [CrossRef]
6. Xu, X.; Liu, J.; Zhuang, D.; Zhang, S. *Spatial-Temporal Dynamic Characteristics and Driving Forces of Forestland Resources in China*; Journal of Beijing Forestry University: Beijing, China, 2004; pp. 41–46.
7. Niu, X.; Hu, Y.; Lei, Z.; Yan, H.; Ye, J.; Wang, H. Temporal and Spatial Evolution Characteristics and Its Driving Mechanism of Land Use/Cover in Vietnam from 2000 to 2020. *Land* **2022**, *11*, 920. [CrossRef]
8. Demissie, F.; Yeshitila, K.; Kindu, M.; Schneider, T. Land use/Land cover changes and their causes in Libokemkem District of South Gonder, Ethiopia. *Remote Sens. Appl. Soc. Environ.* **2017**, *8*, 224–230. [CrossRef]
9. Redo, D.J.; Aide, T.M.; Clark, M.L. The relative importance of socioeconomic and environmental variables in explaining land change in Bolivia, 2001–2010. *Ann. Assoc. Am. Geogr.* **2012**, *102*, 778–807. [CrossRef]
10. Gerhardt, F.; Foster, D.R. Physiographical and historical effects on forest vegetation in central New England, USA. *J. Biogeogr.* **2002**, *29*, 1421–1437. [CrossRef]
11. Schulz, J.J.; Cayuela, L.; Rey-Benayas, J.M.; Schröder, B. Factors influencing vegetation cover change in Mediterranean Central Chile (1975–2008). *Appl. Veg. Sci.* **2011**, *14*, 571–582. [CrossRef]
12. Kanade, R.; John, R. Topographical influence on recent deforestation and degradation in the Sikkim Himalaya in India; Implications for conservation of East Himalayan broadleaf forest. *Appl. Geogr.* **2018**, *92*, 85–93. [CrossRef]

13. Zhang, Q.; Ren, R.; Zhao, L. Spatial distribution pattern and influencing factors of forest in Northeast China. *J. Northeast. Froestry Unversity* **2013**, *41*, 25–28.
14. Zhu, W.B.; Lv, A.F.; Jia, S.F. Spatial distribution of vegetation and the influencing factors in Qaidam Basin based on NDVI. *J. Arid. Land* **2011**, *3*, 85–93. [CrossRef]
15. Spathelf, P.; van der Maaten, E.; van der Maaten-Theunissen, M.; Campioli, M.; Dobrowolska, D. Climate change impacts in European forests: The expert views of local observers. *Ann. For. Sci.* **2014**, *71*, 131–137. [CrossRef]
16. Huang, W.; Deng, X.; Lin, Y.; Jiang, Q. An Econometric Analysis of Causes of Forestry Area Changes in Northeast China. *Procedia Environ. Sci.* **2010**, *2*, 557–565. [CrossRef]
17. Rodríguez García, V.; Caravaggio, N.; Gaspart, F.; Meyfroidt, P. Long- and Short-Run Forest Dynamics: An Empirical Assessment of Forest Transition, Environmental Kuznets Curve and Ecologically Unequal Exchange Theories. *Forests* **2021**, *12*, 431. [CrossRef]
18. Senf, C.; Seidl, R. Mapping the coupled human and natural disturbance regimes of Europe's forests. *BioRxiv* **2020**. [CrossRef]
19. Lambin, E.F.; Meyfroidt, P. Global land use change, economic globalization, and the looming land scarcity. *Proc. Natl. Acad. Sci. USA* **2011**, *108*, 3465–3472. [CrossRef]
20. Singh, M.P.; Bhojvaid, P.P.; de Jong, W.; Ashraf, J.; Reddy, S.R. Forest transition and socio-economic development in India and their implications for forest transition theory. *For. Policy Econ.* **2017**, *76*, 65–71.
21. Zhang, W.; Chong, Z.; Li, X.; Nie, G. Spatial patterns and determinant factors of population flow networks in China: Analysis on Tencent Location Big Data. *Cities* **2020**, *99*, 102640. [CrossRef]
22. Michon, G.; de Foresta, H.; Levang, P.; Verdeaux, F. Domestic forests: A new paradigm for integrating local communities' forestry into tropical forest science. *Ecol. Soc.* **2007**, *12*, 1. [CrossRef]
23. Leblois, A.; Damette, O.; Wolfersberger, J. What has driven deforestation in developing countries since the 2000s? Evidence from new remote-sensing data. *World Dev.* **2017**, *92*, 82–102. [CrossRef]
24. Liu, S.; Yang, Y.; Wang, H. Strategies and countermeasures for the development of plantation management in China: From single-objective management of seeking wood production to multi-objective management of enhancing ecosystem service quality and benefit. *Acta Ecol. Sin.* **2018**, *38*, 1–10.
25. Kull, C.A.; Ibrahim, C.K.; Meredith, T.C. Tropical forest transitions and globalization: Neo-liberalism, migration, tourism, and international conservation agendas. *Soc. Nat. Resour.* **2007**, *20*, 723–737. [CrossRef]
26. Laurance, W.F.; Campbell, M.J.; Alamgir, M.; Mahmoud, M.I. Road Expansion and the Fate of Africa's Tropical Forests. *Front. Ecol. Evol.* **2017**, *5*, 75. [CrossRef]
27. Pfaff, A.; Robalino, J.; Walker, R.; Aldrich, S.; Caldas, M.; Reis, E.; Perz, S.; Bohrer, C.; Arima, E.; Laurance, W.; et al. Road investments, spatial spillovers, and deforestation in the Brazilian Amazon. *J. Reg. Sci.* **2007**, *47*, 109–123. [CrossRef]
28. Asher, S.; Garg, T.; Novosad, P. The ecological impact of transportation infrastructure. *Econ. J.* **2020**, *130*, 1173–1199. [CrossRef]
29. Deng, X.; Huang, J.; Uchida, E.; Rozelle, S.; Gibson, J. Pressure cookers or pressure valves: Do roads lead to deforestation in China? *J. Environ. Econ. Manag.* **2011**, *61*, 79–94. [CrossRef]
30. China Forestry Information Network [EB/OL]. Available online: http://www.lknet.ac.cn/ (accessed on 16 August 2022).
31. Alamgir, M.; Campbell, M.; Sloan, S.; Goosem, M.; Clements, G.R.; Ibrahim-Mahmoud, M.; Laurance, W.F. Economic, Socio-Political and Environmental Risks of Road Development in the Tropics. *Curr. Biol.* **2017**, *27*, 1130–1140. [CrossRef]
32. Breda, N.; Peuffer, M. Vulnerability to forest decline in a context of climate changes: New prospects about an old question in forest ecology. *Ann. For. Sci.* **2014**, *71*, 627–631. [CrossRef]
33. Resources and Environmental Science and Data Center, Chinese Academy of Sciences [EB/OL]. Available online: www.resdc.cn/ (accessed on 25 July 2022).
34. Wang, J.; Xu, C. Geodetector: Principles and Prospects. *Acta Geogr. Sin.* **2017**, *72*, 116–134.
35. Ju, H.; Zhang, Z.; Zuo, L.; Wang, J.; Zhang, S.; Wang, X.; Zhao, X. Driving forces and their interactions of built-up land expansion based on the geographical detector–a case study of Beijing, China. *Int. J. Geogr. Inf. Sci.* **2016**, *30*, 2188–2207. [CrossRef]
36. Wright, S.J.; Muller-Landau, H.C. The Future of Tropical Forest Species1. *Biotropica* **2006**, *38*, 287–301. [CrossRef]
37. Sandel, B.; Svenning, J. Human impacts drive a global topographic signature in tree cover. *Nat. Commun.* **2013**, *4*, 2474. [CrossRef] [PubMed]
38. Sloan, S. Fewer people may not mean more forest for Latin American forest frontiers. *Biotropica* **2007**, *39*, 443–446. [CrossRef]
39. Halas, M.; Klapka, P.; Tonev, P. The use of migration data to define functional regions: The case of the Czech Republic. *Appl. Geogr.* **2016**, *76*, 98–105. [CrossRef]
40. Li, X.; Zhang, Y.; Liu, Y.; Zhao, T. Dynamic Evolution and Future Prediction of Land Use Patterns in the Arid Desert Region of Northwest China from 1990 to 2020. *Forests* **2022**, *13*, 1570. [CrossRef]
41. Lorenzen, M.; Orozco-Ramírez, Q.; Ramírez-Santiago, R.; Garza, G.G. Migration, socioeconomic transformation, and land-use change in Mexico's Mixteca Alta: Lessons for forest transition theory. *Land Use Policy* **2020**, *95*, 104580. [CrossRef]
42. Laurance, W.F.; Peletier-Jellema, A.; Geenen, B.; Koster, H.; Verweij, P.; Van Dijck, P.; Lovejoy, T.E.; Schleicher, J.; Van Kuijk, M. Reducing the global environmental impacts of rapid infrastructure expansion. *Curr. Biol.* **2015**, *25*, 259–262. [CrossRef]
43. Freitas, S.R.; Hawbaker, T.J.; Metzger, J.P. Effects of roads, topography, and land use on forest cover dynamics in the Brazilian Atlantic Forest. *For. Ecol. Manag.* **2010**, *259*, 410–417. [CrossRef]
44. Barber, C.P.; Cochrane, M.A.; Souza, C.M., Jr.; Laurance, W.F. Roads, deforestation, and the mitigating effect of protected areas in the Amazon. *Biol. Conserv.* **2014**, *177*, 203–209. [CrossRef]

45. Hughes, A.C. Understanding and minimizing environmental impacts of the Belt and Road Initiative. *Conserv. Biol.* **2019**, *33*, 883–894. [CrossRef] [PubMed]
46. Caravaggio, N. Economic growth and the forest development path: A theoretical re-assessment of the environmental Kuznets curve for deforestation. *For. Policy Econ.* **2020**, *118*, 102259. [CrossRef]
47. Foster, A.D.; Rosenzweig, M.R. Economic growth and the rise of forests. *Q. J. Econ.* **2003**, *118*, 601–637. [CrossRef]

forests

Article

The Trade-Offs and Synergies of Ecosystem Services in *Pinus massoniana* Lamb. Plantations in Guangxi, China

Rongjian Mo [1,2,†], Yongqi Wang [1,2], Yanhua Mo [1,2,†], Lu Li [1,2] and Jiangming Ma [1,2,3,*]

1 Key Laboratory of Ecology of Rare and Endangered Species and Environmental Protection, Guangxi Normal University, Ministry of Education, Guilin 541006, China; moyanhua2019@mailbox.gxnu.edu.cn (Y.M.)
2 Guangxi Key Laboratory of Landscape Resources Conservation and Sustainable Utilization in Lijiang River Basin, Guangxi Normal University, Guilin 541006, China
3 Guangxi Key Laboratory of Superior Timber Trees Resource Cultivation, Guangxi Forestry Research Institute, Nanning 530002, China
* Correspondence: mjming03@gxnu.edu.cn
† These authors contributed equally to this work.

Abstract: A scientific understanding of the synergistic and trade-off relationships among ecosystem services (ESs) is essential for maintaining the structure, function, and health of plantation forest ecosystems. This understanding facilitates effective ecosystem management practices, and helps identify the types, intensities, and spatial and temporal distribution characteristics of interactions among ESs, which is critical for regional development planning, ecological supplementation, and the maximization of economic benefits. In this study, we used correlation analysis, bivariate spatial autocorrelation, and hot spot analysis to comprehensively analyze the synergistic and trade-off relationships between ESs in *Pinus massoniana* (PM) plantations in Guangxi Paiyang Forest Farm from 2009 to 2018, across both time and space. The study showed that the ESs in PM plantations in Guangxi Paiyang Forest Farm maintained significant positive correlation (synergy), with a mutually reinforcing relationship among services. Notably, the regulating services shifted from weak synergy to weak trade-off relationships over time. From the bivariate spatial autocorrelation analysis, it is clear that the overall trade-off synergistic relationship among the four ESs is basically consistent with the correlation analysis results. From the distribution of multiple ES hot spots, we determined that the number of small groups that can provide three to four service hot spots in Guangxi Paiyangshan Forest Farm is greater. The spatial distribution of cold–hot spots of various ESs varied, and the distribution of cold–hot spots of supply services and regulation services of carbon sequestration and oxygen release was similar.

Keywords: *Pinus massoniana* plantations; forest ecosystem services; trade-off synergy; service cold–hot spot analysis

Citation: Mo, R.; Wang, Y.; Mo, Y.; Li, L.; Ma, J. The Trade-Offs and Synergies of Ecosystem Services in *Pinus massoniana* Lamb. Plantations in Guangxi, China. *Forests* **2023**, *14*, 581. https://doi.org/10.3390/f14030581

Academic Editor: Pedro Álvarez-Álvarez

Received: 10 February 2023
Revised: 12 March 2023
Accepted: 13 March 2023
Published: 15 March 2023

1. Introduction

Forest ecosystems are essential key systems for human beings, providing various services such as timber volume, edible forest by-products, starch and other products, water conservation, biodiversity conservation, and regulation of the atmospheric environment for human life and production [1,2]. Forests serve as the most complete, abundant, and stable carbon reservoir, natural oxygen reservoir, species gene pool, renewable resource reservoir, freshwater resource, and renewable energy reservoir in nature. Forests play a fundamental and critical role in human survival and the realization of sustainable ecological civilization [3]. Additionally, forest ecosystems offer services to humans in terms of medical health care, physical and mental relaxation, tourism, and rest, as well as other cultural experiences. The ecological and economic benefits of forest ecosystems should be comprehensively coordinated, and sustainable management of forest ecosystems should be maintained to achieve a mutually beneficial situation for both humans and nature. Schwaiger, F. et al. [4] used a forest management model to analyze the trade-offs between ESs and biodiversity indicators in two case study areas in Germany, which influenced decision-making in three scenarios of forest management. Langner, A. et al. [5] used simulated scenario analysis and multi-criteria decision analysis methods to quantify trade-offs among four ESs (timber production, carbon storage, biodiversity conservation, and landslide hazard control) in montane forests to make decisions about expected utility in multi-objective forest management. Dai, E. et al. [6] valued four services of timber volume, carbon storage, water production, and soil conservation in the Ganjiang River basin, quantified their trade-offs and synergies at the county scale and sub-basin scale, and detected synergistic relationships between timber volume and carbon storage, soil conservation, and water production. Wu, W. et al. [7] studied the trade-offs and synergistic relationships between two regulating services, soil organic carbon regulation, total nitrogen regulation, and species diversity, as well as their influencing factors in a mixed *Pinus* sylvestris and *Quercus* sylvestris needle-broad forest, a deciduous broad-leaved southern date palm forest, and a broad-leaved evergreen forest of *Cycad* and *Quercus* sylvestris, using secondary forests at different successional stages in the central subtropical region.

Currently, both domestic and international research on forest multifunctionality has primarily focused on the fundamental meaning of forest multifunctionality [8–10], as well as the evaluation of multiple ESs and their spatiotemporal evolution characteristics [11–13]. However, further research is needed to develop and enhance our understanding of forest ES trade-offs and synergies. In this study, we explored the spatiotemporal relationship characteristics among multiple ESs in PM plantation forests by combining statistical analysis methods, spatial mapping, and bivariate spatial autocorrelation methods. This study is a continuation of the previous assessment of ESs in a *Pinus massoniana* plantation. We utilized the value of ESs in PM plantations in Guangxi Paiyangshan Forest Farm as an indicator, and employed correlation analysis to identify the absolute values of correlation coefficients and the positive and negative directions to determine the types and strength characteristics of relationships among ESs. Additionally, we employed the bivariate spatial autocorrelation method to describe the synergistic and trade-off relationships among ESs in space, and applied the hot spot analysis method to identify cold–hot spot areas of ESs. The spatial quantification and mapping of ESs in PM plantation forests can be useful for revealing scale effects and interactions of ESs, and providing a scientific basis for management decisions in PM plantation forests.

2. Materials and Methods

2.1. Study Area

The Guangxi Paiyangshan Forest Farm (Figure 1), located in Ningming County, southwest Guangxi, China, near the border with China and Vietnam is the only large state-owned forestry field directly under the Forestry Department of Guangxi Autonomous Region bordering Vietnam, with a total land area of 28,096.29 hm^2 and a forest land area of 27,462.28 hm^2. The geographical coordinates are 106°30′–107°15′ E, 21°15′–22°30′ N. The southern red loam hilly area in China is a significant timber production region, with approximately 308,000 km^2 of planted forests. PM, one of the primary fast-growing production species, is also a critical afforestation and deforestation species in southern China. It is the most widely distributed and largest area timber species in pine [14,15]. PM has outstanding features, including drought tolerance, ridge thinning resistance, strong growing ability, rapid growth, high quality, and versatility [16]. These qualities have great economic and ecological benefits [17,18], making PM an essential component in forestry production and forest restoration and reconstruction in China [19]. In this study, the Paiyangshan Forest Farm PM plantation is in the northern tropical case study area [20], which has an average annual temperature of 21.8 °C, an average annual rainfall of 1250–1700 mm, a rainy season from May to August, and a relative humidity of 82.5%. The annual number of sunshine hours is 1650.3, and the annual evaporation is 1423.3 mm. After years of afforestation, the native vegetation has been replaced by natural broad-leaved secondary forests and planted forests. The plantation forest tree species are mainly PM, *Eucalyptus robusta*, and *Illicium verum*.

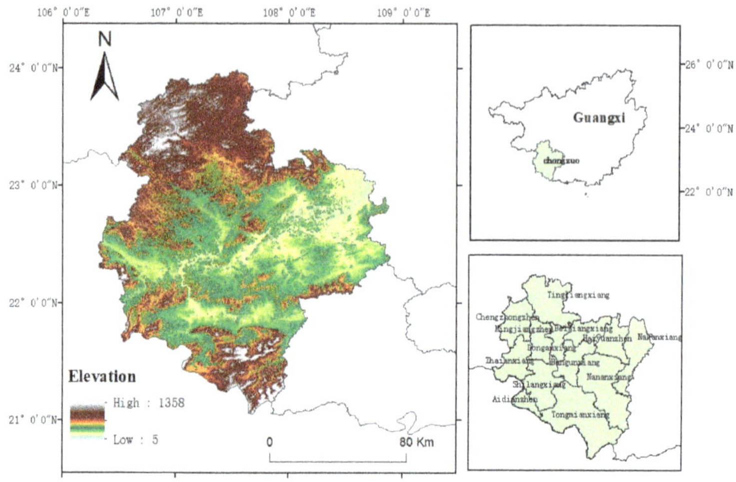

Figure 1. Distribution map of study area.

2.2. Research Methods

In this study, we employed correlation analysis, bivariate spatial autocorrelation analysis, and hot spot analysis to explore the relationship between the ES values of the PM plantations in Guangxi Paiyangshan Forest Farm [20]. The research framework is illustrated in Figure 2.

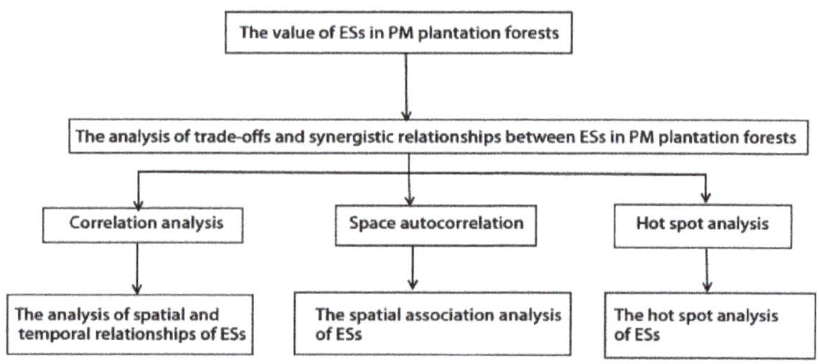

Figure 2. Research framework.

2.2.1. Correlation Analysis

The commonly used statistical analysis methods in ES trade-off studies include correlation analysis, root mean square deviation method, regression analysis, and cluster analysis [21,22].

Correlation analysis is the simplest and most effective method to identify trade-offs between ESs. The magnitude of the coefficients that are significantly correlated between ES pairs and the direction of positive and negative correlation are used to determine the degree and direction of correlation among service pairs. Han, Y. [23] employed simple correlation analysis to calculate the correlation coefficients between production services, water conservation services, carbon sequestration and oxygen release services, environmental purification services, and cultural and recreational services in Xi'an metropolitan area in 1999, 2006, and 2009. The aim was to identify synergies and trade-offs among agricultural ESs. Partial correlation analysis can examine the interrelationships between ES pairs, excluding other factors based on the complex relationships among ESs. Sun, Y. et al. [24] used partial correlation analysis to study the synergistic and trade-off characteristics of water content, soil conservation, net primary productivity of vegetation, and food supply in the Loess Plateau, using Yan'an City as an example. Alamgir, M., et al. [25] used regression analysis and clustering methods to analyze the differences and linkages of multiple ESs among different forest stands.

In this study, a negative correlation coefficient between two ESs that passed the significance test was considered to indicate a trade-off relationship, while a positive correlation coefficient that passed the significance test was considered to indicate a synergistic relationship. Overall, the single services in the study area exhibit spatial heterogeneity, which may lead to both trade-off and synergistic relationships. Therefore, we calculated and tested the correlation coefficients between pairs of four ESs corresponding to small classes in the study area for significance.

In this study, we employed Pearson correlation analysis to investigate the interactions among ESs in PM plantation forests. Pearson correlation analysis is a commonly used method for analyzing relationships among ESs [26,27]. We then used the chart.Correlation function in the R language software to explore the interrelationships among the four major forest services and each sub-forest of the PM plantation, and created a scatter plot matrix. When the correlation coefficient of a pair of ESs is significantly positive, it indicates a synergistic relationship between the services. Conversely, a significantly negative correlation coefficient indicates a trade-off relationship between the services. If the correlation is not significant, then there is spatial compatibility between the services. The strength of the correlation is determined by the absolute value of the correlation coefficient. A correlation coefficient in the range of (0, 0.3) indicates a weak correlation, that in the range of [0.3, 0.5) indicates a moderately strong correlation, and that in the range of [0.5, 1] indicates a strong correlation [28,29].

2.2.2. Bivariate Spatial Autocorrelation

To further examine the spatial distribution of synergistic and trade-off relationships among ecosystem services, we assigned a value to each ES in the vector map of small group units. Next, we imported these data into the GeoDa software to analyze the bivariate spatial autocorrelation between pairs of services in the PM plantation forest using the bivariate spatial autocorrelation module. High–high and low–low agglomerations both indicate synergistic relationships, while high–low and low–high agglomerations indicate trade-off relationships [30].

2.2.3. Hot Spot Analysis

Spatially cold–hot spots refer to areas with either high or low values for ESs provided by PM plantation forests [31]. In this study, we used the hot spot analysis tool in the ArcGIS spatial analysis module and the Getis-Ord Gi* local statistic to identify cold–hot spots for four services [32]. To determine whether a small class was a hot or cold spot, we used the aggregation index degree parameter Z(Gi*) and followed the classification criteria described in the study by Wang, B. et al. [33]: if $Z(Gi^*) > 2.58$, the region was classified as a hot spot; if $1.65 < Z(Gi^*) \leq 2.58$, it was a sub-hot spot; if $Z(Gi^*) < -2.58$, it was a cold spot; if $-2.58 \leq Z(Gi^*) < -1.65$, it was a sub-cold spot; and if $-1.65 \leq Z(Gi^*) \leq 1.65$, it was not considered significant.

3. Results

3.1. Analysis of ES Interactions in PM Plantation Forests

Based on Figure 3, the correlation results of the ecosystem services (ESs) of PM plantations in Guangxi Paiyangshan Forest Farm from 2009 to 2018 can be observed. In 2013, the value per unit area of wood supply, carbon sequestration and oxygen release, and water conservation became more concentrated, while the distribution of soil conservation was more dispersed. In 2018, the data points for water conservation were more concentrated than in 2013, while the data points for other services were more dispersed. From a correlation standpoint, the six groups of services provided by PM plantations in Guangxi Paiyangshan Forest Farm from 2009 to 2018 were all significantly positively correlated, except for carbon sequestration and oxygen release, which were negatively correlated with water conservation in 2018.

Wood supply was strongly positively correlated with carbon sequestration and oxygen release in 2009, and weakly positively correlated with water conservation (correlation coefficient 0.19) and soil conservation (correlation coefficient 0.20). This is similar to the strong synergistic relationship between food provision and carbon sequestration and oxygen release in the value of cropland ESs reported by Tian, Y. [26]. Carbon sequestration and oxygen release were weakly correlated with water conservation (correlation coefficient 0.18) and soil conservation (correlation coefficient 0.17). The correlation between water conservation and soil conservation (correlation coefficient 0.17) was also weak. In 2013, the correlation increased (0.29) and was close to moderate for water conservation and soil conservation, while all other service pairs decreased while maintaining their original correlation. In 2018, the correlation between wood supply and carbon sequestration and oxygen release increased (correlation coefficient 0.68), the correlation between water conservation and soil conservation increased (correlation coefficient 0.48), and the correlation between the other service pairs decreased. There was a negative correlation between carbon sequestration and oxygen release and water conservation, indicating a trade-off between the two. The synergistic relationship between wood supply and carbon sequestration and oxygen release decreased and then increased. The synergistic relationship between water conservation and soil conservation continued to increase, while the synergistic relationship between carbon sequestration and oxygen release and water conservation changed from a synergistic relationship to a trade-off relationship. The synergistic relationship between the other service pairs continued to decrease over the years.

In terms of ES types, from 2009 to 2018, the supply service (wood supply) and regulating service (carbon sequestration and oxygen release) of Guangxi Paiyangshan Forest Farm were in a strong synergistic relationship that first weakened and then strengthened. The synergistic relationship between regulating service (water conservation) and supporting service (soil conservation) changed from weak to moderate synergistic. The synergistic relationship between regulating services (carbon sequestration and oxygen release as well as water conservation) changed from synergistic to trade-off. The relationship between the other service pairs showed a weak synergistic relationship that gradually weakened.

Overall, the service pairs of PM plantation forest at the macroscopic scale of Guangxi Paiyangshan Forest Farm maintained a positive relationship from 2009 to 2018. The supply service and the regulating service showed a strong synergistic relationship, the supply service and the supporting service showed a weak synergistic relationship, and the regulating service and the supporting service showed a weak synergistic relationship. However, the regulating service changed from a synergistic to a weak trade-off relationship among them, and the relationship between water conservation and the supporting service in the regulating service changed from a weak synergistic to a moderate synergistic relationship.

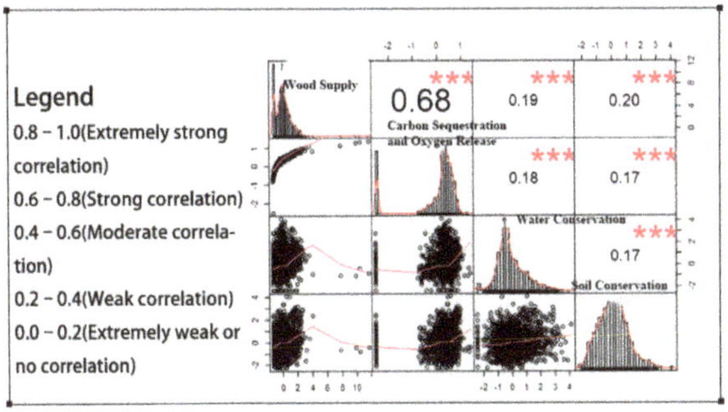

(**a**)

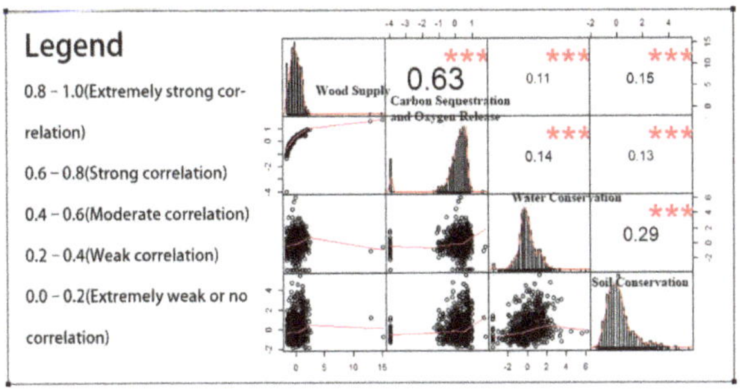

(**b**)

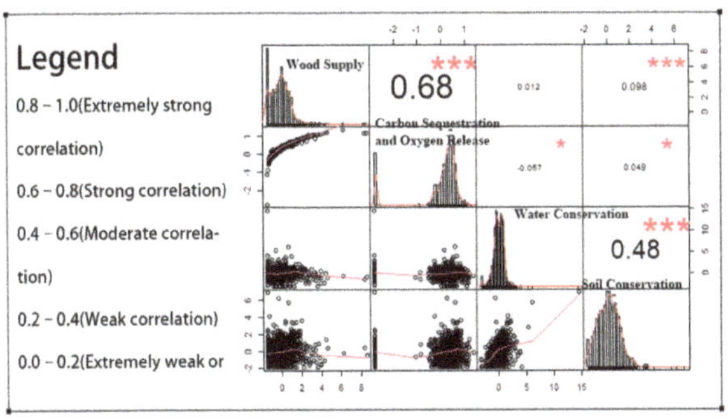

(**c**)

Figure 3. A scatter plot matrix was used to show the results of the ES correlation analysis of PM plantation forests in Guangxi Paiyangshan Forest Farm from 2009 to 2018. Pearson correlation coefficients range from −1 to 1, indicating negative and positive correlations, respectively. The strength of correlation increases as the absolute value of the Pearson correlation coefficient approaches 1. Statistical significance is indicated by asterisks, with * representing significance at the 0.05 level, and *** at the 0.001 level. The scatterplot matrices for 2009, 2013, and 2018 are presented as (**a**), (**b**), and (**c**), respectively.

3.2. *Spatial Association of ESs in PM Plantation Forests*

After analyzing the temporal and spatial correlation of ESs, we gained initial insights into the synergistic and trade-off relationship between the ESs of PM plantations in Guangxi Paiyangshan Forest Farm. We determined that the synergy was greater than the trade-off at the scale of the whole forest. Bivariate spatial autocorrelation analysis was conducted on the ESs of PM plantations in Guangxi Paiyangshan Forest Farm, and the spatial distribution of "HH" type areas was determined to have high spatial correlation of high value of PM plantation function, indicating enhanced synergy with the ESs of PM plantations in adjacent small groups. The "LL" type area is the area with a high spatial correlation of low value of PM plantation function, indicating weakened synergy with the adjacent small group PM plantation ES pairs. The "LH" and "high–low" type areas are small classes with high and low spatial correlation of ES pairs, respectively, indicating a trade-off between ES pairs in adjacent small classes, whereby one service is enhanced or weakened in a small class unit while another function is weakened in an adjacent small class. In other words, one service is enhanced or weakened while another function is weakened or enhanced in the adjacent small group. Due to space limitations, we have only analyzed the spatial correlations between wood supply services and other services here.

Based on Table 1 and Figure 3, the analysis results are significant at a level higher than 95%. As seen in Table 1, the trade-off synergistic relationship of each service pair is generally consistent with the results of the correlation analysis (Figure 3). However, the local autocorrelation index, Moran I, between carbon sequestration and oxygen release and water conservation in 2018 is negative, indicating a weak trade-off, whereas other service pairs are primarily synergistic with each other. Specifically, the synergistic relationship between wood supply and carbon sequestration and oxygen release, water conservation, and soil conservation is spatially robust, with the strongest synergistic relationship being between wood supply and carbon sequestration and oxygen release. The synergistic relationship gradually decreases from 2009 to 2018 with other services. The synergistic relationship between carbon sequestration and oxygen release and soil conservation is initially strong but then decreases. The synergistic relationship between carbon sequestration and oxygen release and water conservation shows a weak synergistic relationship from 2009 to 2013 and a weak trade-off relationship in 2018. The overall spatial relationship between water conservation and soil conservation is weakly synergistic, increasing first and then decreasing. The overall synergistic relationship between multiple service pairs shows a decreasing trend from 2009 to 2018.

Table 1. Bivariate local autocorrelation Moran I index of ESs in PM plantations in Paiyangshan Forest Farm.

ESs Service Pairs	2009	2013	2018
Wood Supply–Carbon Sequestration and Oxygen Release	0.4248	0.2992	0.3369
Wood Supply–Water Conservation	0.1844	0.1442	0.0393
Wood Supply–Soil Conservation	0.1720	0.1204	0.0705
Carbon Sequestration and Oxygen Release–Water Conservation	0.1710	0.1409	−0.0295
Carbon Sequestration and Oxygen Release–Soil Conservation	0.1094	0.1160	0.0266
Water Conservation–Soil Conservation	0.1716	0.2298	0.1291

The analysis of spatial autocorrelation, with wood supply as the central indicator, and other ESs of neighboring small groups as the corresponding indicators, revealed significant spatial heterogeneity in the trade-off synergistic relationship between wood supply and other services in the PM plantation.

From 2009 to 2018, the synergistic and trade-off areas between wood supply and carbon sequestration and oxygen release were more synergistic than trade-off areas (Figure 4). In 2009, the synergistic enhancement areas were mainly located in the Beishan branch, Gongwu branch, a small portion of small classes in Honghu branch, and the northwestern part of Pucheng branch. However, these areas decreased significantly by 2013. Specifically, the synergistic enhancement areas in Beishan branch shifted to the north, and the synergistic enhancement areas in Gongwu branch and Pucheng branch decreased. By 2018, the synergistic enhancement area of Gongwu branch decreased significantly, while the synergistic enhancement area of Nianke branch and Honghu branch increased slightly. Additionally, the synergistic weakening area was mainly concentrated in the west and east of Gongwu branch, the middle of Nianke branch, Honghu branch, Nalai branch, and Dawangshan branch in 2009. The synergistic weakening area in the western part of Dawangshan branch, Nalai branch, and Gongwu branch decreased, but Beishan branch increased. The number of synergistic weakening areas increased in Gongwu branch and Pucheng branch and decreased in Beishan branch in 2018, with little change in the number of synergistic weakening areas from 2009 to 2018. Moreover, the weighted areas were mainly distributed in Gongwu branch and Honghu branch in 2009. However, the weighted areas were sporadically distributed in major sub-fields in 2013. In 2018, the weighted areas increased, mainly distributed in Honghu branch and Nianke branch. The number of "high–low" and "low–high" balance areas remained close over the years.

From 2009 to 2018, the synergistic and trade-off areas between wood supply and water conservation were more synergistic than trade-off areas (Figure 5). In 2009, the areas of synergistic enhancement were mainly located in the southern Gongwu branch and the western and eastern parts of the Pucheng branch. However, the synergistic enhancement area significantly reduced in 2013, particularly in the Gongwu branch. In 2018, the synergistic enhancement area of the Pucheng branch decreased while the southern Gongwu and Beishan branches saw a significant increase. Conversely, the areas where the synergy weakened in 2009 were mainly concentrated in the Honghu branch, Nianke branch, the northwestern part of the Nalai branch, and the Dawangshan branch. This area significantly reduced in 2013 and only remained in the Beishan branch, southwestern part of Nianke branch, northwestern part of Honghu branch, and Gongwu branch. However, the synergistic weakening area increased again in 2018, particularly in the Honghu branch and Dawangshan branch, while it decreased in the Beishan branch. Over the years, the number of regions showing synergistic changes decreased and then increased. In 2009, the "low–high" trade-off area was mainly located in the southern part of Gongwu branch, Honghu branch, and the southern part of Pucheng branch, while the "high–low" trade-off area was mainly located in the Honghu branch, Beishan branch, the northwest end of Gongwu branch, and Nianke branch. In 2013, the number of trade-off areas decreased, with "high–low" trade-off areas mainly in the northwestern part of the forest and "low–high" trade-off areas in the southeastern part. In 2018, the number of weighing areas increased, with the "high–low" weighing areas mainly existing in the northern part of the entire forest, such as Honghu branch and Dawangshan branch, while the "low–high" weighing areas were located in the southern part of Beishan branch, Gongwu branch, and the southern part of Pucheng branch. The number of trade-off areas has fluctuated over the years, and the number of "low–high" trade-off areas has consistently exceeded the number of "high–low" trade-off areas.

From 2009 to 2018, the synergistic and trade-off areas between wood supply and water conservation were more synergistic than trade-off areas (Figure 6). In 2009, the synergistic enhancement areas were mainly located in the northwestern part of the Gongwu and Pucheng branches, but this area significantly reduced in 2013 and was only distributed in the northwestern part of the Pucheng branch. However, in 2018, the number of synergistic enhancement areas increased, particularly in the northwestern part of the Pucheng, Gongwu, Nianke, and northern part of Daeangshan branches. The areas of weakened synergy were distributed across all major branches in 2009 but decreased in 2013, mainly in the Beishan and Nianke branches, the western part of the Honghu branch, and the northeastern part of the Dawangshan branch. However, in 2018, the area of weakened synergy increased, mainly in the Honghu, Nianke, and Dawangshan branches. In 2009, the "low–high" weighing area was distributed across all branches, while the "high–low" weighing area was mainly located in the Honghu, Beishan, and Gongwu branches. The number of weighing areas decreased in 2013, and the "low–high" weighing areas were mainly located in the northern and southern parts of Gongwu and Pucheng branches, while the "high–low" weighing areas were located in the western and middle parts of the forest. However, in 2018, the number of weighing areas increased, and the "low–high" weighing areas were mainly located in the Gongwu, Dawongshan, and Pucheng branches, while the "high–low" weighing areas were located in the Gongwu, Honghu, and Pucheng branches. The number of trade-off areas decreased and then increased over the years, and the "high–low" trade-off area was consistently higher than the "low–high" trade-off area.

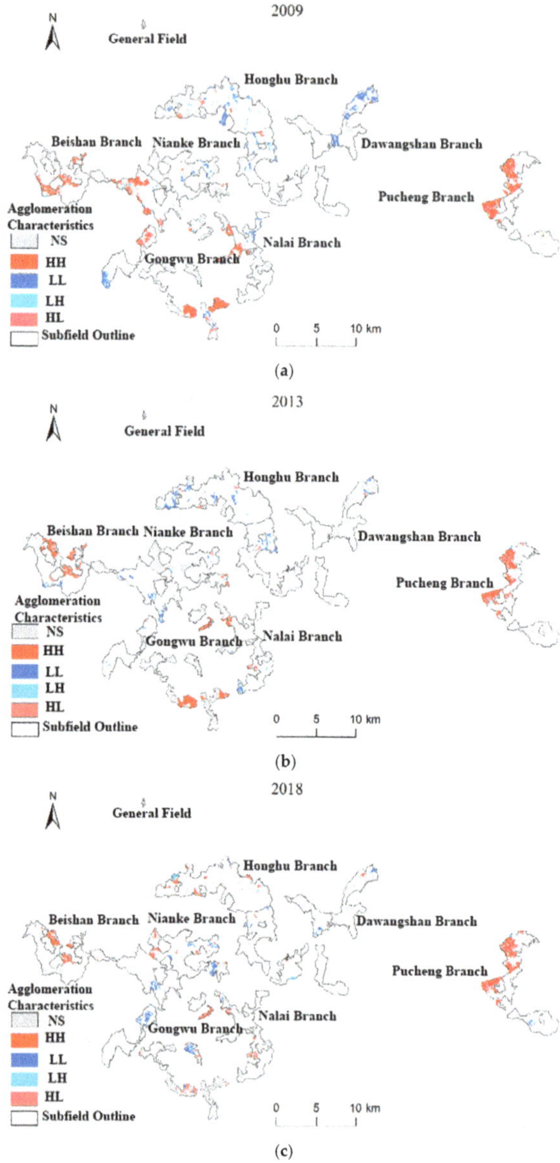

Figure 4. This is a Local LISA plot that shows the relationship between wood supply and carbon sequestration and oxygen release services in the PM plantations of Guangxi Paiyangshan Forest Farm from 2009 to 2018. The analysis used the following indicators: NS (not significantly clustered), HH (high–high clustering), LL (low–low agglomeration), LH (low–high agglomeration), HL (high–low agglomeration). The plot includes the following LISA maps: (**a**) wood supply and carbon sequestration and oxygen release services in 2009; (**b**) wood supply and carbon sequestration and oxygen release services in 2013; and (**c**) wood supply and carbon sequestration and oxygen release services in 2018.

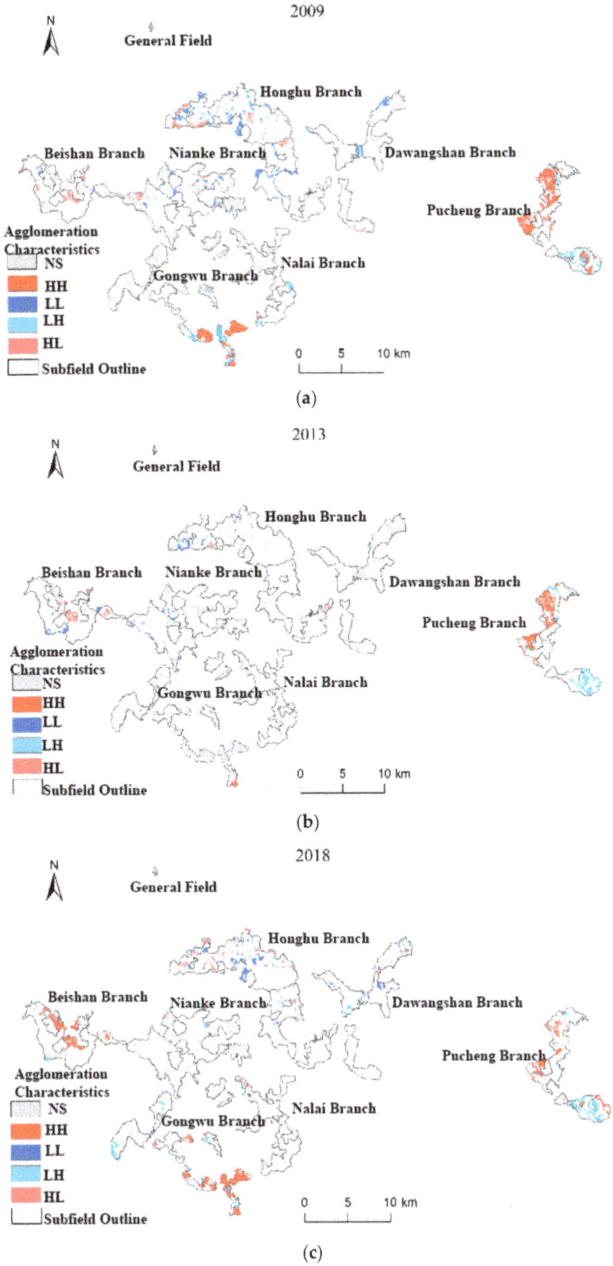

Figure 5. The Local LISA map shows the relationship between wood supply and water conservation services in PM plantations in Paiyangshan Forest Farm from 2009 to 2018. The map is categorized into NS (not significantly clustered), HH (high–high clustering), LL (low–low agglomeration), LH (low–high agglomeration), and HL (high–low agglomeration). We present three LISA maps: (**a**) the map of wood supply and water conservation services in 2009, (**b**) the map of wood supply and water conservation services in 2013, and (**c**) the map of wood supply and water conservation services in 2018.

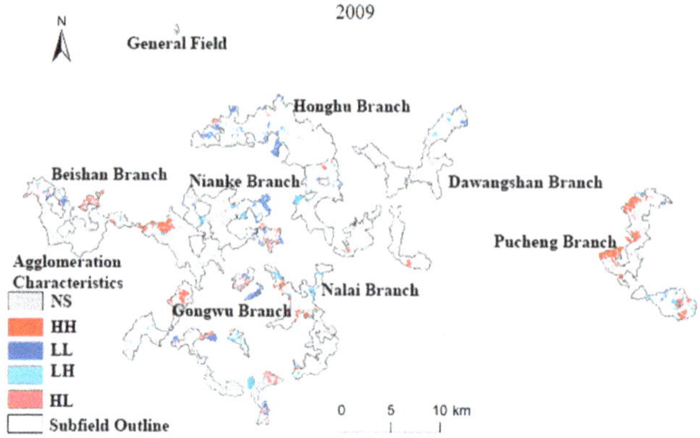

(a)

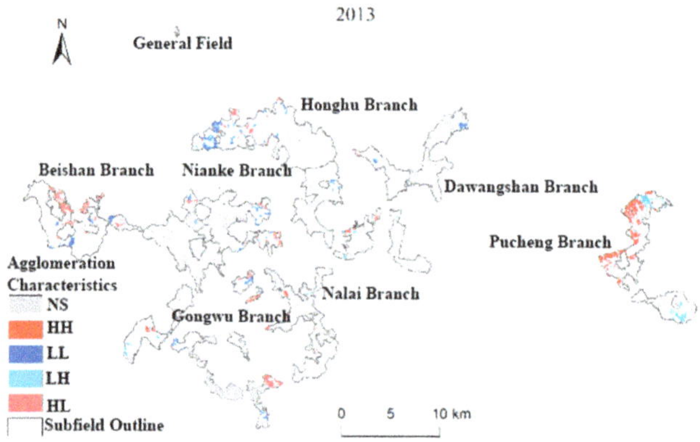

(b)

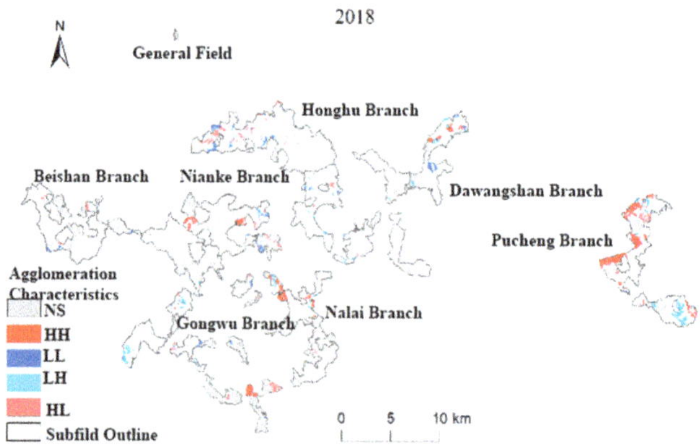

(c)

Figure 6. These Local LISA plots show the relationship between wood supply and soil conservation services in PM plantations in Paiyangshan Forest Farm from 2009 to 2018. The plots are categorized into NS (not significantly clustered), HH (high–high clustering), LL (low–low agglomeration), LH (low–high agglomeration), and HL (high–low agglomeration). The LISA maps for wood supply and soil conservation services are presented for three years: (**a**) 2009, (**b**) 2013, and (**c**) 2018.

3.3. Single ES Space Cold Hot Spot Analysis

The spatial variation and clustering characteristics of ES values were analyzed using hot spot analysis. The number and proportion of small groups of cold and hot spots were counted for each service value in PM plantation forests.

According to Table 2, the percentage of cold and hot spots of the four ESs in Guangxi Paiyangshan Forest Farm changed from 2009 to 2013. The service with the largest percentage of hot spots and sub-hot spots in small groups shifted from carbon sequestration and oxygen release to water conservation. The service with the smallest percentage of cold spots and sub-hot spots from 2009 to 2018 remained carbon sequestration and oxygen release. Among the four services, carbon sequestration and oxygen release were the most prominent in Guangxi Paiyangshan Forest Farm.

Table 2. Statistics on the number of small classes of cold hot spots of ESs in PM plantations in Guangxi Paiyangshan Forest Farm in 2009–2018.

ES			Wood Supply	Carbon Sequestration and Oxygen Release	Water Conservation	Soil Conservation
2009	Hot Spots and Sub-hot Spots	Number	621	894	532	674
		Proportion	28.33%	40.78%	24.27%	30.75%
	Cold Spots and Sub-cold Spots	Number	722	489	1042	629
		Proportion	32.94%	22.31%	47.54%	28.70%
2013	Hot Spots and Sub-hot Spots	Number	450	499	412	423
		Proportion	31.80%	35.27%	29.12%	29.89%
	Cold Spots and Sub-cold Spots	Number	441	349	541	510
		Proportion	31.17%	24.66%	38.23%	36.04%
2018	Hot Spots and Sub-hot Spots	Number	486	583	752	517
		Proportion	28.45%	34.13%	44.03%	30.27%
	Cold Spots and Sub-cold Spots	Number	396	338	649	486
		Proportion	23.19%	19.79%	38.00%	28.45%

By combining Table 2 and Figure 7, we can observe that in 2009, the number of hot and cold spots of wood supply service value in small groups was 621 and 722, respectively. Hot spots and sub-hot spots were more concentrated in spatial distribution, mainly in the central and eastern part of Beishan branch, Gongwu branch, northern and western part of Pucheng branch. Cold spots and sub-cold spots were mainly distributed in the central and western part of Gongwu branch, Nianke branch, Honghu branch and Dawangshan branch. In 2013, the number of hot and cold spots of small classes of wood supply service value amounted to 450 and 441, respectively, with an increase in the proportion of hot spots and a decrease in the proportion of cold spots. Compared with 2009, the number of hot spots decreased mainly in Gongwu branch, while it increased in Beishan branch. The number of small groups of cold spots mainly increased in Gongwu branch as well as Pucheng branch and decreased in Nianke branch. The number of small classes of hot and cold spots of wood supply service value quantity in 2018 was 486 and 396, respectively. Hot spots and sub-hot spots were mainly distributed in the northern part of Beishan branch, some small classes in the northern part of Honghu branch as well as the northern part of Pucheng branch, and the hot spot distribution area was reduced. Cold spots and sub-cold spots were more dispersed, mainly distributed in the southwest end of Beishan branch, the middle of Gongwu branch, the south end of Honghu branch, the south of Pucheng branch, and the west of Dawangshan branch. The distribution of cold spots significantly reduced compared with 2013. From 2009 to 2018, the percentage of hot spots of wood supply services first increased and then decreased, the percentage of cold spots continued to decrease, the area of lower wood supply services continued to decrease, and the distribution pattern of the whole forest value volume changed to high east–west and low middle.

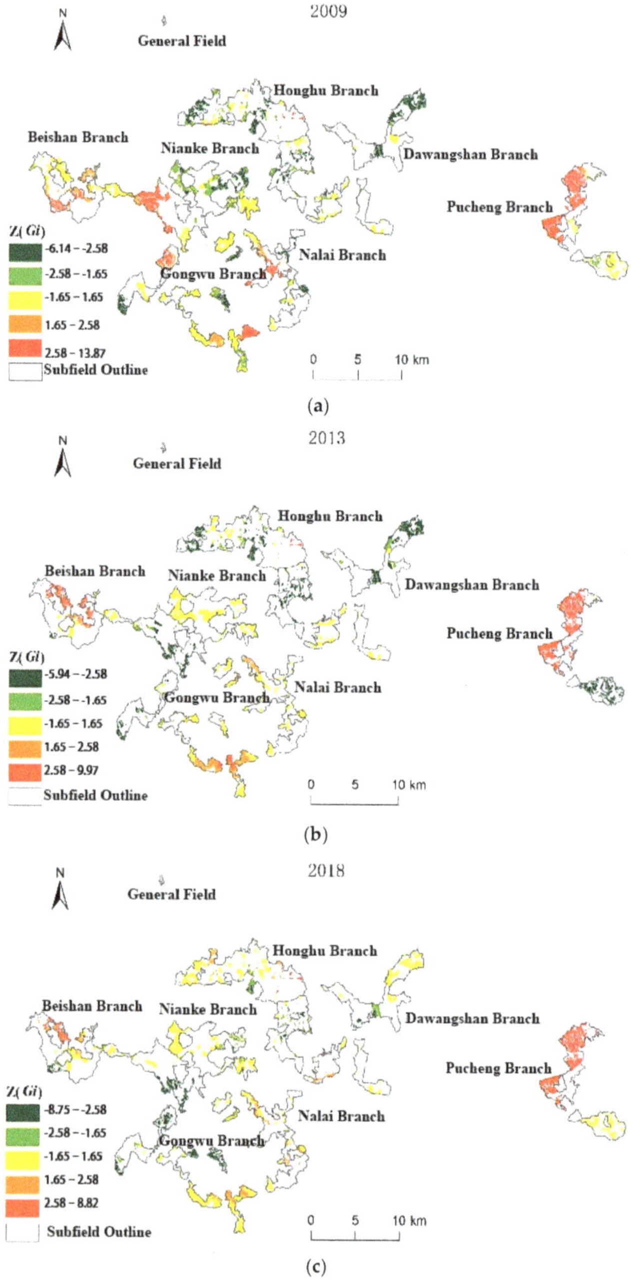

Figure 7. This figure shows the distribution of cold–hot spots of wood supply for PM plantations in Paiyangshan Forest Farm from 2009 to 2018, with Z(Gi*) as the aggregation index degree parameter. The color scheme used is red for hot spot regions, orange for sub-hot spot regions, yellow for insignificant regions, dark green for cold spot regions, and light green for sub-cold spot regions. Panel (**a**) displays the distribution of cold–hot spots in 2009, panel (**b**) displays the distribution in 2013, and panel (**c**) displays the distribution in 2018.

By combining Table 2 and Figure 8, we can observe that in 2009, there were 894 hot spots and 489 cold spots in the number of small groups with service values of carbon sequestration and oxygen release. The hot spots and sub-hot spots were concentrated in the central part of Beishan branch, the northwestern and eastern parts of Gongwu branch, and the northern part of Pucheng branch. Meanwhile, the cold spots and sub-cold spots were mainly distributed in the southern and southwestern ends of Gongwu branch, the northeastern part of Dawangshan branch, the northwestern part of Honghu branch, and Nalai branch. In 2013, the number of small groups of hot–cold spots for the value of carbon sequestration and oxygen release services decreased to 499 and 349, respectively, with the percentage of hot spots decreasing and the percentage of cold spots increasing. Compared with 2009, the hot spots decreased in the southern part of Beishan branch but increased in the northern part of Beishan branch and the northwestern part of Gongwu branch. Meanwhile, the number of small groups of cold spots mainly in the western part of Gongwu branch and Dawangshan branch decreased, but that in the southern part of the Gongwu branch increased. In 2018, there were 583 hot–cold spots of the value of carbon sequestration and oxygen release services, with 338 being cold spots. The percentage of hot and sub-hot distribution areas decreased, but the distribution range changed. The hot spot area of the value of carbon sequestration and oxygen release appeared in Honghu branch, and the distribution range of cold spots and sub-cold spots changed significantly. In 2013, the percentage of small classes of cold spots was mainly concentrated in the eastern part of Gongwu branch and Nianke branch, and there were no cold spots in Honghu branch and Beishan branch. However, the percentage of cold spots for carbon sequestration and oxygen release services increased and then decreased from 2009 to 2018, while the percentage of hot spots continued to decrease. Overall, from 2009 to 2018, the distribution pattern of the entire forest value volume changed to high east–west and low middle.

By combining Table 2 and Figure 9, we can observe that in 2009, the number of hot–cold spots of water conservation service value in small groups was 532 and 1042, respectively. The distribution of cold–hot spots was relatively concentrated, with hot spots and sub-hot spots mainly distributed in the east and south of PM plantations in Guangxi Paiyangshan Forest Farm (south of Gongwu branch), and cold spots and sub-cold spots mainly distributed in the northwest of PM plantations in Guangxi Paiyangshan Forest Farm (north of Beishan branch and Nianke branch, northwest of Narai branch and Honghu branch, and west and northeast of Dawangshan branch). In 2013, there were 412 and 541 small classes of hot–cold spots in the value volume of water conservation services. The percentage of hot spots increased while the percentage of cold spots decreased. Compared with 2009, the hot spots were mainly located in the south of Gongwu branch. The number of small groups of cold spots mainly decreased in the northwest of Gongwu branch, Nianke branch, the northwest of Nalai branch, and Dawangshan branch. In 2018, there were 752 and 649 small classes of hot–cold spots in the value volume of water conservation services, and the percentage of hot spots continued to increase while the percentage of cold spots continued to decrease. Hot spot areas appeared in the southwest of Beishan branch and Gongwu branch, and hot spot areas of water conservation value volume appeared. The distribution range of cold spots and sub-cold spots changed considerably, with the Beishan branch transforming into a hot spot area, the northwestern part of the Gongwu branch not being significant, and cold spots appearing in the Dawangshan branch. The percentage of hot spots for water conservation services continued to increase while the percentage of cold spots continued to decrease from 2009 to 2018. The distribution pattern of cold–hot spots of the entire forest value shifted from high in the southeast and low in the northwest to high in the west of the southeast and low in the north of the center.

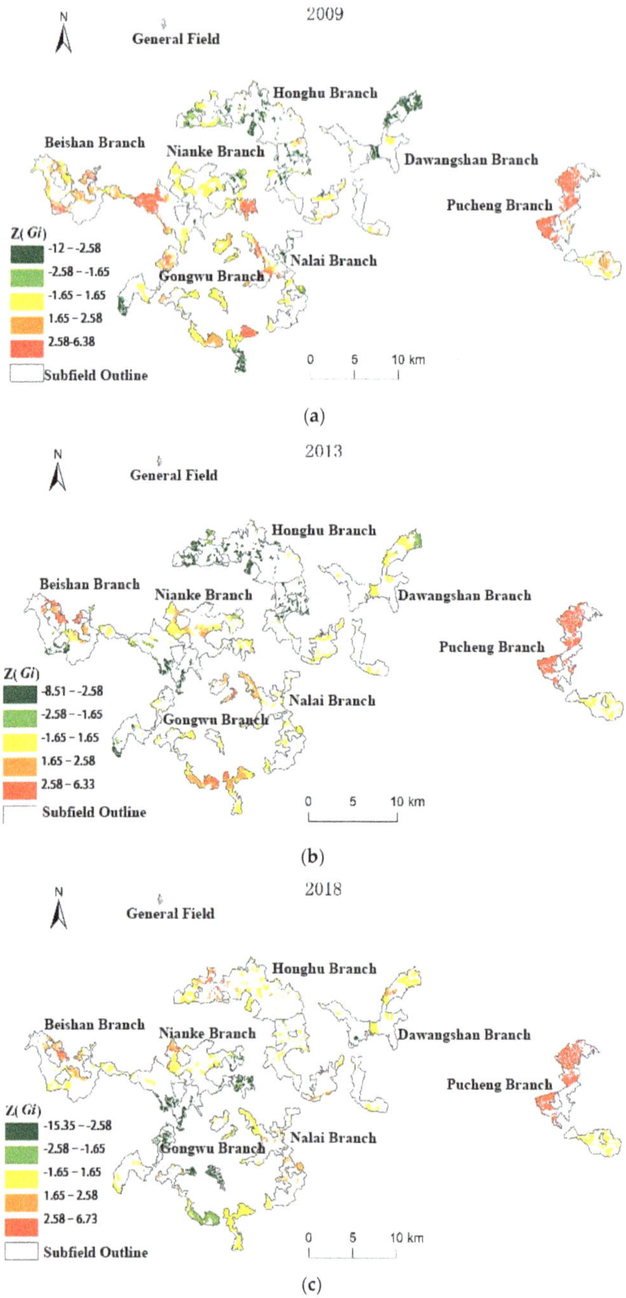

Figure 8. Distribution of cold–hot spots of carbon sequestration and oxygen release in PM Plantations in Paiyangshan Forest Farm from 2009 to 2018. Z(Gi*) is the aggregation index degree parameter. Red indicates a hot spot region, orange indicates a sub-hot spot region, yellow indicates an insignificant region, dark green indicates a cold spot region, and light green indicates a sub-cold spot region. The distribution of cold–hot spots of carbon sequestration and oxygen release is shown for the years 2009 (**a**), 2013 (**b**), and 2018 (**c**) in the following figures.

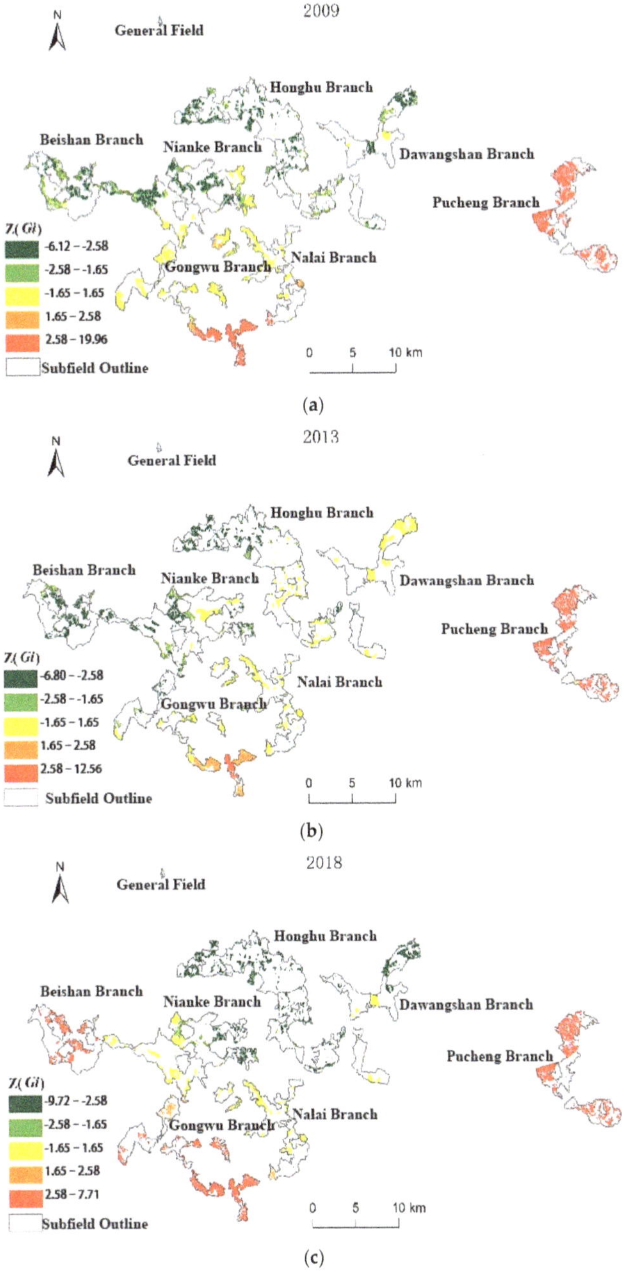

Figure 9. Distribution of cold–hot spots of water conservation in PM Plantations at Paiyangshan Forest Farm, 2009–2018. The aggregation index degree parameter Z(Gi*) was used to analyze the spatial pattern of water conservation services. Red indicates hot spot regions, orange indicates sub-hot spot regions, yellow indicates insignificant regions, dark green indicates cold spot regions, and light green indicates sub-cold spot regions. The following maps show the distribution of cold–hot spots of water conservation services in 2009 (**a**), 2013 (**b**), and 2018 (**c**), respectively.

By combining Table 2 and Figure 10, we can observe that in 2009, the number of small groups of hot–cold spots for soil conservation service value was 674 and 629, respectively. Hot spots and sub-hot spots were mainly distributed in the northwestern and central parts of Gongwu branch, the northwestern part of Nalai branch, and Pucheng branch, while cold spots and sub-cold spots were mainly distributed in the northern part of Beishan branch, the eastern and northwestern end of Nianke branch, the central and southern parts of Gongwu branch, and Honghu branch. In 2013, the number of hot and cold spot small classes of soil conservation service was 423 and 510, respectively, with the percentage of hot spots decreasing and the percentage of cold spots increasing. Compared with 2009, hot spots were mainly in the northwestern and central part of Gongwu branch and the northwestern part of Nalai branch, while their numbers increased in the central part of Nalai branch and Pucheng branch. The number of small groups of cold spots increased mainly in the eastern part of Beishan and Dawangshan branch, and decreased in Gongwu branch. In 2018, the number of hot–cold spot small classes for the value of soil conservation services was 517 and 486, respectively, with an increase in the percentage of hot spots and a decrease in the percentage of cold spots. The hot spot distribution area changed less, and decreased in Nalai branch. The cold spot distribution area was widely reduced, distributed in the eastern part of Nianke branch and Honghu branch, and other former cold spot areas became hot and cold insignificant areas. The proportion of hot spots in soil conservation services decreased and then increased, while the proportion of cold spots increased and then decreased from 2008 to 2019. The distribution pattern of cold–hot spots of the entire forest value shifted from an overall staggered pattern of cold–hot spots to high in the east and central part and low in the north.

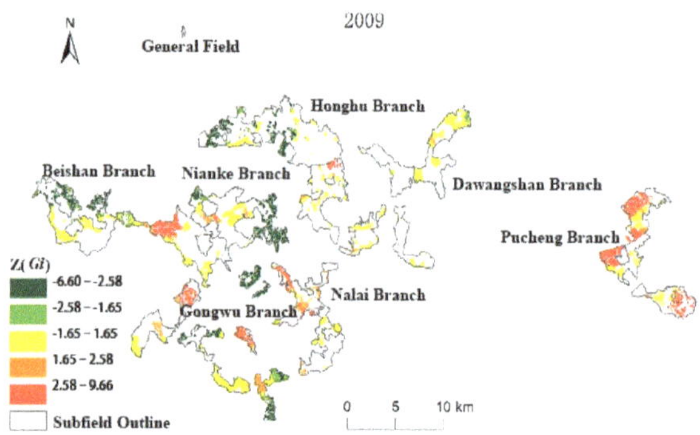

(a)

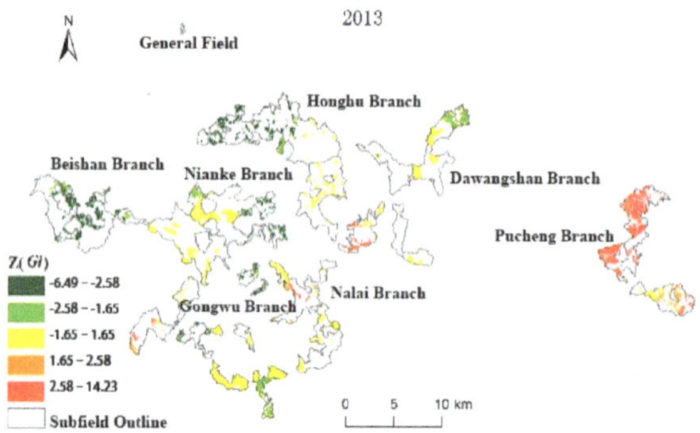

(b)

Figure 10. *Cont.*

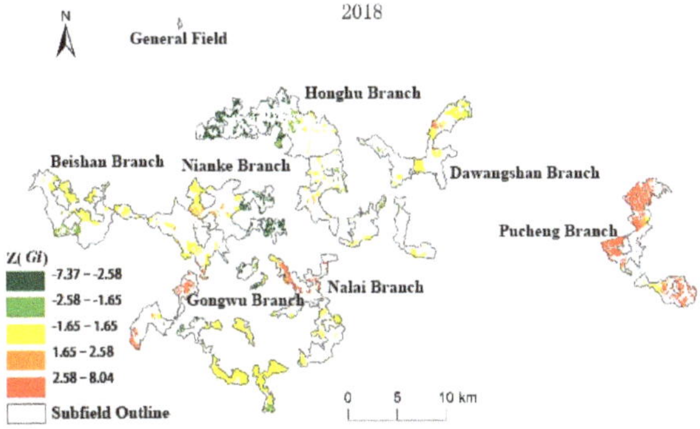

(c)

Figure 10. The distribution of cold–hot spots of soil conservation for PM plantations in Paiyangshan Forest Farm from 2009 to 2018 is presented in this figure. The aggregation index degree parameter is denoted by Z(Gi*). The color red represents hot spot regions, orange represents sub-hot spot regions, yellow indicates insignificant regions, dark green denotes cold spot regions, and light green indicates sub-cold spot regions. Panel (**a**) displays the distribution of cold–hot spots of soil conservation in 2009; panel (**b**) shows the distribution in 2013; and panel (c) depicts the distribution in 2018.

3.4. Multiple ESs Cold Hot Spot Analysis

The distribution and extent of cold–hot spots of individual ESs have been explored in detail earlier. The same service provides different amounts of value in different regions, and different regions can also provide multiple ESs. However, the regions provide different service capacities, both high and low, specifically reflected in this paper by the amount of service value per unit area [20]. Furthermore, the cold–hot spots of four ES value amounts are spatially superimposed to identify the cold–hot spots of multiple ESs in different regions.

The overall spatial distribution of the number of multiple hot spot ESs in Guangxi Paiyangshan Forest Farm over the years shows a pattern of high in the east and west and low in the middle (Figure 11). The gray small classes indicate that there are no hot spot services in the small mark. The western and northern part of Pucheng branch is the distribution area of four hot spot ES supply, while the southern part of Gongwu branch only provides one hot spot ES. Among these, the number of hot spot ESs provided by small classes of PM plantation forest in Gongwu branch is more complicated. In 2013, the number of hot spot ESs increased in Beishan branch and the southern part of Gongwu branch, while it decreased in the northwestern and eastern part of Gongwu branch. In 2018, the number of hot spot ESs decreased in the southern part of Gongwu branch and increased in Beishan branch, the central part of Gongwu branch, Honghu branch, Nalai branch, and Dawangshan branch. The largest number of small groups was able to provide three to four hot spot services from 2008 to 2019.

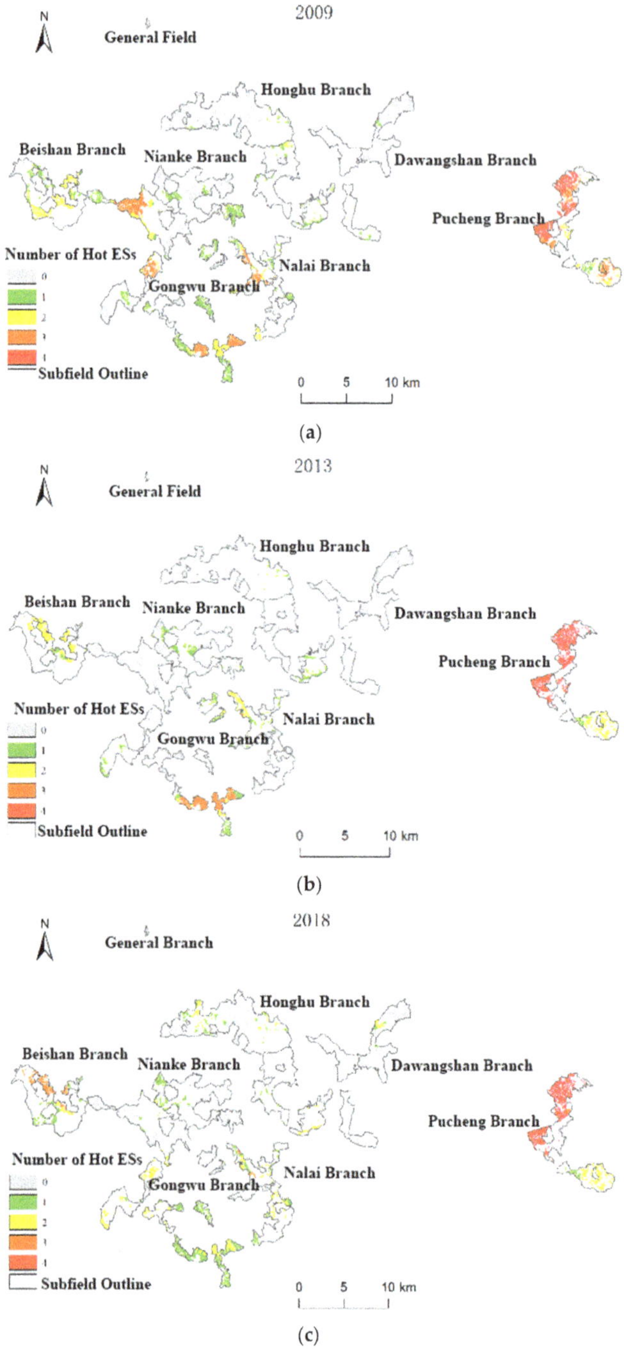

Figure 11. Number of hot spot ESs in PM plantations in Guangxi Paiyangshan Forest Farm from 2009 to 2018: (**a**) 2009; (**b**) 2013; (**c**) 2018.

4. Discussion

Most studies investigating the interconnections of various ESs have focused on administrative units such as provinces, basins, watersheds, and counties, but few have examined the spatial associations between ES pairs at the scale of forestry units. In this study, we quantified the synergistic and trade-off relationships among four ESs—wood supply, carbon sequestration and oxygen release, water conservation, and soil conservation—in Guangxi Paiyangshan Forest Farm, and explored the spatial autocorrelation distribution and hot spot distribution of ES pairs. Correlation analysis can provide a good overall perspective for qualitatively analyzing trade-off or synergistic relationships of ES pairs, but it cannot fully capture the spatial information of these relationships. Therefore, spatial autocorrelation analysis is a good complement to the analysis. Our results indicate that the relationships among ESs in PM plantations, like other forest ecosystems [34,35], exhibit a variety of synergistic and trade-off relationships, with obvious spatial heterogeneity and a predominance of synergistic relationships.

Different spatial scales can influence the interactions among ESs, and the synergies and trade-offs among them can also change [36]. By comparing the correlation coefficients among the four ES functions in PM plantations in Guangxi Paiyangshan Forest Farm, we identified the important relationships of ESs at the camping unit scale. We detected a significant positive correlation between the ES functions of the PM plantation, indicating a strong synergistic relationship between wood supply, carbon sequestration and oxygen release services, and a weak synergistic relationship between wood supply, water conservation, and soil conservation, which is consistent with the findings of Dai, E. et al. [6]. We also observed a weak synergy between water conservation services and the other three services (water conservation and soil conservation increased from weak to moderate synergy). Carbon sequestration and oxygen release had a weak synergistic relationship with water conservation, and carbon sequestration and oxygen release had a weak synergistic relationship with soil conservation services. Furthermore, we identified an important trade-off between carbon sequestration and oxygen release services and water conservation services. Based on the previously conducted ES assessment of PM plantation forests [20], it is known that the wood supply capacity, as well as carbon sequestration and oxygen release capacity, of PM plantation forests increase with the accumulation of stand volume and productivity with the increase in forest age. At the same time, soil erosion slows down under the effect of vegetation cover, and the soil conservation capacity of PM plantation forest increases. However, the water conservation capacity is mainly influenced by both rainfall and evaporation, with the increase in vegetation cover, evaporation increases and water conservation capacity decreases. Therefore, forest management should focus on the holistic theory of ESs and balance the relationship between provisioning services and regulating and supporting services according to local environmental problems to improve the overall benefits, rather than the economic benefits of a single service.

Overall, the ESs of PM plantation forests are dominated by synergistic relationships. There is a strong synergy between provisioning services and regulating services (carbon sequestration and oxygen release), a weak synergy between provisioning services and regulating services (water conservation), a weak synergy between provisioning services and supporting services, and a weak synergy between regulating services and supporting services. This indicates that the presence and growth of PM plantations contribute to the positive growth of these three services, providing a basis for multi-objective management operations in PM plantations. However, if the wood supply capacity is realized solely as timber output capacity, it will create a trade-off with the other three services. The output of timber volume will weaken the stand stock and ultimately affect the ESs of the whole stand [37,38].

There are four types of spatial relationships among ES pairs in PM plantations: "HH" synergistic relationship, "LL" synergistic relationship, "HL" trade-off relationship, and "LH" trade-off relationship. Understanding the overall synergy and trade-offs is important, but it is also necessary to understand the spatial distribution of these relationships in order to develop strategies that can be tailored to local conditions.

The analysis of ES hot and cold spots in PM plantations revealed that the relationship between ES pairs affects the differences in service provisioning capacity among small groups of plantation forests in the region. The probability of multiple service hot spots occurring in small classes with synergistic changes in service pairs is higher, and their spatial distribution is increasingly similar. This finding is consistent with the results of Tian, Y.'s study [26]. On the other hand, the hot and cold spots provided by small plantation classes showing trade-off relationships overlap more, and the likelihood of hot spots appearing in the same area is lower, resulting in greater spatial variability. Furthermore, while some small plantation classes may provide multiple hot spot services, others may only provide one or no hot spot services.

5. Conclusions

This study utilized correlation analysis, bivariate spatial autocorrelation, and hot spot analysis to synthesize the synergistic and trade-off relationships over time and space of ESs in PM plantations in Guangxi Paiyangshan Forest Farm from 2009 to 2018, and revealed the following main findings:

- There were significant correlations among the four pairs of ESs in Guangxi Paiyangshan Forest Farm. A strong synergistic relationship was observed between wood supply and carbon sequestration and oxygen release services, whereas a weak synergistic relationship existed between wood supply and both water conservation and soil conservation. Additionally, a weak synergistic relationship was observed between water conservation and the other three services. Notably, the synergistic relationship between water conservation and soil conservation improved from weak to moderate, while the synergistic relationships between carbon sequestration and oxygen release and both water conservation and soil conservation were weak. Overall, supply services had a strong synergy with regulating services (carbon sequestration and oxygen release), a weak synergy with regulating services (water conservation), a weak synergy with supporting services, and regulating services had a weak synergy with supporting services.
- There were significant spatial differences in the bivariate spatial autocorrelations among the four ES pairs, and the overall spatial trade-off and synergistic relationships among the service pairs were generally consistent with the results of the correlation analysis. Wood supply showed synergistic relationships with carbon sequestration and oxygen release, water conservation, and soil conservation, with the strongest synergistic relationship observed between wood supply and carbon sequestration and oxygen release services. The synergistic relationship between wood supply and other services weakened over time.
- The spatial distribution of cold–hot spots for each ES in Guangxi Paiyangshan Forest Farm varies. There are both similarities and significant differences in the spatial distribution of cold–hot spots of different types of ESs. The distribution of cold–hot spots for provisioning services and regulating services (carbon sequestration and oxygen release) was similar, while the distribution of cold–hot spots for carbon sequestration and oxygen release, as well as water conservation services, which are also regulating services, differed significantly.
- Once the trade-offs and synergistic relationships between service pairs in time and space are understood, the decision-making process for PM plantation management should consider the spatial heterogeneity of different ecological processes and their links to different services. It is important to recognize the interrelationships among ESs, clarify their spatial and temporal characteristics, and identify the driving mechanisms behind them. Coordination and optimization of multiple services should be prioritized, and one service should not be enhanced at the expense of others.

- This study has some limitations. For example, while analyzing the trade-offs and synergies of ESs for the entire PM plantation in Guangxi Paiyangshan Forest Farm, the study lacked research on the scale effects of trade-offs and synergies of ESs [39]. The research on the value of ESs and the factors influencing the synergy and trade-off relationship between multiple services is not deep enough, and the study only focused on the ecosystem supply, regulation, and support services while neglecting the cultural services of PM plantations. In the future, we will continue to study the synergy and trade-off relationship between multiple ecosystem services in depth.

Author Contributions: R.M. and Y.M. conceptualized the framework, acquired the funding and supervised the overall project; L.L. and Y.W. collected the data; R.M. and Y.M. analyzed the data; R.M. and Y.M. wrote the manuscript; J.M., L.L. and Y.W. provided modification comments; J.M., L.L. and Y.W. reviewed the final manuscript. All authors have read and agreed to the published version of the manuscript.

Funding: This work was supported by the National Natural Science Foundation of China (32260387), the Open Research Fund of Guangxi Key Laboratory of Superior Timber Trees Resource Cultivation (2019-B-04-01), the Guangxi Key Research and Development Program (GuikeAB21220057), and the Guangxi Innovation-Driven Development Project (GuikeAA20161002-1 and GuikeAA17204087-7).

Institutional Review Board Statement: Not applicable.

Informed Consent Statement: Not applicable.

Data Availability Statement: The data are contained within the article.

Acknowledgments: We thank Zhangqi Yang of Guangxi Forestry Science Research Institute for his guidance on experimental methods, the state-owned Paiyang Mountain Forest Farm of Guangxi Zhuang Autonomous Region for help in field sampling, and graduate students from the School of Life Sciences of Guangxi Normal University for their assistance in indoor experiments.

Conflicts of Interest: The authors declare no conflict of interest.

References

1. Zhu, Y.; Lv, J. A review of domestic forest ecosystem service valuation methods and indicators. *For. Econ.* **2015**, *37*, 74–84.
2. Ou, Z.; Sun, Y.; Deng, Z.; Feng, D. Forest ecosystem service trade-offs: Perceptions, methods and drivers. *Soil Water Conserv. Sci. China* **2020**, *18*, 150–160.
3. Cademus, R.; Escobedo, F.J.; McLaughlin, D.; Abd-Elrahman, A. Analyzing trade-offs, synergies, and drivers among timber production, carbon sequestration, and water yield in Pinus elliotii forests in southeastern USA. *Forests* **2014**, *5*, 1409–1431. [CrossRef]
4. Schwaiger, F.; Poschenrieder, W.; Biber, P.; Pretzsch, H. Ecosystem service trade-offs for adaptive forest management. *Ecosyst. Serv.* **2019**, *39*, 100993. [CrossRef]
5. Langner, A.; Irauschek, F.; Perez, S.; Pardos, M.; Zlatanov, T.; Öhman, K.; Eva-Maria, N.; Lexer, M.J. Value-based ecosystem service trade-offs in multi-objective management in European mountain forests. *Ecosyst. Serv.* **2017**, *26*, 245–257. [CrossRef]
6. Dai, E.; Wang, X.; Zhu, J.; Xi, W. Quantifying ecosystem service trade-offs for plantation forest management to benefit provisioning and regulating services. *Ecol. Evol.* **2017**, *7*, 7807–7821. [CrossRef]
7. Wu, W.; Xiang, W.; Gou, M.; Xu, C.; Ouyang, S.; Fang, X. Trade-offs and synergies of ecosystem services in three secondary forest species in Central Asia. *J. For. Environ.* **2019**, *39*, 256–264.
8. Zhao, T.; Ouyang, Z.; Zheng, H.; Wang, X.; Miao, H. Evaluation of forest ecosystem service functions and their values in China. *J. Nat. Resour.* **2004**, 480–491.
9. Jin, F.; Lu, S.; Yu, X.; Rao, L.; Niu, J.; Xie, Y.; Zhang, Z. Evaluation of forest ecosystem service functions and their values in China. *J. Appl. Ecol.* **2005**, 1531–1536.
10. Yu, X.; Lu, S.; Jin, F.; Chen, L.; Rao, L.; Lu, G. Valuation of forest ecosystem service functions in China. *J. Ecol.* **2005**, 2096–2102.
11. Wang, B.; Ren, X.; Hu, W. Assessment of forest ecosystem service functions and their values in China. *For. Sci.* **2011**, *47*, 145–153.
12. Wang, H.; Liu, J.; Jing, Q.; Cui, X. Exploration of forest resource value accounting system. *For. Econ.* **2019**, *41*, 62–68.
13. Han, C.; Chen, N.; Sun, S.; Zhao, C. Advances in research on hydrological regulation functions and mechanisms of forest ecosystems. *J. Ecol.* **2019**, *38*, 2191–2199.
14. Deng, L.; Li, M. Current status and outlook of Horsetail pine plantation forest research. *Anhui Agric. Sci.* **2009**, *37*, 2968–2971.
15. Qin, X.; Ding, G. Characteristics of soil organic carbon and its relationship with nutrients in Sargassum pine plantation forests of different ages. *Zhejiang For. Sci. Technol.* **2012**, *32*, 12–17.

16. Qin, Q.; Tang, J.; Deng, X.; Song, X.; Qin, Z. Research on soil fertility evaluation of Horsetail pine plantation forest in Guangxi. *For. Surv. Plan.* **2017**, *42*, 16–21, 32.
17. An, N.; Ding, G.; Chen, H.; Nong, Z.; Huang, B. Selection of high-yielding tallow trees and cultivation of high-yielding tallow stands in Horsetail pine. *Guizhou Agric. Sci.* **2015**, *43*, 118–122.
18. Xie, W.; Ye, S.; Yang, M.; Zhao, L. Biomass and distribution pattern of Sargassum pine plantation forest in southeastern Guizhou hilly area. *J. Beihua Univ. (Nat. Sci. Ed.)* **2009**, *10*, 68–71.
19. Lei, L.; Xiao, W.; Zeng, L.; Huang, Z.; Gao, S.; Tan, B. Effects of different forestry measures on soil respiration in Sargassum pine forests. *For. Sci. Res.* **2015**, *28*, 713–719.
20. Mo, R.; Wang, Y.; Dong, S.; Ma, J.; Mo, Y. Ecosystem Service Evaluation and Multi-objective Management of *Pinus massoniana* Lamb. Plantations in Guangxi, China. *Forests* **2023**, *14*, 213. [CrossRef]
21. Dai, E.; Wang, X.; Zhu, J.; Gao, J. Progress and trends in ecosystem service trade-off/synergy research. *Adv. Earth Sci.* **2015**, *30*, 1250–1259.
22. Yang, W. Research methods for trade-offs of forest ecosystem services. *Green Technol.* **2018**, 14–19.
23. Han, Y. *Study on the Trade-Offs and Synergistic Relationships of Agro-Ecosystem Services and Their Driving Forces in Xi'an Metropolitan Area*; Shaanxi Normal University: Xi'an, China, 2016.
24. Sun, Y.; Ren, Z.; Hao, M.; Duan, Y. Spatial and temporal variation of ecosystem service trade-offs and synergies in the Loess Plateau and the factors influencing them: The case of Yan'an City. *J. Ecol.* **2019**, *39*, 3443–3454.
25. Alamgir, M.; Turton, S.M.; Macgregor, C.J.; Pert, P.L. Ecosystem services capacity across heterogeneous forest types: Understanding the interactions and suggesting pathways for sustaining multiple ecosystem services. *Sci. Total Environ.* **2016**, *566*, 584–595. [CrossRef] [PubMed]
26. Tian, Y. *Synergistic and Trade-Off Relationships and Management Strategies for Ecosystem Services in Cropland*; Zhejiang University: Hangzhou, China, 2018.
27. Peng, J.; Liu, Z.; Liu, Y.; Chen, X.; Zhao, H. Evaluation of multifunctionality of arable landscapes in counties of Beijing, Tianjin and Hebei regions. *J. Ecol.* **2016**, *36*, 2274–2285.
28. Wang, X.; Ma, X.; Feng, X.; Zhou, C.; Fu, B. Spatial and temporal characteristics of ecosystem service trade-offs and synergistic relationships in key fragile ecoregions. *J. Ecol.* **2019**, *39*, 7344–7355.
29. Jopke, C.; Kreyling, J.; Maes, J.; Koellner, T. Interactions among ecosystem services across Europe: Bagplots and cumulative correlation coefficients reveal synergies, trade-offs, and regional patterns. *Ecol. Indic.* **2015**, *49*, 46–52. [CrossRef]
30. Wang, C.; Liu, C.; Wu, Y.; Liu, Y. Spatial patterns of ecosystem services and trade-offs and synergistic relationships in loess hilly areas—A case study of Yuzhong County. *J. Ecol.* **2019**, *38*, 521–531.
31. Song, J.; Wen, L.; Wang, F.; Li, K.; Wu, C.; Zhang, H.; Zhang, X. Spatial and temporal dynamics of ecosystem service values in the Ulaanbaatar Desert. *J. Ecol.* **2021**, *41*, 2201–2211.
32. Zhang, S.; Zhang, K. Study of Moran index and G-factor of spatial autocorrelation local indicators. *Geod. Geodyn.* **2007**, 31–34.
33. Wang, B.; Zhao, J.; Hu, X. Spatial pattern analysis of ecosystem services in the Heihe River Basin based on the InVEST model. *J. Ecol.* **2016**, *35*, 2783–2792.
34. Lan, J.; Lei, X.; Zhang, Y. Multifunctional trade-offs and synergistic relationships in snowy spruce forests of the central Tianshan Mountains. *For. Sci.* **2019**, *55*, 9–18.
35. Van der Plas, F.; Ratcliffe, S.; Ruiz-Benito, P.; Scherer-Lorenzen, M.; Verheyen, K.; Wirth, C.; Zavala, M.A.; Ampoorter, E.; Baeten, L.; Barbaro, L.; et al. Continental mapping of forest ecosystem functions reveals a high but unrealised potential for forest multifunctionality. *Ecol. Lett.* **2018**, *21*, 31–42. [CrossRef]
36. Burkhard, B.; Kroll, F.; Nedkov, S.; Müller, F. Mapping ecosystem service supply, demand and budgets. *Ecol. Indic.* **2012**, *21*, 17–29. [CrossRef]
37. Zhu, J.; Dai, E.; Zheng, D.; Wang, X. Characteristics of trade-offs between wood production and carbon sequestration functions in plantation forests under the influence of logging: An example from the Huitong Forest Ecology Experiment Station, Hunan. *J. Geogr.* **2018**, *73*, 152–163.
38. Rong, J.; Lei, X.; Zhang, H.; Feng, Q. Forest management planning with carbon storage and timber production objectives in mind. *J. Northwest For. Acad.* **2012**, *27*, 155–162.
39. Zhang, L.G.; Bao, B.F.; Dong, L. Spatial and temporal pattern evolution of grain yield and spatial heterogeneity of driving factors in Poyang Lake eco economic zone. *Econ. Geogr.* **2018**, *38*, 154–161.

forests

MDPI

Article

Effects of Forestry Transformation on the Ecosystem Level of Biodiversity in Poland's Forests

Ewa Referowska-Chodak [1,*] and Bożena Kornatowska [2]

1 Institute of Forest Sciences, Warsaw University of Life Sciences, Nowoursynowska 159, 02-776 Warszawa, Poland
2 Institute of Environmental Protection—National Research Institute, Słowicza 32, 02-170 Warszawa, Poland; bozena.kornatowska@ios.edu.pl
* Correspondence: ewa_referowska_chodak@sggw.edu.pl

Abstract: This paper presents the results of an analysis of the effects of Poland's forest management evolution over the last 75 years on forest biodiversity at the ecosystem level. Forest biodiversity changes in the two politically and economically different eras (socialism and democracy) are interpreted based on four indicators used in assessments of forest stands (naturalness; habitat diversity; forest management system; forest stand age structure). In the era of socialism (1945–1989), there were dynamic increases in the area of semi-natural forests as well as in the proportion of the most fertile habitats, whilst the proportion of the poorest habitats decreased quite dynamically. Then, the clearcutting management system was regularly implemented, with adverse impacts on forest spatial structure diversity. The proportion of old/mature tree stands and the stand average age increased at relatively slow rates. In the era of democracy (1990–2020), there were comparatively more dynamic increases observed in the area of forests undisturbed by man, as well as in the proportions of mixed broadleaved and wetland forest habitats. At the same time, the proportion of old/mature stands and stand average age kept increasing at relatively fast rates. The area of forests managed with the use of the shelterwood system increased and the area of forest plantations substantially decreased. On the other hand, irrespective of the era under study, there occurred a noticeable not-so-favourable decreasing trend in the proportion of the youngest forest stands. All in all, during the analysed period of more than seven decades, the evolution of forest management practice implemented in Poland's forests by State Forests National Forest Holding led to the restoration of/an increase in biodiversity at the ecosystem level. Yet, there have remained unsolved issues, as regards the following aspects: organisational (the assurance of further reconstruction of forest stands, and the restoration of water profiles), political (a lack of up-to-date national forest policy), and financial (the costs of protecting/restoring biodiversity vs. State Forests' self-financing), as well as conceptual (old-growth stands in managed forests, and controversy over clearcutting) and natural/anthropogenic (climate change, and the eutrophication of forest habitats) issues. The solutions may require measures outside the limits of Poland's forestry, if not far beyond national borders.

Keywords: forest spatial structure; habitat diversity; management practices; naturalness; SFM indicator; stand age structure

Citation: Referowska-Chodak, E.; Kornatowska, B. Effects of Forestry Transformation on the Ecosystem Level of Biodiversity in Poland's Forests. *Forests* **2023**, *14*, 1739. https://doi.org/10.3390/f14091739

Academic Editors: Chao Wang, Fan Zhang and Wei Liu

Received: 29 June 2023
Revised: 22 August 2023
Accepted: 25 August 2023
Published: 28 August 2023

1. Introduction

Forest management, dependent upon historical, political, economic and social conditions, determines the shape, sustainability and biodiversity of forest ecosystems [1]. Thus, changes in the goals and principles of forest management result in alterations of forest ecosystems and in their ability to provide services related to biodiversity and its conservation [2,3]. The examination the aforesaid relations at the large spatial and temporal scales is important and informative. In this study, we examined the relationship between changing forest management and the state of forest ecosystems using the example of Poland's forests

(Central Europe), taking into consideration the period from the end of World War II (1945) to the present.

In order to obtain a better understanding of the changes that occurred during the analysed period, it is necessary to go slightly back in time. Of many phases of forest use in Poland, the most significant one for the diversity of forest ecosystems were those which began about 250 years ago. Starting from 1772, Poland was split between the bordering empires (Russia, Prussia and Austria) for over 120 years. At that time, the country's environment was intensively exploited, and in the 19th century, forest coniferous tree species were propagated; this also occurred all over Europe [4–9]. Modelled on agriculture, single-species, single-generation and single-layer spatially ordered forest stands were cultivated, leading to the elimination of the ecosystem and spatial diversity of forests almost completely [10,11].

After Poland regained its independence in 1918, works on a modern model of multifunctional forestry began to emerge. Established at that time (1924) was the Polish State Forests Enterprise who has to date continued functioning, and is now known as State Forests National Forest Holding (the State Forests or the SF, hereafter also referring to Poland's forests managed by this enterprise). Currently, the SF is the EU's largest specialised entity managing national forests [12]. In 1939, the outbreak of World War II abruptly terminated the initial stage of Poland's forestry development and brought on the destruction of national forests due to warfare, military battles and over-exploitation by the occupying forces until 1945 [13,14]. Nevertheless, subsequent generations of Polish foresters have continued to pursue the initial course and to build on the achievements gained before the war, although the extent of the use of former knowledge has changed over time.

On account of historical changes in Poland in the last decades, we divided the period under study (1945–2020) into two stages, which are quite different in terms of Poland's political and economic conditions, i.e., the era of socialism (1945–1989) and the era of democracy (since 1990). The two eras have made their important marks on the shape of forest management in Poland.

Extensive nationalisation and industrialisation were characteristic of Poland under socialist conditions [15,16], resulting in severe environmental pollution, which was every so often followed by the extinction of natural habitats, and which was also recorded in the case of forests [17,18]. The economy was centrally planned and regulated, devoid of market mechanisms, and national forestry was implemented in line with a resource-based economic model [4,7]. Forest management was chiefly focused on industrial wood production and the achievement of commercial goals [19], whereas ecological principles and forest sustainability were of secondary importance [7]. Nonetheless, the managers of state-owned forests undertook a number of measures so as to reduce ongoing biodiversity loss one way or another.

Appropriate measures had a chance to be implemented on a considerable scale only after the transformation of Poland's political (and economic) system into a democratic one which brought about the free market (starting from 1989). This led to the implementation of the concept of sustainable forest management (SFM), with a strong emphasis being put on the protection and restoration of forest ecosystems [20–23], including ecosystem-level biodiversity. A very important factor for the maintenance direction of reforms was to continue the state ownership of forests under the conditions of the transformed economic system [3]. In the meantime, Poland joined several international processes on the protection of, inter alia, forest biodiversity (e.g., the Convention on Biological Diversity, and the Ministerial Conference on the Protection of Forests in Europe—now FOREST EUROPE), and started implementing jointly developed recommendations/guidelines on its own forest industry and management practice.

Two milestone actions undertaken during the era of democracy have boosted the protection of biodiversity in Poland's forests (especially those managed by the SF), i.e., the accession to the European Union (2004) and the implementation of the Birds and Habitats

Directives [24]. The ratification of the Aarhus Convention (2003) [25] resulted in greater-than-before public participation in shaping the rules and ways of forest management, especially in terms of forest nature conservation.

The aforementioned historical, political, economic and social changes (together with the increased awareness of environmental issues) during the period under study have considerably influenced the state of forest biodiversity in Poland. So far, these aspects have been studied based on selected indicators, in relation to biodiversity at the landscape level. The results of Referowska-Chodak and Kornatowska [26] show that in the era of socialism, there were prompt increases in total forest cover, wood resources (total growing stock) and the total area of protective forests essential for biodiversity conservation. In the era of democracy, average growing stock density increased intensively and forest management has put a greater emphasis on reducing forest fragmentation and clearcut logging. Likewise, there was an average increase observed in protected forests' areas in both eras under study. At the crossing point, the area of protected forest in Poland increased most vividly. The changes throughout the study period were considered positive for the forest landscape. Yet, numerous problems and challenges in this have still remained. These include, inter alia, unnecessary beneficial alterations in the ownership structure of Poland's forests; outdated national forest policies; insufficient funding for nature conservation; uneven distribution and ongoing forest fragmentation across the country; climate change impacts (e.g., extreme weather events) [26]. For a more complete picture of the consequences of the evolution of forest management in Poland, there is a need for further comprehensive analyses as regards other than landscape biodiversity levels, including but not limited to the ecosystem biodiversity level.

Considering the above, the main aim of the present study was to evaluate the effects of forestry evolution under Poland's conditions on the state of forest biodiversity at the ecosystem level, in reference to selected indicators. In the perspective of 75 years (1945–2020), the specific objectives were distinguished: (1) to determine the direction and dynamics of changes in forest ecosystems, (2) to identify drivers of the observed changes and their effects, and (3) to identify threats to forest ecosystems and the direction of further actions for the benefit of forest biodiversity at the ecosystem level. The study period comprised the two political eras in Poland (socialism and democracy) with differently formulated forest management objectives. In the era of socialism, forest management was focused on industrial wood production and the achievement of commercial goals whereas in the era of democracy, the emphasis has been put on the implementation of the concept of sustainable forest management (SFM) and the restoration of forest ecosystems. The regulations and guidelines for forest management developed in the eras under study are quite different but in both cases have directly influenced the state and structure of forest ecosystems.

The results of a comprehensive ana"ysis'of relationships between forest management and forest biodiversity at the ecosystem level, carried out in consideration of conceivable conflicts, as well as the long-term perspective and the large spatial scale, can be useful for enhancing sustainable forest management, for example in the regions with historical conditions or forest structures analogous to those in Poland. On a broader scale, such studies can serve as forms of documentation of changes occurring under the impact of human activity on the surrounding nature.

2. Materials and Methods

2.1. Indicators

The indicators used to measure the state of forest biodiversity, such as the set of criteria and indicators (C&I for SFM) developed by FOREST EUROPE [27] often reflect the condition of more than one level of biodiversity. In order to assess forestry evolution effects on different levels of forest biodiversity in Poland's forests (State Forests), we used the criteria/indicators that reflect the studied aspects to the greatest extent.

In the present study, the widely used biodiversity indicators, such as species richness and abundance as well those related to alpha, beta, gamma diversity, were not used as they merely refer to a species level. Our aim was to focus on analysing complex systems composed of various factors and forest management undertakings that shape forest ecosystems—not only in terms of species characteristics, but also spatial and temporal diversity. Therefore, in regard to the ecosystem level, the following proved useful: selected C&I for SFM [27]; selected factors listed by Keller [28] as determinants of biodiversity related to forest ecosystem stability; selected indicators used by Mederski et al. [7] to analyse the effect of "ecological" forestry policy on the landscape change.

The influence of the evolution of Polish forestry on forest biodiversity at the ecosystem level was therefore studied based on the following four indicators:

1. Naturalness—C&I for SFM: Indicator 4.3, *naturalness* [27]. The indicator refers to a degree of alteration of a certain ecosystem classified in the following classes [3]: forests undisturbed by man ("*the natural forest development cycle persists or was restored and show characteristics of natural tree species composition, natural age structure, deadwood component and natural regeneration and no visible signs of human activity*"); semi-natural forests ("*neither undisturbed by man nor plantations but displaying some characteristics of natural ecosystems*"); plantations ("*usually representing ecosystems on their own, established artificially by planting or seeding, often with introduced tree species, and intensively managed*"; "*completely distinct from the original ecosystem*"). In this study, consistent with the methodology adopted in Poland, forests undisturbed by man were considered forests under strict protection in forest nature reserves (area per ha). Forests protected in national parks were not analysed, as institutions other than State Forests are responsible for their management [26]. In the case of forest plantations, we took into account those established for increased timber production in shorter production cycles (ha). Seed plantations (focused on the conservation of gene resources) and Christmas tree plantations (usually established under power lines) were not included in the analysis. The area of semi-natural forests (ha) was calculated as the difference between the total area of forests administered by the SF over the study years [26] and the sum of the areas of other forest categories examined, i.e., forests undisturbed by man and plantations;
2. Habitat diversity: *diversity of habitat conditions* is, as stated by Keller [28], one of the factors that influence forest ecosystem functioning. In the present study, we focused on 4 basic types of forest habitats in Poland (the term considered to represent the physical conditions of forest sites). The habitats under this study were distinguished by forest site fertility from the poorest to the most fertile, i.e., those of (1) coniferous forests, (2) mixed coniferous forests, (3) mixed broadleaved forests and (4) broadleaved forests. Based on available data, we calculated the respective habitat area proportions (%) in each of the study years. In further descriptions, we referred to the habitat moisture gradient through the analysis of changes in the proportion (%) of the wettest habitats in the forests under study;
3. Forest management system: indicator *methods of final felling* [7]. This indicator refers to general principles of forest use and regeneration, as well as to the implementation of specific activities in time and space so as to ensure that the intended production goal is achieved [11]. The management system directly influences forest ecosystem characteristics and *spatial structure*, which, in accordance with Keller [28], is a factor with significant effects on ecosystem stability. In this study, depending on data availability, there are presented figures on forest areas (ha) covered by specific treatments under each analysed management system in a given study year or the total forest area (ha) planned for the application of a given management mode in the long-term perspective;
4. Forest stand age structure: *the age structure of stands*, consistent with Keller [28], influences the formation of specific conditions for an ecosystem's development and its richness, and thus has effects on long-term ecosystem stability. This is also an indicator for SFM (C&I, Indicator 1.3, *age structure and/or diameter distribution* [27]), originally considered in terms of economic aspects and the contribution of forest resources to global carbon cycles [3,27].

2.2. Scope of Analyses

It was assumed that the analyses would include forest areas managed by the State Forests National Forest Holding (SF), and references to all forests in Poland would be made only if information regarding the SF was not available or the analysed data were relevant for the interpretation of the obtained results. This approach was based on the following: (1) almost 77% of the area of all Polish forests is currently under management of the SF [29]; (2) a standardised forest management practice has been implemented in all the forests administered by the SF [12]—this confirms the consistency of the presented results and conclusions drawn; (3) more and more detailed and reliable data (especially historical) concerning the SF are available compared to those on other forests in Poland; (4) State Forests is the entity managing state land, and therefore, it is principally influenced by pressures exerted by the government and decisions on land management policy.

The time scale of the analysis was the period of 1945–2020 (75 years). The influence of the evolution of Poland's forestry on forest biodiversity at the ecosystem level is presented based on data pertaining to 10-year intervals (Figure 1).

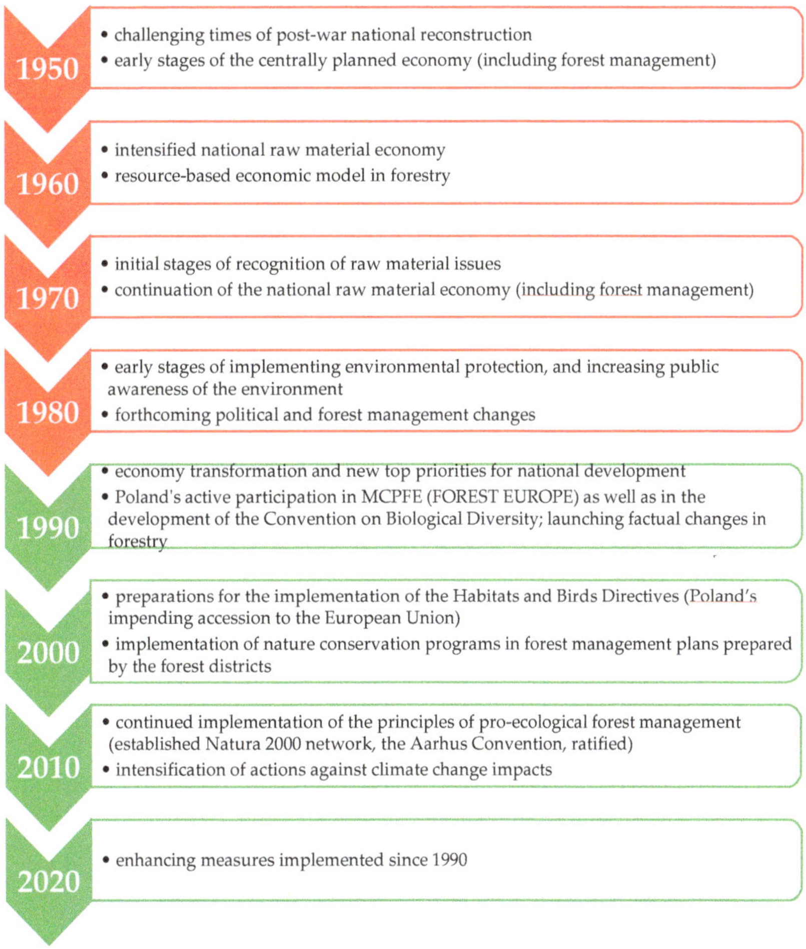

Figure 1. Timeline of developments related to Poland's forestry in 10-year intervals of the period under study (the era of socialism—red; the era of democracy—green).

The year 1990 was assumed as the breakthrough year (the sanctioned start of forest management transformation). The data used for the analyses represented the status at the end of the year given in the tables/figures. In the case of data being unavailable for a given study year, if possible, data from contiguous years were used. Relevant information was then indicated in the footnotes to the tables/figures. Due to the different reporting mode (a marketing year covering the fourth quarter of the previous year and the first three quarters of the following year), the results for the area under different types of forest management systems in the year 1960 was calculated as the sum of $\frac{3}{4}$ of the parameter value for the 1959/1960 marketing year and $\frac{1}{4}$ of the value for 1960/1961.

The substantive scope of the study included the following: (1) evaluating information/data compiled in terms of the characteristics of the analysed feature, taking into account differentiation over time, as well as the direction and dynamics of changes; (2) providing a comprehensive analytical commentary; (3) identifying threats to forest ecosystems and indicating the directions of beneficial actions.

2.3. Sources and Analysis of Information

Once the indicators were selected and the scope of the study was defined, pertinent data for the study years were compiled. The presented numerical data came from reports published by State Forests National Forest Holding (Poland) that included financial/economic information or that regarding the state of forests [30–33]; the Forest Data Bank [34]; statistical yearbooks on forestry and environmental protection published mainly by Statistics Poland (GUS); relevant available monographs and articles. Due to the lack of pertinent studies/reports for some study years (especially, for those at the beginning of the study period), several statistics were not available; nevertheless, the trend and dynamics of changes could still be revealed. For each analysed indicator, the aforesaid trend and dynamics of changes in indicator elements were determined both for the entire study period and the periods of the political eras under study. For the purpose of the discussion of the results, including the description of factors that influence changes in the analysed parameters, we used the results of several articles from the Scopus (Elsevier) database (keyword: Polish forestry), as well as those found using a snowballing approach.

3. Results and Discussion

3.1. Naturalness

The classification of forests according to the degree of their naturalness illustrates the distance between the current and potential natural status of a particular forest and reflects the history and intensity of human interventions in forest ecosystems [3]. In Poland, forests with preserved primeval features (thus with the minimum aforesaid distance) appear very sparsely [11], mainly in the case of outside forests managed by the SF. This is mainly due to forest management being carried out for centuries, which was particularly exaggerated during the period when Poland's territory was split between Prussia, the Habsburg monarchy, and Russia (1795–1918). As a result, there is now a dominating presence of forests with a relatively greater distance between the current and potential natural status, and this distance depends upon local conditions and history; therefore, it is differentiated across the country.

At the beginning of the period under study—just after World War II ended (1945)—there existed no structured network of forest (or other) protected areas [26]. At that time, fast-growing tree plantations were not implemented [35,36]. Therefore, the entire forest area under the administration of State Forests (5408 thousand ha [26]) can be classified as semi-natural forests. In the course of time, forests began to diversify in some measures (Table 1). The values presented in the table below are the result of a long search for the most reliable data or of data interpolation as historical information (especially as regards the era of socialism) were often unavailable or divergent.

Table 1. Different degrees of naturalness in forests under the management of State Forests *.

Year	1950	1960	1970	1980	1990	2000	2010	2020
Forests undisturbed by man (thousand ha)	0.02	<5.0 [1]	<5.9 [2]	<8.7 [2]	1.0	0.9	1.6	10.0
Semi-natural forests (thousand ha)	<5740	6125	6493	~6700	6787	6943	7066	7108
Plantations (thousand ha)	n.a. (>0)	10.0	8.6 [3]	n.a. (>0)	17.0 [4]	8.9 [5]	4.1 [6]	2.5

* sources: [26,35–44]; [1] area of all forest reserves, [2] area of all strict reserves, [3] based on data for 1965 and 1972, [4] based on data for 1987 and 1989, [5] based on data for 1998 and 2003, and [6] data for 2009; n.a.—data not available.

The dynamics of changes in the area of forests of different naturalness degrees. During the period under study, the trends of changes in the area of the examined forests varied depending on the forest category. There was an increase in the area of forests undisturbed by man (which increased in the era of socialism by 1 thousand ha, and by 9 thousand ha in the era of democracy). The area of semi-natural forests also increased (by 1379 thousand ha in the era of socialism, and by 321 thousand ha in the era of democracy). In the case of plantations, after a period of increase in their area (an increase of 17 thousand ha under socialism), a decrease was observed (a decrease of 14.5 thousand ha under democracy). The dynamics of the aforesaid changes varied. The increase in the area of semi-natural forests was more dynamic in the era of socialism (on average 30.6 thousand ha/year) when compared to that in the era of democracy (on average over 10.7 thousand ha/year). This was contrary to the changes in the area of forests undisturbed by man; in the era of socialism the increase rate was on average 22 ha/year, whereas in the era of democracy, apart from an initial downward trend, it was 300 ha/year. The area of plantations was quite variable In the era of socialism (depending on the projects implemented and their success), whereas in the era of democracy, the area of newly established plantations was smaller compared to that of already existing plantations converted into forests with the use of earlier management/reforestation practices.

Forests undisturbed by man. Currently, at any rate, practically every forest in Poland is under indirect pressures, such as industrial emissions [11]. Therefore, it is conceivable to refer only to forests with ceased direct pressures. Polish forests strictly protected under the Nature Conservation Act of 16 April 2004 [45] can be categorised as "forests undisturbed by man" and assessed by means of Indicator 4.3, naturalness [27]. In forests managed by the SF, the area of "undisturbed" forests increased bit by bit (Table 1) and in 2020, it was 0.14% of the total area of State Forests. The low share of strictly protected forest ecosystems was emphasised as an important problem by the Polish State Council for Nature Conservation as early as 2007, and later in 2016 [46,47]. The Council pointed out, inter alia, that undisturbed forest sites can serve as reference plots for improving forest management practice [46]. They can be used to determine the natural stand type characteristic of a region under given habitat conditions [11], as well as to observe and better understand ecological processes [3]. Although the area under strict protection in forest nature reserves designated in the SF has been increasing since 2007 and then since 2016 (Table 1), the area of undisturbed protected forest ecosystems could be considerably enhanced. The value of undisturbed forests is greater the larger and more consistent their area is, as this allows natural ecosystem dynamics to occur [3]. On the scale of European countries, a particularly high proportion of forests undisturbed by man occur in Georgia (17.7%), Bulgaria (18.1%), and Liechtenstein (22.4%). However, the average for Europe as a whole is 2.2%, and in Northern Europe region it reaches a maximum of 3.9% [3].

Semi-natural forests. The majority of Poland's forests were strongly changed due to the long-lasting implementation of the resource-based economic model of forest management [48,49], in the time when the country was partitioned (1795–1918) and later, in the era of socialism (1945–1989). Focusing on timber production turned forests into unstable formations with seriously diminished species, ages and spatial structures. Yet, the forest ecosystem is a living, dynamic natural entity, with great potential for regeneration. Nowadays, the degree of naturalness could be improved if forests were not affected by human interference [11]—as referred to in the paragraph above—or through silvicultural practices.

Already in the early 1980s, the Polish scientific community highlighted a need for the implementation of a "close-to-nature silviculture model", allowing for restoring complexity and increasing the stability of forest ecosystems [48]. However, this concept had a chance to gradually materialise only after the political, economic and social transformation which took place in the late 1980s and early 1990s. Hence, the results presented in Table 1 show data on forests categorised as semi-natural in the era of socialism whose features were considerably less natural when compared to those in the era of democracy (and especially, when compared to present forests). Even so, the classification adopted in this study is fit for its purpose, because forest plantations—if not intensively managed for an extensive period of time—can be considered semi-natural forests [3]. Notwithstanding the resource management approach, "regular" Polish forests were not plantations, and even during the era of socialism, there was harvested no more wood than its annual increase [26].

Plantations. The concept of fast-growing tree plantations flourished in State Forests during the era of socialism. Poplar (*Populus* sp.) plantations were established in the 1950s, with the goal of establishing 50 thousand ha by the year 2000. Over time, the cost of growing poplar plantations on forest lands turned out to be too high, so the project was discontinued in the 1980s, as was the case with failed willow (*Salix* sp.) plantations in the early 1970s. Then, 100,000 ha of spruce (*Picea* sp.), larch (*Larix* sp.), birch (*Betula* sp.) and Douglas-fir (*Pseudotsuga menziesii*) plantations was to be established in 1970–1990, of which only a little over 4000 hectares was realised. In the first decade of the 21st century, there were efforts undertaken to establish plantations of tree species such as bird cherry (*Prunus avium*) and black locust (*Robinia pseudoacacia*) but since 2010, projects of this kind have been given up by State Forests [36]. Even though the current Forest Management Instruction [50] still specifies "fast-growing tree plantations", current plantations (especially in the case of poplar) are usually established on agricultural lands with poor soils, so as to secure and improve subsoil quality and provide income for private land owners [42,51], e.g., from wood sale to paper factories. In view of the above, as far as Poland's forests (State Forests) are concerned, the present role of plantations is quite minor, comparable to that across Europe (3.8% on average [3]), and significantly lesser than, for example, that in the United Kingdom, Ireland and Belgium, where plantations dominate in the total forest area (in proportions of 89.2%, 86.2% and 68.2%, respectively [3]). However, due to the growing demand for timber, plantations of fast-growing trees on agricultural lands (outside of the SF) may be necessary [42].

Naturalness and biodiversity. It is interesting to note that in regard to forest ecosystems, naturalness and biodiversity are not always correlated, as a forest habitat located in environments affected by strong limiting factors can have a considerably high level of naturalness, even if its biodiversity is not high [52,53]. For instance, in Poland, the process of ageing of unmanaged Carpathian beech forests has been accompanied by a decrease in the diversity of the tree and herb layers [54]. Based on data from permanent plots (established in 1936), Brzeziecki [55–57] reports considerable changes in tree species composition and abundance, as well as the low recruitment rates of formerly dominant species, in spontaneously developing tree stands strictly protected for about 100 years in the Białowieża Forest (a unique temperate forest in eastern Poland). There are also observed cases when high biodiversity is observable in less natural forests. For example, the outstanding species composition and physiognomy of Euro-Siberian steppic woods with *Quercus* spp. (Natura 2000 priority habitat 91I0) are attributable to, inter alia, human-dependent cattle grazing [58].

3.2. Habitat Diversity

The potential natural vegetation in Poland was classified as temperate lowland forest (predominant) and mountain forest [59]. Presumably, lowland forests in fertile habitats would be dominated by oak-hornbeam communities [60]. However, this pattern was disrupted as a result of centuries of a gradual takeover of fertile soils for agriculture, leaving forests mainly on poorer soils [61,62]. This essentially translated into then existing structure of forest habitat types. On the other hand, it is important to be aware that forest management per se also affects habitat quality, by changing basic parameters—fertility, water content and pH [1,63]. The planting of Scots pine (*Pinus sylvestris*) stands on fertile soils, which was carried out in Poland in the 1800s, and later, in the era of socialism, had adverse effects on soil fertility, enzymatic activity and physicochemical properties [64,65]. The evolution of the structure of habitats in State Forests in terms of their fertility during the study period is shown in Figure 2.

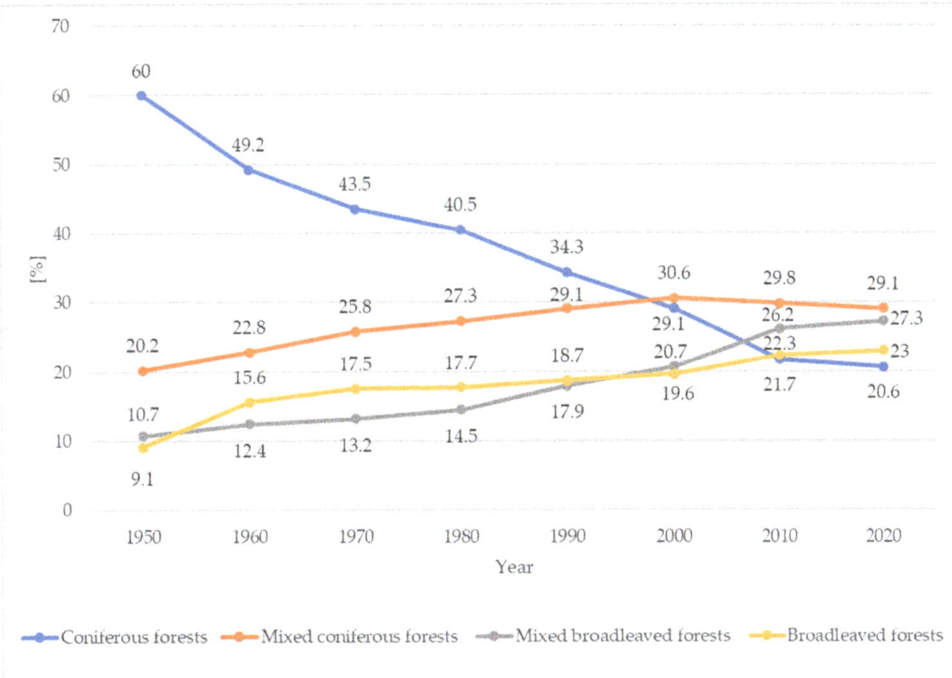

Figure 2. Proportions (%) of the main forest habitat types (distinguished by site fertility) in State Forests. Sources of information: [17,30–33,66–68]. Notes: 1950—data for 1953; 1960—data for 1961; 1970—data for 1968; 1980—data for 1978.

The dynamics of changes in the share of forest habitats of different fertility levels. Over the course of the analysed 75 years (in particular, the 67 years for which data are presented in Figure 2), the structure of forest habitat types underwent a significant transformation: the share of the least fertile habitats decreased three-fold, and the proportion of the most fertile and fertile ones increased two-fold. In the case of the poorest forest habitats (coniferous forests), the decrease rate was higher in the era of socialism (until the year 1990, the average annual decrease was 0.69% of total forest area) when compared to that in the era of democracy (average annual decrease: 0.46% of total forest area). In 2020, the share of mixed coniferous forests in the total forest area was 144% of the value recorded in 1953. A stable increase in the share of mixed coniferous forests was recorded in the era of socialism (average annual increase: 0.24%), whereas in the era of democracy, after an initial increase, the share of coniferous forests decreased to the level recorded in the beginning of this era. More remarkable increases in forest habitat shares were observed in the case of mixed broadleaved forests and broadleaved forests—in 2020, their shares were 255% and 252% of the initial values (1953), respectively. However, a considerable difference can be noted between the eras under study—the increase in the share of mixed broadleaved forests was more dramatic in the era of democracy when compared to that in the era of socialism (an annual increase of 0.31% and 0.19% in total forest area, respectively). This was not the case of broadleaved forest habitats; their share increased annually by 0.26% in the era of socialism and only by 0.14% during the era of democracy.

Causes of changes in habitat fertility. We identified four causes of changes in forest habitat fertility. Firstly, the changes may be due to habitat eutrophication resulting from the deposition of airborne nitrogen compounds [62], which under Poland's conditions is one of the highest in Europe [3]. Even though the total nitrogen content in forest soils in Poland is not high as that in Europe, it is gradually increasing—only in 2009/2012–2015 it increased by 0.1 g/kg, the same value of increase as that in the pH of these soils (by 0.1) [3]. It should be noted that emissions of nitrogen and sulphur are a potential threat to Polish forest ecosystems as the permissible levels of acid depositions are often exceeded [69]. On the other hand, it should be emphasised that the magnitude of air pollution is a factor that is decisively independent of forest management, and depends, inter alia, on national policy in regard to the environment and the country's development.

Secondly, changes in habitat structure within forests managed by the SF may be a result of forest land reclamation, which was carried out since 1955 at a rate of 20–75 thousand ha/year [37,40,70], with a much higher rate in the years 1970–1980 (>100–200 thousand ha/year [68]). As part of forestry undertakings, especially during the era of socialism, mineral fertilisation was applied to forest soils, which directly affected forest resources and related ecosystem services [2]. Over time, however, this form of interference in forest habitats was scaled down, and in the last decades (democracy), fertilisation was applied to soils only in very small areas—for example, in 2010 it was scaled down to 41 hectares of forests under SF management and in 2020, it was scaled down to 15 hectares [41,70]. In addition to fertilisation, certain tree species were planted to enhance habitat fertility [11,64]. The influence of stand species and spatial composition on soil trophic properties was also confirmed by Augusto et al. [71]. An analogous effect was—and still is—achieved via the conversion of stands from conifer monocultures established in too-fertile habitats into mixed broadleaved and broadleaved forests (carried out by SF for about 50 years [7,17]). Admittedly, the crowns of deciduous tree species increase the pH of throughfall precipitation [72].

Thirdly, changes in the habitat structure may be related to afforestation realised in Poland since the end of World War II, which was especially intensive in the period until the late 1970s [26,61,62]. Much of reforested/afforested land was characterised by greater fertility than that found in the forests managed by SF, especially in the period just after the war ended. This was due to the fact that areas with the most fertile soils were formerly deforested for agriculture. For this reason, in post-agricultural areas designated for afforestation, Brunic Arenosols (65.8%) were dominant on the scale of SF. In southern Poland, even more fertile soils were observed—Cambisols and Luvisols. The dominant forest habitat types assigned to these areas include mesotrophic fresh mixed coniferous forest (35.9%) and fresh mixed broadleaved forest (26.5%), covering a total of 62.4% of afforested post-agricultural land [73]. Therefore, the gradual inclusion of afforested land in the acreage of SF has undoubtedly had an effect on the overall structure of forest habitats, causing the proportion of the poorest habitats to decrease (Figure 2).

Fourthly, the documented changes in habitat structure may be a result of the increasingly accurate identification of a habitat's current and potential productive capacity [74]. Pertinent soil physical and chemical analyses and studies on spontaneous natural vegetation, including index plants, have been carried out over the last 20 years [75], with the purpose to improve the precision of forest habitat classification. In this context, pioneer solutions have been implemented in the promotional forest complexes (PFCs) established by State Forests during the era of democracy [76]. PFCs are an original Polish concept of implementing and promoting sustainable forest management.

The share of humid and wet habitats. The grid of forest habitat types in Poland's forests (State Forests) used for forest management planning also takes into account the division of forest habitats listed in Figure 2 in terms of, inter alia, their degree wetness (dry/fresh/humid/marsh) [77]. In 1972, the share of forested humid/marsh (wetland) habitats in the area of state-owned forests was 11.4% [68]. This was the time (the post-war period—socialism) of the implementation of land drainage projects to make wet habitats (forested and non-forested) productive [78,79]. It is worth noting that the draining of wetlands to turn them into pastures and arable fields was carried out not only in Poland, but also in large areas in other parts of Europe [80]. In Poland, a new approach to forest land improvement was assumed after the transformation of the political system from socialism into democracy, which no longer required almost every hectare of land administered by State Forests to be made productive [11,78,79]. As a result of the remedial measures taken (see Section 3.5), the share of forested moist and wet/wetland habitats in SF has increased to 15.8% (calculated based on [33]). This is a positive development in the perspective of preserving biodiversity, as well as increasing the population of trees [81].

3.3. Forest Management Framework

The method of management directly influences forest spatial diversity—the number of forest stories/layers, tree density, and tree size (e.g., differentiated DBHs) as well as the shape of crowns [82–86], the microclimate, insolation that reaches the forest floor at a given stage of stand development, soil moisture and nutrient contents [87]. Forest organisms respond to the specific living conditions shaped by humans; those that can endure can adapt [84].

Polish forestry cultivates high forests almost exclusively [11], with four main management systems being currently in use: clearcutting, shelterwood cutting, selection cutting and special (functional) cutting. The first three systems apply to commercial forests and the last one applies to those with important protective functions [11,50,88]. Under the aforesaid forest management systems, the following can be used: clearcuts, step cuts, group selection cuts and single-tree selection cuts [11,88]. The changes in the way forests have been managed since the end of World War II are shown in Table 2; notably, during the study period, the methods for reporting relevant information/data were modified and data were not always available.

Table 2. Management systems in Poland's forests administered by State Forests *.

Year	1950	1960	1970	1980	1990	2000	2010	2020
Clearcutting system (thousand ha)	n.a.	n.a.	n.a.	n.a.	n.a.	n.a.	3219.0	3190.6
Clearcuts (thousand ha)	16.9	47.8	40.8	34.6	33.5	29.5	26	30.3
Shelterwood cutting system (thousand ha)	n.a.	n.a.	n.a.	n.a.	n.a.	n.a.	3137.5	3378.5
Various types of cuts (e.g., step cuts, group selection cuts) (thousand ha)	58.4	8.3	11.4	8.3	n.a.	n.a.	n.a.	n.a.
Selection cutting system (single-tree selection cuts) (thousand ha)	–	–	–	–	n.a.	n.a.	174.4	110.1
Special (functional) cutting system (thousand ha)	–	–	–	–	n.a.	n.a.	541.6	442.2

* sources: [34,61,62,68,89]; n.a.: data not available; —: no such cutting system.

The dynamics of changes in different forest management methods. Irrespective of the political era under study (socialism vs. democracy), two basic forest management systems dominate in the SF: clearcutting and shelterwood cutting. However, there is a fundamental difference between their dynamics, depending on the examined era: the forest area under the shelterwood cutting system has been increasing, and that under the clearcutting system has been decreasing in the era of democracy (Table 2), even though the total area of forests administered by SF has been augmented [26]. The system of selection cutting has as yet been of minor importance, and was established only in the era of democracy, when Polish forestry put a greater emphasis on biodiversity conservation. The outcome of the new approach to forestry in the democratic era is also the method of management dedicated to forests of special importance for, among others, nature conservation purposes (special management—the special cutting system). Over the last decade, selection cutting and special cutting systems have been used in areas decreasing in size.

Clearcutting system/clearcuts. This management method has a strong impact on the spatial structure of forest ecosystems, as an entire mature stand is removed at once, and a succeeding stand—resulting from forest regeneration over a wide-open area—is even-aged (single-generation), single-storey, and most often comprises pioneer or post-pioneer species [11,90]. Due to the dominant share of the Scots pine [33], which is a pioneer species, as well as because of a considerable area of poor habitats being occupied with forests in Poland (see Section 2.2), the clearcutting method was and still is widely applied as a forest management practice (Table 2; [11,91]). It is worth noting that single-generation forests, being a result of the use of the clearcutting management system, have not been a problem only for Poland's forest ecosystems, but also for those in Europe, where about 75 percent of the total forest area is covered by even-aged forests [3]. In Poland, besides being used in coniferous forests, the clearcutting system is currently used in parts of alder forests [11,50,88]. Two aspects of clearcutting management are worth pointing out: the total and unit clearcut area. The total clearcut area has gradually decreased over the years, regardless of the political era under study (Table 2), despite the fact that the overall area of forests administered by SF has increased [26]. On the other hand, however, there were recorded increases in the total area of clearcutting (e.g., in the year 2020—Table 2). In most cases, this was due to the necessity to remove trees because of large-scale wind damage or forest dieback caused by the impacts of biotic factors [2,61,62,92]. Nevertheless, in 2020 (democracy), only 20% of all timber harvested came from clearcutting. In comparison, in the marketing year 1959/1960 (socialism), 55% of timber mass was harvested under the clearcutting management system [26]. In the era of socialism, the unit area of clearcutting was usually 6 ha. This was justified on practical grounds in single-generation Scots pine stands, because the method was the least expensive and time-consuming, which translated into higher income from the forest [7,48]. Until the 1960s, even if ill-chosen, the clearcutting management system was also used in large areas covered by mountain beech forests [84]. In the era of democracy, clearcutting in 6 ha areas has been used only on certain sites, and the forest areas of smaller size are routinely planned for harvesting with the use of this method [22,88], which is considered a manifestation/evidence of the "greening" of Poland's forestry practices [22,61].

Shelterwood cutting system/various types of cuts. This management method interferes with forest spatial structure to a moderate degree; for a certain period of time, two generations of trees coexist, afterwards an older generation is harvested, so forest cover continuity is maintained [11,90]. Currently, this system is used in broadleaved forests and in alder forests not managed under the clearcutting system [11,50,88]. The implementation of the shelterwood management system has become particularly important in the era of democracy, being used as a measure with the purpose of adapting to the changes in the fertility of forest habitats (see Section 2.2). Concurrently, as can be seen in the historical data (Table 2), in the early stages of socialism, the area of forests under the shelterwood management system was much larger than that under the clearcutting system. At that time, foresters attempted to bring back the species composition and spatial structure of Poland's forests—altered in earlier times—to a more natural state. However, due to the soon-to-be-implemented central planning of the economy (characteristic of socialism), as well as various adverse impacts/overexploitation in forests, these attempts had to be abandoned [93,94].

Selection cutting system/single-tree selection cuts. Under selection management—along with forest cover continuity— multi-generation forests have been maintained [11,90]. This kind of management is currently recommended for fir (*Abies alba*) forests and mountain Norway spruce (*Picea abies*) forests [11,50,88]. Forests managed in this manner currently account for only 1.5% of the total forest area administered by SF (the value derived from Table 2). In comparison, in Switzerland, about 8% of the total forest area is managed this way [95]. On a European scale, about $\frac{1}{4}$ of forest area is covered by multi-generation forests [3]. Thus, evidently, in Poland's forests managed by SF, the proportion of unevenly-aged forests is comparatively low, with a noticeable decreasing trend in recent years (Table 2). The downward trend may be due to the process of degradation and resultant reconstruction of artificial spruce monocultures in mountainous areas, which have been proven to have little resistance to adverse abiotic and biotic factors [6].

Selection management as well as various cuts under the shelterwood cutting system have beneficial effects on the diversification of the spatial structure of forests [11,48,54,90]. At a stand level, the multi-layered forest structure allows for a more efficient use of light, water and soil nutrients by forest vegetation [11,86]. Also, the provision of ecosystem services, such as biodiversity maintenance, is generally more advantageous compared to the establishment of single-generation forests, especially young ones [3,48,90,96]. On the other hand, however, the selection cutting system is quite expensive and difficult to implement [11]. It also does not provide solutions to all dilemmas and issues concerning nature. Noted was an increased share of ruderal plant species and increased homogenisation of the undergrowth when this practice was applied [1,54,97]. Nonetheless, analogous phenomena are also observed in unmanaged forests [54].

Special management system. Special management is related to forests with marginalised or excluded commercial functions (e.g., designated nature reserves, and parts of protective forests [11,50,88]). Although such types of forests existed in the era of socialism [26], a special mode of their management has become a part of management planning only in the era of democracy. In the last decade, there has been observed a downward trend in the area of forests under the special management system (Table 2). Nevertheless, in the same decade, the overall area of forests in nature reserves and that of protective forests generally increased [26]. The indicated inconsistency can be explained by the less-restricted nature conservation measures undertaken in some protective forests (under the management of SF). Another reason behind this could be the relinquishment of some of constraints—initially introduced under requirements related to forest management certification. In light of the imperative to respect Poland's nature conservation law, less-restricted nature conservation measures would be unlikely in the case of nature reserves [45,98].

3.4. *Age Structure of Forest Stands*

Forest biodiversity is determined, inter alia, by the age structure of stands [3,28]. Forest age structure can be considered at a stand level (to which a reference was made in Section 3.3: a single- or multi-generation stand, depending on the forest management system used at the site), or at the scale of the entire country. Under the present study, forest biodiversity was determined at the scale of national forests managed by SF (Table 3). It should be noted that the co-presence of stands of different ages—combined with open spaces—shapes a mosaic of the environments valuable for maintaining the diversity of different groups of organisms, for instance bats [10,91,99].

Table 3. The proportion of area covered by stand age classes in forests managed by SF *.

Year	1950 [1]	1960	1970 [2]	1980 [3]	1990	2000	2010	2020
Age class I (1–20 years old) (%)	23.2	29.4	24.4	21.6	14.2	12.0	10.8	10.9
Age class II (21–40 years old) (%)	22.8	23.1	21.6	21.2	24.7	20.1	15.0	13.4
Age class III (41–60 years old) (%)	18.5	18.0	19.1	21.5	20.5	22.7	24.4	19.2
Age class IV (61–80 years old) (%)	13.1	13.6	14.6	16.0	18.1	19.4	19.1	20.8
Age class V and older (>80 years old) (%)	14.3	13.4	14.1	15.2	18.1	20.9	23.0	24.0
Stands under the selection cutting system (uneven-aged) and those in the restocking class (with distributed age classes) (%)	n.a.	n.a.	3.0	3.1	3.3	4.1	6.2	9.5
Felling sites, blanks and irregularly stocked open stands, and not reforested/afforested forest land (%)	8.1	2.5	3.2	1.5	1.1	0.8	1.5	2.2

* Sources: [30–33,75,89,94]. [1] data for 1948; [2] data for 1967; [3] data for 1978; n.a.—data not available.

The dynamics of changes in the age structure of forest stands. In the period 1948–2020, the age structure of Poland's forests administered by SF meaningfully changed (Table 3). The proportion of the area covered by younger stands (of up to 40 years old) decreased from 46.0% to 24.3%, with a lower decrease rate in the era of socialism (approx. 0.16%/year) compared to that in the era of democracy (approx. 0.49%/year). As for the other age classes, the proportions of their areas increased over the entire period under study—with differentiated dynamics: lower proportions for 41–80-year-old stands (middle-aged, age classes III and IV) and higher proportions for those that were older (age class V and higher; tree stands under the selection cutting system and those in the restocking class). The area covered by middle-aged stands, in age classes III and IV, increased from 31.6% to 40.0%. The increase rate was lower in the era of democracy (about 0.05%/year) than that in the era of socialism (about 0.16%/year). In contrast, the area of forest stands containing older trees increased from 14.3% to 33.5%, with a lower increase rate (approx. 0.17%/year) in the era of socialism compared to that observed in the era of democracy (approx. 0.40%/year). A natural consequence of the evolution described above is the change in the average age of forest stands. During the period under study, the average age of forest stands increased by about 20 years. Initially, reaching the average increase of 10 years took about 45 years (the era of socialism), whereas the duration of the next average increase of 10 years was 30 years (the era of democracy).

Young stands (1–40 years old). The initial dominance of forest stands classified in younger age classes (in the era of socialism until the early 1980s) was related to forest restoration activities, which were undertaken for the regeneration of the forests over-exploited during World War II, as well as actions carried out with the purpose of intensive post-war reforestation/afforestation [17,26,62,100]. Nowadays, at the European level, only in Iceland, there is a relatively large proportion of forests in the regeneration phase being recorded [3]. In Poland, due to the shrinking of reforestation/afforestation areas, and also restrictions regarding clearcutting management introduced in the era of democracy (including the reduction of the size of clearcut areas—see Section 3.3)—the proportion of area covered by younger stands has considerably decreased, particularly that covered by the youngest stands (1–20-year-old trees). In this regard, it is worth noting that in Table 3, the area covered by this generation of trees is also included in the category "Tree stands under the selection cutting system..."; for the most part, these stands appear in multi-generation forests [33]. A continuous reduction in the area covered by the youngest stands may pose a future risk to forest sustainability and the proper age class structure [61,62]. The diversification of forest age structure is a key driver of biodiversity, as it enriches habitats with a range of species associated with each stage of the successional forest cycle [10]. In Poland, the arrangement of the areas covered by trees in certain age classes (a comparative level, yet not necessarily optimal) is considered "normal" if stands of the age classes I and II cover about 18% (each) of the forest area [33]. The values presented in Table 3 evidently differ from those.

Middle-aged stands (41–80 years old). The slowed-down afforestation/reforestation process has influenced a successive increase in the proportion of the area covered by middle-aged forests (with various fluctuations in age class III forests). At the turn of the 20th and 21st centuries, age classes II and III showed a dominant trend, and since 2010, age classes III and IV have prevailed (41–80 years old) (Table 3). The current proportion of stands of this age slightly exceeds the average. This is considered "normal" in the system of age classes in Poland's forests, where stands in age classes III and IV should each cover about 18% of the forest area [33]. Available data show that in the majority of European countries, forests in the intermediate development phase also prevail and cover considerably large areas [3]. It is worth noting that forest stands composed of species such as pine, spruce, birch (common in Poland's forests) in the intermediate development stage (50–100 years old) are characteristic of the highest biomass production, as was reported in Sweden [101].

Old stands (>80 years old; tree stands under the selection cutting system and those in the restocking class). After the collapse of Eastern European socialism, old-growth forests were defined as threatened by the major socio-economic restructuring processes that occurred in the transition of the economy from state-led to market-oriented [2]. However, this has not been the case of Poland's forests as they have mostly remained state property [102,103]. Currently, 81–100-year-old tree stands constitute 15.2% of the forest area under the administration of State Forests. Under the "normal" arrangement of the age classes, they should account for approx. 18% [33]. Stands older than 100 years of age (tree stands under the selection cutting system and those in various stages of regeneration) constitute another 18.3%, even though their assumed share under the "normal" arrangement of the age classes amounts to approx. 9% [33]. Thus, a noteworthy overrepresentation of the oldest stands in relation to the aforesaid "normal" pattern is observed. The increase in the proportion of older stands in managed forests, related to the process of management transformation towards SFM has been observed not only in Poland, but also throughout Central Europe [49,63,104]. In several European countries (the Netherlands, Norway and Portugal), the proportion of forest stands in the mature phase of development is considerably higher (in terms of the area covered) compared to that of stands in other developmental phases [3]. This is related, for example, to an increase in the rotation age of forest tree species [63]. Currently, in Poland's forests managed by SF, depending on the geographical region and habitats, the rotation ages are 70–140 years for Scots pine (typically 110–120 years); 80–130 years for Norway spruce; 90–140 years for fir (*Abies* sp.); 100–140 years for beech (*Fagus* sp.); 120–240 years for oak (*Quercus* sp.) [50]. Apart from the adopted rotation age, the presence of older stands may be achievable thanks to the designation of many protected areas throughout forests administered by SF [26,98,105]. In general, mature forests provide ample ecosystem services related to, inter alia, biodiversity, which are more favourable compared to the presence of young forests [3,10,101]. Older forests are characterised by greater spatial diversity and the presence of specific microhabitats (which is an element of diversity at the ecosystem level) and support the existence of numerous species, often appearing in greater abundance than they do in younger forests [10,96,101,106]. Of these species, some appear only in older stands [87]. The presence of old trees (even individual specimens) has great aesthetic value, as they shape the diversity of forest landscapes [11].

Average age of stands. The gradual increase in the average age of forest stands managed by SF (more intensive in the era of democracy—Figure 3) reflects the general direction of changes in Poland's forests.

The average stand age (64 years) in forests under SF management is higher than that that in privately-owned forests (58 years [107]). The value of the calculated average age of privately-owned forests results from, among other factors, the afforestation of private lands at the beginning of the era of democracy. As a result, considerably large areas covered by young forests contributed to the lowering of the calculated average value. At the same time, the average stand age of forests managed by SF is of a lower value than that determined for protected forests in national parks (92 years [107]). This is due to long-term strict protection (for example, via taking them out of use). With the current average age of 64 years (Figure 3), older stands in state-owned forests (those administrated under SF) are dominated by fir *Abies* sp. (which are on average 81 years old), hornbeam *Carpinus* sp. (77 years old) and beech *Fagus* sp. (73 years old).A lower rotation age (40–80 years) results in an age below the average for all SF-administered forests (64 years—Figure 3) for poplar *Populus* sp. (52 years) as well as alder *Alnus* sp. and birch *Betula* sp. (58 years) [33,50]. It is worth emphasising that in the period 1950–2010, the average age of stands in Poland's forests administered by SF increased by 18 years (Figure 3), whilst in Europe as a whole, it decreased by 7 years [108].

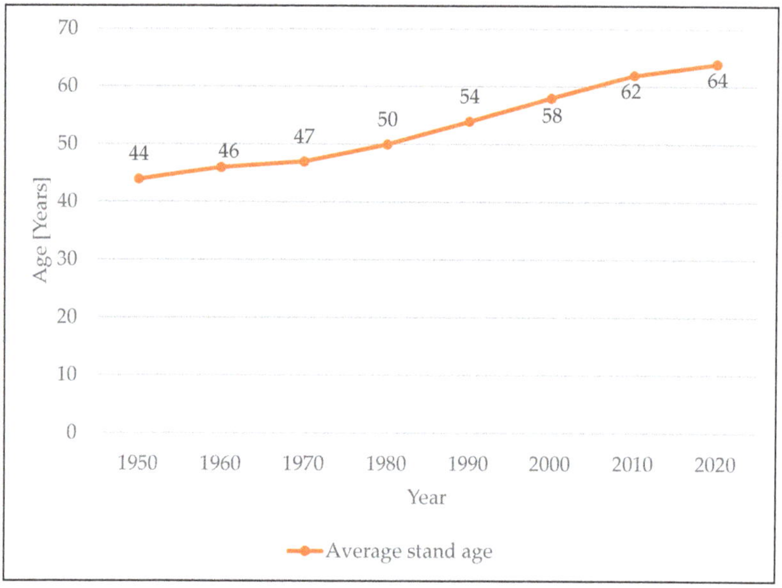

Figure 3. Average stand age (years) within state forests. Sources of information: [30–33,75,89,94]. Notes: 1950—data for 1948; 1970—data for 1967; 1980—data for 1978.

3.5. Issues and Directions of the Protection of Forest Ecosystems in Poland

In the field of the protection of biodiversity at the ecosystem level in Poland's forests, there still remain problems and dilemmas to be solved (Figure 4).

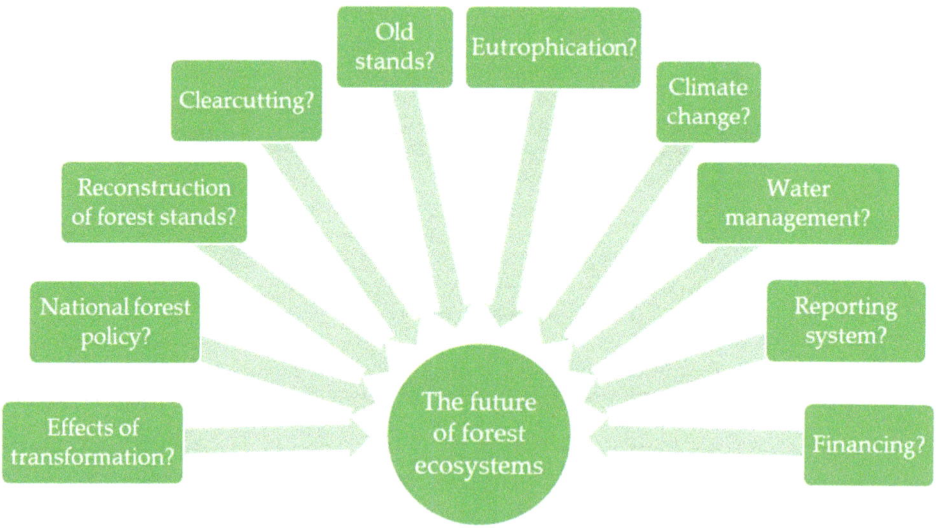

Figure 4. Problems and dilemmas of enhancing forest biodiversity at the ecosystem level in Polish forests/forests administrated under State Forests.

Effects of transformation of forest management model. Forest management objectives, adopted in Poland at the beginning of the era of democracy and thereafter amended, have rendered the realisation of forest production functions conditional on the needs of ecosystem preservation being taken into account [109]. This is a meaningfully different approach to that implemented in the era of socialism, where the production function played a distinctive, superior role resulting in considerable anthropogenic changes in forest ecosystems [48]. Not without significance for the course of changes in Polish forestry is the fact that SF, as the organisation managing the majority of the country's forests, has been the most important forest policy maker at a national level [76]. Many positive changes have already been achieved; nevertheless, the full effect of forest management transformation on the enhancement of forest ecosystems is still to be seen. Considering the now-existing alteration of some forest ecosystems, the model of a semi-natural forest with a diversified species and spatial structure [11,48,49] is not always possible to implement in the time of one generation of a stand [48]. In view of this, it is necessary to ensure future, legal, organisational and financial measures, for continuing the instigated processes, as they are necessary in boosting the diversity and security of Poland's forests.

Lack of revised national forest policy. In Poland, the policy on forests and forestry in force was adopted in the year 1997 [21]. This means, inter alia, that there have been no solutions included that were developed afterwards, as part of the process of FOREST EUROPE (e.g., the resolution "Conserving and Enhancing Forest Biological Diversity in Europe", MCPFE 2003 Vienna Resolution 4), or those related to Poland's accession to the European Union (2004) and the implementation of a network of nature protection under Natura 2000 [110]. The decisions/recommendations adopted under the framework of these processes address the issues of the protection, conservation and management of forest ecosystems in practice. Even though in the forests administered by SF or those protected in national parks (designated as state-owned lands) many consecutive endorsements have been already implemented [50], there are forests under other forms of ownership, where the situation is not as promising. This is because of the lower requirements for forest management documentation [20] as well as unsolved organisational management issues. Consequently, in view of safeguarding the richness of all forest ecosystems in the country, a national forest program (NFP) should be legitimately established, which is not the case in Poland, even though comprehensive works on the NFP proposal were carried out in 2013–2016, with a wide range stakeholders participating. One of the issues debated was the better recognition of the necessity to safeguard forest biodiversity.

The reconstruction of tree stands. The first, initial component of the forest management transformation model, implemented as early as in 1970s (under socialism) [7,17], considered a gradual replacement of coniferous stands artificially planted on fertile forest sites in the 1800s and in the era of socialism, by mixed and broadleaved stands made of species indigenous to a given forest habitat [5]. Therefore, in the years 1948–2020, the proportion of forest area covered by coniferous tree species decreased from the initial 87.1% to 75.6% [33,94]. This approach enabled the application of more complex forest management systems (see Section 3.3), which enhanced the differentiation of the forest's spatial structure [11]. On the other hand, however, recovering the spatial (and species) structure of forest ecosystem may not be welcomed by society, with expressions of dissatisfaction and low acceptance due to their individuals' own preferences when it comes to the forest landscape. The results of the study by Edwards et al. [111] revealed that from the perspective of the surveyed persons, the forest should be full of light, dry and of low density. Dense understorey vegetation is not appreciated, as it reduces visibility and causes a loss of the sense of safety. Poland's State Forests are in general accessible to the public, who contributes to shaping the forest ecosystem through the mechanism of public consultations on draft forest management plans [112]. In consideration of the conflict of interests between forest management practices implemented towards forest restoration and public expectations, as for forest appearance, it is important to increase public awareness of the reasoning behind the naturalisation of forest spatial (and species) structure, e.g., through educational campaigns carried out by SF.

Dilemmas of the clearcutting forest management system. The clearcutting management system (which in Poland concerns a forested area of over 3 million ha—see Section 3.3) is considered to strongly and negatively interfere with the forest ecosystem [11,113]. The assumption is that the greater the clearcut area, the greater the adverse impact [96,114]. Clearcutting results in local climate changes followed by, e.g., the fast decomposition of forest litter and the leaching of plant nutrients from the soil [11,90]. At the same time, the pioneer stage of forest regeneration is consolidated, and forest age, species and layer structures are most often in decline [11,48], which decreases forest resistance to damage from biotic and abiotic factors [11]. In order to limit the negative consequences of clearcutting, since 1995, it has been recommended (by State Forests) to leave behind old-growth tree groups with all the lower layers [11,22]. Apart from the above, the clearcutting system remains one of the elements of forest management which evokes negative emotions in Polish society [115]. However, it should be noted that there are a number of forest species for which the presence of open spaces is essential for proper functioning. For example, studies on bat activity carried out on Poland's Scots pine stands showed the beneficial effects of clearcuts (and 1 m high young forests as well) on bats [91], which was confirmed via the study results obtained outside Poland [116,117]. Likewise, a considerable increase in undergrowth vegetation diversity after cutting trees in coppice stands was noted by Decocq et al. [87]. Hence, the use of clearcuts should not be completely abandoned, especially if the restrictions regarding their maximum area are respected.

Large areas covered by old stands. Apart from the beneficial effects of old-growth forests (see Section 3.4), the ageing of forests may be associated with certain problems of forest management. An increased stand age and volume leads to a greater susceptibility of stands to disturbances [118,119], resulting in damage due to abiotic factors [62], such as strong winds [11,120], biotic factors such as insect outbreaks [120] and anthropogenic factors, such as pollution [121]. The ageing of beech stands, for example, may lead to top soil acidification, lower nutrient concentrations, and consequently, to plant species homogenisation in the undergrowth, which means that older stands do not always generate better conditions for the conservation of species richness [63,84]. In view of forest management, wood from old-growth trees is not profitable enough as it depreciates with time, thus implying economic losses [61,62]. The profitability of wood production is an important issue. If the optimal economic rotation age was taken into account, then, e.g., for Scots pine (the dominant species in forests managed by SF), it would range between 75 and 91 years, rather than 110 and 120 years—as is currently the case in Poland's forests [122,123]. Decreasing the rotation age was postulated by the authors of the above calculations, in order to meet legal requirements as regarding SF's self-financing [123]. It is important to note that the large area of old-growth forests will at some point require regeneration; thus, there will be established young forests. Young forests do not yield good profits from large-size timber and require considerable maintenance expenditures. In Poland, there is very high demand for wood, which is first of all due to its advanced wood processing industry [29]. Mature wood is very valuable for the furniture industry (juvenile wood is used by the paper industry). Timber and its products manufactured in Poland have been exported to European and non-European markets [29], so a reduction in the supply of wood from state forests will result in lesser wood availability not only in Poland, but also in other countries. The presented issues/dilemmas require further analysis and the development of pertinent procedures in order for the optimal solution to be chosen for a specific place, time and in relation to the needs of people and nature now and in the future.

The gradual eutrophication of forest habitats. The risk of eutrophication concerns about 60% of Poland's area, mainly its central part, but also the north-eastern part (the latter is a region with a low level of atmospheric pollution) [62]. A soil nitrogen surplus can lead to nutrient imbalances, growth reduction, nitrogen leaching and groundwater pollution [3]. Gradual eutrophication (see Section 3.2) may therefore pose a threat to the sustainability of forest ecosystems [61,62] and lead to forest homogenisation [124]. Negative changes in the undergrowth can be partially reduced by controlling the species composition of the stand [125]. Eutrophication poses a particular threat to rare habitats on sites with unfavourable fertility conditions, e.g., the Central European lichen Scots pine forests (communities of *Cladonio-Pinetum* association) that appear in Poland and that are protected under the Natura 2000 program (code 91T0) [126]. Solving the eutrophication problem can be very challenging due to the nature of its causes which are beyond the control of forest managers, and therefore require the comprehensive, interdisciplinary cooperation among many different institutions.

Ecosystem diversity and climate change. One of the most important contemporary problems of biodiversity protection is climate change [127]. The habitats most dependent on water resources are particularly vulnerable to climate change impacts [128]. In Poland, amongst the forests in marshy and wet habitats, floodplain forests comprise the richest ecosystems of broadleaved forests [81]. Higher mean temperatures (these every so often being extremely high) and recurrence of droughts increase the risk of habitat overdrying [80]. As a result, forest trees are more and more vulnerable and threatened by dieback [62,127,129]. Global warming also poses the risk of frequent fires [29,53]. Furthermore, every so often, forests experience extreme weather events and their impacts [127]. In response to it all, some counteractive actions have been implemented by SF, with a scope and scale that changes over time. In the whole period under study, there have been fire protection measures undertaken, in light of the expanding scale, especially in the 21st century, due to a greater fire risk attributable to prolonged drought periods caused by climate change. In addition to technical solutions that allow a reduction in the risk of fire/fire damage, such as the forest fire detection system that enables the relevant actors to take quick actions towards fire control [62,70,80,130], over the years, there was a fire hazard map for forests (that is open-access) that was developed and is updated twice a day in the period from 1 April to September 30 [131].With the use of this tool, some forest areas can be periodically restricted or closed for use by the general public [20] so as to protect forest ecosystems against accidently set fire. All these solutions should also be implemented in the future. Faced with uncertainty about the conditions for the further development of a given forest, in Poland's forests there is a complex cutting system that has been implemented [11,48], and that includes shelterwood or selection cuttings. These forest management methods contribute to the improvement of forest spatial structure. The diverse and rich structure of forest ecosystems increases forest continuity, resilience and stability [48,54,84,132], as well as compensating for the mortality observed after the perturbations [86]. Since the transformation of Poland's political and economic system, complex cuttings have been more frequently implemented compared to the era of socialism (see Section 3.3), which has contributed to the shaping of multi-aged forests, as well as natural ones [63]. The process of transformation from a regular into an irregular form of forests develops gradually [11,132] and should be continued in the future. An element supporting the diversification of forest spatial structure (also biodiversity) is the planting of trees which form a lower layer of crowns as well as plants that form an underlying layer of vegetation (understorey) [11]. In the case of the understorey, the extent of planting has been quite small in the last two decades [29,75]. This may be related to the lower level of needs for such activities (many tasks have been already accomplished), and also to the requirements of the Natura 2000 program. Planting additional trees and shrubs may lead to the degradation of protected habitats, such as Central European lichen Scots pine forests (code 91T0), thermophilous oak forests *Potentillo albae-Quercetum* (91I0) or habitats of the most valuable and threatened plant and animal species. Therefore, care should be taken when introducing additional vegetation. Supporting forest ecosystems under climate change includes enhancing natural water retention in forests (so-called small water retention) which has been implemented for some time in the era of democracy and needs to be continued, bearing in mind that Poland's forests were subject to destructive drainage operations in the 1800s and 1900s.

Water management in forests. In forests managed by State Forests, the estimated area of forest drainage was about 13% [11], i.e., 850 thousand ha [79]. Polish foresters tried to avert this practice [133], as wherever drainage works were carried out, large, sometimes irreversible hydrological changes were observed in forest ecosystems, especially those situated in forest swamps and peat bogs [134], resulting in biodiversity loss [135]. A decrease in water resources in forest habitats is particularly hazardous. In view of long periods of drought due to ongoing climate change, bearing in mind the above considerations, as well as the transformed priorities of forest management [11,78,79], in the era of democracy, as early as in the second half of the 1990s, the first works began towards the restoration of water resources in forests [79]. The devices used included gates, river bars, dikes, fords, overspills, fish ladders, ditches and small retention reservoirs. Other solutions involved the implementation of soft engineering elements, such as the introduction of woody, shrubby, and/or herbaceous vegetation to enhance water retention potential and to contribute to the protection of riverbanks [80,81]. In the 21st century, these works were intensified and financially supported by the EU's funds [61,62,81]. The total level of retention achieved as a result of the actions taken in 2007–2015 exceeded that planned at approx. 32.3 million m^3 of water, and the actions taken in the period 2016–2020 resulted in the retention of another 2.5 million m^3 of water [70,136]. Notwithstanding the unquestionable successes at the scale of State Forests, it should be noted that the situation in Poland is still quite unfavourable—Poland ranks as one of the last countries in Europe in terms of available water resources [80,81]. Therefore, it is necessary to continue the activities towards water retention, which will benefit both forest ecosystems and non-forest ecosystems spatially and hydrologically. These are first and foremost water bodies, swamps and peat bogs. Urgent works related to water retention should also be undertaken outside the area administered by State Forests. The described above measures concerning small retention are especially recommended due to their low cost [81]. It is important to stress that restoring water retention largely contributes to the protection and restoration of ecosystems, especially those dependent on hydrogenic habitats, and supports the implementation of the Sustainable Development Agenda in terms of, inter alia, halting biodiversity loss (Goal 15, [137]). On the other hand, it should be noted that so far achievements in water retention in Poland's forests (and beyond) have been in part due to the work of beavers, the population of which in Poland has considerably increased from approx. 200 specimens in 1960 [37] to about 142,000 specimens in 2020 [43].

Reporting systems. Poland has been reporting on the state of forests under the framework of FOREST EUROPE. In the submitted reports, Polish forests that are strictly protected, compliant with the Nature Conservation Act [45], are referred to as "forests undisturbed by man". Yet, this approach does not take into account all the nuances of Polish forest management and its impact on the naturalness of forest ecosystems. It is estimated that at the beginning of 2020, at least 530 000 ha of forests managed by State Forests (7.5% of the SF area) was completely excluded from timber harvesting. Apart from forest nature reserves, this area comprised other types of strictly protected forests, the protection zones for selected species, xylobiont refuges, wetland habitats, some protective forests and FSC reference forests, all designated in line with national law or internal instructions and good practices, and applicable only within SF [138]. In comparisons with the average percentage of forests undisturbed by man in Europe (2.2% [3]), the aforesaid value of 7.5% of the SF area excluded from logging seems to be a good result. This value reflects the reality of Poland's forests undisturbed by man better than that obtained by means of the method used in the reporting system of FOREST EUROPE. The method for determining the area of forests undisturbed by man could be verified in the future. This might improve the social perception of forest management in a situation where more and more people expect foresters to focus on the protective functions of forests (and not those related to wood production). It should be emphasised that at least 23% of Poles believe that the primary task of State Forests should be nature conservation [139].

Financing the protection of biodiversity. The main source of income for State Forests is sale of felled timber [12,136]. Supporting the non-productive functions of forests, in particular the function of biodiversity safeguarding, can increase the costs of management and result in lower profits [140]. Currently, State Forests management is profitable, which should allow for the maintenance of the course of positive changes in regard to biodiversity, recorded step by step since the end of World War II. This trend may also be influenced by the considerably strong social pressure focused on protecting the country's natural resources that arose along with transformation of Poland's political system and economy. Increasing the area of Natura 2000 sites that are under strict protection, as planned in the EU Biodiversity strategy for 2030 [141], is also relevant. A consequence of expanding the area of protected forests may be a lower supply of timber, as well as a reduction in employment in the forest wood sector. Now, it is of great social importance and provides employment for almost 0.5 million people in Poland [29]. It is necessary to anticipate a scenario in which profitability comes to a halt, so as not to lose the so-far-attained effects of foresters' work in the field of biodiversity protection and restoration in Polish national forests. Within this context, the aforesaid social considerations are not without significance.

4. Conclusions

The effects of the evolution of management of Poland's forests carried out by State Forests National Forest Holding (SF) in the period 1945–2020 on forest biodiversity at the ecosystem level were analysed with the use of selected indicators (naturalness; habitat diversity; forest management system; forest stand age structure). Regardless of the deficiency of precise data or any statistics for the years considered in this study, in some situations, it was possible to determine the trends and dynamics of changes in particular parameters related to ecosystem diversity in Polish forests. Such results were interpreted on the basis of a comprehensive literature review, even if not all possible sources of information were analysed. Nonetheless, the sources used allowed us to look at the analysed parameters and problems from different angles.

Even though the period of 75 years does not seem as a long time in view of a forest ecosystem's lifespan and functioning, the obtained results indicate improvements in the ecological condition of ecosystems in the examined forests which, over the study period, were intensively influenced by political, economic and social changes at a national level. During the time of a centrally planned socialist economy (the era of socialism: 1945–1989), there were dynamic increases observed in the area of semi-natural forests and in the share of broadleaved forest habitats. The proportion old forest stands, as well as the stand average age, increased at a relatively slow rate. The proportion of poor coniferous forest habitats dynamically decreased. The clearcutting forest management system prevailed, and as a result the spatial structure of the forest was depleted. Along with a transition from a centrally planned economy to a market economy (the era of democracy—ongoing since 1990) and forest management transformation, in Poland's forests, there have been observed increases in the area of forests undisturbed by man, and the shares of mixed broadleaved forest habitats as well as wet and wetland/swamp forest habitats. The proportion of older stands and the average age of stands have also increased at a relatively fast rate. The area of forests managed under the shelterwood cutting system has expanded, which can help maintain biodiversity within the forest ecosystem through boosting a forest's spatial structure. The area of forest plantations has considerably decreased. In general, regardless of the era under study, there has been a decreasing trend observed in the proportion of the youngest stands, which is particularly unfavourable in view of forest sustainability.

All in all, the evolution of forest management in Poland's forests/those administered under State Forests during the whole study period has led to the restoration of/an increase in forest biodiversity at the ecosystem level, which was also perceptible in the era of socialism, when Poland's forestry had to face barriers related to the centrally planned and regulated economy. Thanks to the gradual change in forest management priorities towards treating forest as an ecological system performing multiple functions as well as the implementation of closer-to-nature forestry, it was possible to enrich the species, ages and spatial structures of forests and enhance their diversity and sustainability. Further studies are needed on long-term relationships between forest management and forest plant communities, which will require a compilation of comprehensive data (not available at the moment). Likewise, in the future, it would be useful to carry out analogous smaller-scale (regional-level) studies on the diversity of habitats and other analysed parameters, so as to capture their specific trends, the threats they face and the need to counter the risks. Problematic, however, is the lack of comprehensive information related to forest management development in Poland right after 1945. Additionally, changes in the administrative boundaries of individual regional units of the state forests introduced over time affect the consistency of the study's results.

Even with the accomplishments referred to in the present study, in the field of the protection of the ecosystem-level biodiversity in Poland's forests, there still remain unsolved problems, such as organisational (ensuring the further reconstruction of forest stands, and improving water relations), political (revising the national policy on forests), financial (bringing together costs of biodiversity protection/restoration and State Forests' self-financing), conceptual (addressing the issue of old-growth stands, and the pros and cons of clearcutting) and natural and anthropogenic (related to climate change and the eutrophication of habitats) issues. In some cases, actions beyond Poland's forestry, and even beyond the country's borders, will be necessary to carry out meet these challenges.

In times of urgent necessity to protect biodiversity on a local, national, continental and global scale, the method of forest management, adopted objectives, priorities and solutions are of great importance. The basis for their arrangement should be, inter alia, long-term experience and comprehensive knowledge gained worldwide, which—under similar or changing external conditions—will make it possible to avoid evident mistakes or provide the right solution. The results and their interpretation presented in this paper provide useful information that can be used to improve forest management with regard to its impact on biodiversity at the ecosystem level.

Author Contributions: Conceptualisation, E.R.-C. and B.K.; methodology, E.R.-C.; formal analysis, E.R.-C.; investigation, E.R.-C. and B.K.; writing—original draft preparation, E.R.-C.; writing—review and editing, B.K.; visualisation, E.R.-C.; project administration, E.R.-C. All authors have read and agreed to the published version of the manuscript.

Funding: This research received no external funding.

Data Availability Statement: Publicly available datasets were analysed in this study. These data can be found through the following links: https://stat.gov.pl/obszary-tematyczne/rolnictwo-lesnictwo/lesnictwo (accessed on 10 November 2022), https://stat.gov.pl/obszary-tematyczne/srodowisko-energia/srodowisko (accessed on 30 October 2022), https://www.bdl.lasy.gov.pl/portal/publikacje (accessed on 5 November 2022, https://www.bdl.lasy.gov.pl/portal/tworzenie-zestawienia-ru (accessed on 5 November 2022).

Acknowledgments: We would like to thank very much the reviewers of our work for their efforts and valuable comments that greatly helped us to improve the manuscript.

Conflicts of Interest: The authors declare no conflict of interest. The funders had no role in the design of the study; in the collection, analyses, or interpretation of data; in the writing of the manuscript, or in the decision to publish the results.

References

1. Van Calster, H.; Baeten, L.; De Schrijver, A.; De Keersmaeker, L.; Rogister, J.E.; Verheyen, K.; Hermy, M. Management Driven Changes (1967–2005) in Soil Acidity and the Understorey Plant Community Following Conversion of a Coppice-with-Standards Forest. *For. Ecol. Manag.* **2007**, *241*, 258–271. [CrossRef]
2. Griffiths, P.; Kuemmerle, T.; Baumann, M.; Radeloff, V.C.; Abrudan, I.V.; Lieskovsky, J.; Munteanu, C.; Ostapowicz, K.; Hostert, P. Forest Disturbances, Forest Recovery, and Changes in Forest Types across the Carpathian Ecoregion from 1985 to 2010 Based on Landsat Image Composites. *Remote Sens. Environ.* **2014**, *151*, 72–88. [CrossRef]
3. FOREST EUROPE. *State of Europe's Forests 2020*; Ministerial Conference on the Protection of Forests in Europe FOREST EUROPE; Liaison Unit Bratislava: Bratislava, Slovakia, 2020.
4. Rykowski, K. Forest Policy Evolution in Poland. *J. Sustain. For.* **1997**, *4*, 119–126. [CrossRef]
5. Węgiel, A.; Jaszczak, R.; Rączka, G.; Strzeliński, P.; Sugiero, D.; Wierzbicka, A. An Economic Aspect of Conversion of Scots Pine (*Pinus sylvestris* L.) Stands in the Polish Lowland. *J. For. Sci.* **2009**, *55*, 293–298. [CrossRef]

6. Keeton, W.S.; Crow, S.M. Sustainable Forest Management Alternatives for the Carpathian Mountain Region: Providing a Broad Array of Ecosystem Service. In *Ecological Economics and Sustainable Forest Management: Developing a Trans-Disciplinary Approach for the Carpathian Mountains*; Soloviy, I., Keeton, W.S., Eds.; Ukrainian National Forestry University Press: Lviv, Ukraine, 2009; pp. 109–127, ISBN 978-966-397-109-0.
7. Mederski, P.; Jakubowski, M.; Karaszewski, Z. The Polish Landscape Changing Due to Forest Policy and Forest Management. *iForest* **2009**, *2*, 140–142. [CrossRef]
8. Nienartowicz, A.; Lewandowska-Czarnecka, A.; Ortega, E.; Deptuła, M.; Filbrandt-Czaja, A.; Kownacka, M. Afforestation of Heathlands and Its Influence on the Land Cover, Accumulation of Plant Biomass and Energy Flow in the Landscape: An Example from Zaborski Landscape Park. *EQ* **2015**, *21*, 91–99. [CrossRef]
9. Dietze, E.; Brykała, D.; Schreuder, L.T.; Jażdżewski, K.; Blarquez, O.; Brauer, A.; Dietze, M.; Obremska, M.; Ott, F.; Pieńczewska, A.; et al. Human-Induced Fire Regime Shifts during 19th Century Industrialization: A Robust Fire Regime Reconstruction Using Northern Polish Lake Sediments. *PLoS ONE* **2019**, *14*, e0222011. [CrossRef]
10. Brockerhoff, E.G.; Barbaro, L.; Castagneyrol, B.; Forrester, D.I.; Gardiner, B.; González-Olabarria, J.R.; Lyver, P.O.; Meurisse, N.; Oxbrough, A.; Taki, H.; et al. Forest Biodiversity, Ecosystem Functioning and the Provision of Ecosystem Services. *Biodivers. Conserv.* **2017**, *26*, 3005–3035. [CrossRef]
11. Jaworski, A. *Sposoby Zagospodarowania, Odnawianie Lasu, Przebudowa I Przemiana Drzewostanów [Management Practices, Regeneration, Conversion and Transformation of Forest Stands]*, 2nd ed.; Hodowla lasu [Silviculture]; PWRiL: Warsaw, Poland, 2018; Volume 1, ISBN 978-83-09-01113-2. (In Polish)
12. Directorate-General of the State Forests. *Sprawozdanie Finansowo-Gospodarcze Za 2021 Rok [Report on Financial and Economic State of State Forests in 2021]*; DGLP: Warsaw, Poland, 2022; p. 51. (In Polish)
13. Karlikowski, T. (Ed.) *Lata Wojny i Okupacji [Years of War and Occupation]*, 1st ed.; Z dziejów Lasów Państwowych i leśnictwa polskiego 1924–2004 [The history of the State Forests and the Polish Forestry 1924-2004]; CILP: Warsaw, Poland, 2006; Volume 2, ISBN 83-88478-93-1. (In Polish)
14. Banach, J.; Skrzyszewska, K.; Skrzyszewski, J. Reforestation in Poland: History, Current Practice and Future Perspectives. *REFOR* **2017**, *3*, 185–195. [CrossRef]
15. Bałtowski, M. *Gospodarka Socjalistyczna w Polsce: Geneza—Rozwój—Upadek [The Socialist Economy in Poland: Genesis—Development—Decline]*, 1st ed.; Wydawnictwo Naukowe PWN: Warsaw, Poland, 2009; ISBN 978-83-01-16044-9. (In Polish)
16. Swadźba, S. System Gospodarczy Polski w Latach 1918–2018 [Poland's Economic System from 1918 to 2018]. *Optimum* **2019**, *1*, 19–31. [CrossRef]
17. Rozwałka, Z. (Ed.) *Las w Liczbach [Forest in Numbers]*; ARW A. Grzegorczyk: Warsaw, Poland, 1997; ISBN 83-86902-11-6. (In Polish)
18. Szaro, R.C.; Bytnerowicz, A.; Oszlányi, J.; Godzik, B. (Eds.) *Effects of Air Pollution on Forest Health and Biodiversity in Forests of the Carpathian Mountains*; NATO science series; IOS: Amsterdam, The Netherlands; Washington, DC, USA; Ohmsha: Tokyo, Japan, 2002; ISBN 978-1-58603-258-6.
19. Siry, J.P.; Newman, D.H. A Stochastic Production Frontier Analysis of Polish State Forests. *For. Sci.* **2001**, *47*, 526–533. [CrossRef]
20. The Forest Act, 1991 [Dz. U. 1991.101.444 as amended, in Polish]. Available online: http://isap.sejm.gov.pl/isap.nsf/download.xsp/WDU19911010444/U/D19910444Lj.pdf (accessed on 4 March 2020).
21. Ministerstwo Ochrony Środowiska, Zasobów Naturalnych i Leśnictwa. *Polityka Leśna Państwa [National Forest Policy]. Document Approved by the [Polish] Council of Ministers on 22 April 1997*, 1st ed.; MOŚZNiL: Warsaw, Poland, 1997.
22. Director-General of the SF. Zarządzenie Nr 11 Dyrektora Generalnego Lasów Państwowych z Dnia 14 Lutego 1995 r. w Sprawie Doskonalenia Gospodarki Leśnej Na Podstawach Ekologicznych [Order No. 11 of the Director General of the State Forests of 14 February 1995 on Improving Forest Management on an Ecological Basis]. Signature: ZZ-710-13/95 1995. Available online: https://www.wroclaw.lasy.gov.pl/documents/21578137/0/Zarzadzenie+Nr+11.pdf/3b10b05b-1803-46b9-9461-d76d2bb20c5f (accessed on 15 March 2022).
23. Director-General of the SF. Zarządzenie [Order] Nr 11a Dyrektora Generalnego Lasów Państwowych z Dnia 11 Maja 1999 r., Zmieniające Zarządzenie Nr 11 Dyrektora Generalnego Lasów Państwowych z Dnia 14 Lutego 1995 Roku w Sprawie Doskonalenia Gospodarki Leśnej Na Podstawach Ekologicznych. Signature: ZG -7120-2/99. 1999. Available online: https://www.wroclaw.lasy.gov.pl/documents/21578137/0/Zarz_11A_DGLP_11.05.1999_r.pdf/4edb8ccd-1e91-4166-ae9d-07adc7e1cbf0 (accessed on 15 March 2022).
24. Czerepko, J.; Geszprych, M.; Gołos, P. Basic Assumptions for Forest Management and Nature Conservation from Axiological, Legal, and Economic Perspective. *Folia For. Pol.* **2017**, *59*, 68–78. [CrossRef]
25. Aarhus Convention: Convention on Access to Information, Public Participation in Decision-Making, and Access to Justice in Environmental Matters, Adopted on 25 June 1998 in Aarhus. Polish Version: 2003 [Dz. U. 2003.78.706]. Available online: http://isap.sejm.gov.pl/DetailsServlet?id=WDU20030780706 (accessed on 10 November 2021).
26. Referowska-Chodak, E.; Kornatowska, B. Effects of Forestry Transformation on the Landscape Level of Biodiversity in Poland's Forests. *Forests* **2021**, *12*, 1682. [CrossRef]
27. MCPFE Updated Pan-European Indicators for Sustainable Forest Management. Annex 1 to Madrid Ministerial Declaration. In Proceedings of the 7th Ministerial Conference, Madrid, Spain, 20–21 October 2015.

28. Keller, W. Vermehrt die Waldbewirtschaftung die Biodiversität? In *Erhaltung der Biodiversität—Eine Aufgabe für Wissenschaft, Praxis und Politik: Publikation zur Tagung "Forum für Wissen" vom 1. Februar 1995 an der WSL in Birmensdorf*; Forum für Wissen; Schlaepfer, R., Eidgenössische Forschungsanstalt für Wald, Schnee und Landschaft, Eds.; Bibliothek WSL: Birmensdorf, Switzerland, 1995; pp. 33–38, ISBN 978-3-905620-42-9.
29. Leśnictwo [Forestry]. Statistical Yearbook of 2020. 2021. Available online: https://stat.gov.pl/obszary-tematyczne/roczniki-statystyczne/roczniki-statystyczne/rocznik-statystyczny-lesnictwa-2021,13,4.html (accessed on 15 March 2022).
30. BULiGL. *Wyniki Aktualizacji Stanu Powierzchni Leśnej i Zasobów Drzewnych w Lasach Państwowych Na Dzień 1 Stycznia 1991 r. [Results of the Update of Forest Area and Timber Resources in the State Forests as of 1 January 1991] 1991*; BULiGL: Raszyn, Poland, 1991.
31. BULiGL. *Wyniki Aktualizacji Stanu Powierzchni Leśnej i Zasobów Drzewnych w Lasach Państwowych Na Dzień 1 Stycznia 2001 r. [Results of the Update of Forest Area and Timber Resources in the State Forests as of 1 January 2001] 2001*; BULiGL: Raszyn, Poland, 2001.
32. BULiGL. *Wyniki Aktualizacji Stanu Powierzchni Leśnej i Zasobów Drzewnych w Lasach Państwowych Na Dzień 1 Stycznia 2011 r. [Results of the Update of Forest Area and Timber Resources in the State Forests as of 1 January 2011] 2011*; BULiGL: Raszyn, Poland, 2011.
33. BULiGL. *Wyniki Aktualizacji Stanu Powierzchni Leśnej i Zasobów Drzewnych w Lasach Państwowych Na Dzień 1 Stycznia 2021 r. [Results of the Update of Forest Area and Timber Resources in the State Forests as of 1 January 2021] 2022*; BULiGL: Raszyn, Poland, 2022.
34. BULiGL Bank Danych o Lasach [Forest Data Bank]—A Database on Resources and Condition of Polish Forests. Available online: https://www.bdl.lasy.gov.pl (accessed on 11 November 2022).
35. Zajączkowski, K. *Hodowla lasu. T. 4[a], Plantacje drzew szybkorosnących [Silviculture. Vol. 4[a], Plantations of Fast-Growing Trees]*; PWRiL: Warszawa, Poland, 2013; ISBN 978-83-09-01143-9.
36. Zajączkowski, K. Nowe Bazy Surowcowe: Plantacje Drzew Poza Lasem Oraz Zadrzewienia [New Resource Bases: Tree Plantations Outside the Forest and Afforestation]. In *Lasy i Gospodarka Leśna Jako Instrumenty Ekonomicznego i Społecznego Rozwoju Kraju [Forests and Forest Management as Instruments of Economic and Social Development of the Country]: Materiały Piątego Panelu Ekspertów w Ramach Prac Nad Narodowym Programem Leśnym Rozwój, Sękocin Stary, 17 Września 2014 Roku*; Kaliszewski, A., Rykowski, K., Eds.; Instytut Badawczy Leśnictwa: Sękocin Stary, Poland, 2015; pp. 290–312, ISBN 978-83-62830-44-2.
37. *Rocznik Statystyczny Leśnictwa i Przemysłu Drzewnego 1961 [Statistical Yearbook of Forestry and Wood Industry 1961]*; PWRiL: Warsaw, Poland, 1963. (In Polish)
38. *Rocznik Statystyczny Leśnictwa i Gospodarki Drewnem 1979 [Statistical Yearbook of Forestry and Wood Economy]*; GUS: Warsaw, Poland, 1979. (In Polish)
39. Ochrona Środowiska [Environment]. *Statistical Yearbook of 1990*; GUS: Warsaw, Poland, 1991.
40. Leśnictwo [Forestry]. *Statistical Yearbook of 2000*; GUS: Warsaw, Poland, 2001.
41. Leśnictwo [Forestry]. *Statistical Yearbook of 2010*; GUS: Warsaw, Poland, 2011.
42. Boruszewski, P.; Laskowska, A.; Jankowska, A.; Klisz, M.; Mionskowski, M. Potential Areas in Poland for Forestry Plantation. *Forests* **2021**, *12*, 1360. [CrossRef]
43. Ochrona Środowiska [Environment]. Statistical Yearbook of 2020. Available online: https://stat.gov.pl/obszary-tematyczne/srodowisko-energia/srodowisko/ochrona-srodowiska-2021,1,22.html (accessed on 15 March 2022).
44. GDOŚ Centralny Rejestr Form Ochrony Przyrody [Central Register of Nature Protection Forms]—A Database on Nature Protection Form in Poland. Available online: https://crfop.gdos.gov.pl/CRFOP (accessed on 11 November 2022).
45. The Nature Conservation Act, 2004 [Dz. U. 2004.92.880 as amended, in Polish]. Available online: http://isap.sejm.gov.pl/isap.nsf/download.xsp/WDU20040920880/U/D20040880Lj.pdf (accessed on 20 March 2020).
46. Polish State Council for Nature Conservation (PROP). Najważniejsze Problemy Ochrony Przyrody w Polsce [The Main Problems of Nature Conservation in Poland]; Polish State Council for Nature Conservation (PROP): 2007. Available online: http://www.salamandra.org.pl/component/content/article/36-prawo/170-problemy-ochrony-przyrody.html?directory=177 (accessed on 30 March 2022).
47. Polish State Council for Nature Conservation (PROP). Opinia w Sprawie Najpilniejszych Wyzwań Dotyczących Ochrony Przyrody w Polsce w Roku 2016 [Opinion on the Most Pressing Challenges for Nature Conservation in Poland in 2016]; Polish State Council for Nature Conservation (PROP): 2016. Available online: https://otop.org.pl/uploads/media/prop-16-04_najpilniejsze-wyzwania-ochrony-przyrody.pdf (accessed on 11 July 2022).
48. Bernadzki, E. Cele hodowli wczoraj i dziś [Silvicultural objectives yesterday and today]. *Sylwan* **1997**, *141*, 23–31.
49. Klocek, A.; Zając, S. Teoria portfela w leśnictwie—Optymalizacja składu gatunkowego drzewostanów [Portfolio theory in the forestry—Optimization of the species composition]. *Sylwan* **2018**, *162*, 971–979. [CrossRef]
50. Święcicki, Z. (Ed.) *Instrukcja Urządzania Lasu [Forest Management Planning Instruction]*; CILP: Warsaw, Poland, 2012; Volume 1, ISBN 978-83-61633-69-3. (In Polish)
51. Czarnecki, A.; Lewandowska-Czarnecka, A. Potential of Poplar Plantation for Enhancing Polish Farm Sustainability. In *Proceedings of the Geo-Environment and Landscape Evolution II: Monitoring, Simulation, Management and Remediation*; WIT Press: Rhodes, Greece, 2006; Volume 1, pp. 241–249.
52. European Environment Agency. *Developing a Forest Naturalness Indicator for Europe: Concept and Methodology for a High Nature Value (HNV) Forest Indicator*; European Environment Agency: Copenhagen, Denmark, 2014; ISBN 978-92-9213-478-5.
53. European Environment Agency. *European Forest Ecosystems: State and Trends*; European Environment Agency: Copenhagen, Denmark, 2016; ISBN 978-92-9213-728-1.

54. Durak, T. Long-Term Trends in Vegetation Changes of Managed versus Unmanaged Eastern Carpathian Beech Forests. *For. Ecol. Manag.* **2010**, *260*, 1333–1344. [CrossRef]
55. Brzeziecki, B. Wieloletnia dynamika drzewostanów w Puszczy Białowieskiej (w warunkach ochrony ścisłej) [Multi-year stand dynamics in the Bialowieza Forest (under strict protection)]. In *Stan Ekosystemów Leśnych Puszczy Białowieskiej [State of the Forest Ecosystems of the Bialowieza Forest]*; Wikło, A., Ed.; CILP: Warsaw, Poland, 2016; pp. 45–58, ISBN 978-83-63895-38-9. (In Polish)
56. Brzeziecki, B. Puszcza Białowieska jako ostoja różnorodności biologicznej [Białowieża Forest as a biodiversity hotspot]. *Sylwan* **2017**, *161*, 971–981. [CrossRef]
57. Brzeziecki, B.; Hilszczański, J.; Kowalski, T.; Łakomy, P.; Małek, S.; Miścicki, S.; Modrzyński, J.; Sowa, J.; Starzyk, J.R. Problem masowego zamierania drzewostanów świerkowych w Leśnym Kompleksie Promocyjnym "Puszcza Białowieska" [Problem of a massive dying—off of Norway spruce stands in the 'Białowieża Forest' Forest Promotional Complex]. *Sylwan* **2018**, *162*, 373–386. [CrossRef]
58. Bubel, K.; Reczyńska, K.; Pech, P.; Świerkosz, K. Secondary Serpentine Forests of Poland as a Refuge for Vascular Flora. *Diversity* **2021**, *13*, 201. [CrossRef]
59. Mayer, H. *Die Wälder Europas*; Gustav Fischer Verlag: Stuttgart, Germany, 1984.
60. Matuszkiewicz, W.; Faliński, J.B.; Kostrowicki, A.S.; Matuszkiewicz, J.M.; Olaczek, R.; Wojterski, T. *Potencjalna Roślinność Naturalna Polski. Mapa Przeglądowa 1:300 000. Arkusze 1-12 [Potential Natural Vegetation of Poland. Survey Map 1:300 000. Sheets 1–12] 1995*; IGiPZ PAN: Warszawa, Poland, 1995.
61. Forest Research Institute (Poland); Jabłoński, M.; Jabłoński, T.; Kowalska, A.; Małachowska, J.; Piwnicki, J. *Raport o Stanie Lasów w Polsce 2010 [Report on the Condition of Forests in Poland in 2010]*; CILP: Warsaw, Poland, 2011.
62. Zajączkowski, G.; Jabłoński, M.; Jabłoński, T.; Szmidla, H.; Kowalska, A.; Małachowska, J.; Piwnicki, J. *Raport o Stanie Lasów w Polsce 2020 [Report on the Condition of Forests in Poland in 2020]*; CILP: Warsaw, Poland, 2021; p. 162. (In Polish)
63. Durak, T.; Holeksa, J. Biotic Homogenisation and Differentiation along a Habitat Gradient Resulting from the Ageing of Managed Beech Stands. *For. Ecol. Manag.* **2015**, *351*, 47–56. [CrossRef]
64. Szymański, S. Siedlisko jako podstawa planowania hodowlanego [Site as base of silvicultural planning]. *Sylwan* **1985**, *129*, 13–21.
65. Błońska, E.; Januszek, K. Wpływ Składu Gatunkowego Drzewostanów Na Aktywność Enzymatyczną i Właściwości Fizykochemiczne Gleb Leśnych [The Influence of Tree Species on Enzyme Activity and Physical-Chemical Properties of Forest Soils]. *Rocz. Glebozn.—Soil Sci. Annu.* **2010**, *61*, 5–14.
66. Trampler, T. Produkcyjność krain i dzielnic przyrodniczo-leśnych [Productivity of natural-forest lands and districts]. *Biul. Inst. Badaw. Leśnictwa* **1954**, *21*, 420–429.
67. *Rocznik Statystyczny Leśnictwa 1945–1967 [Forestry Statistical Yearbook 1945–1967]*; GUS: Warsaw, Poland, 1968. (In Polish)
68. *Rocznik Statystyczny Leśnictwa i Gospodarki Drewnem 1981 [Statistical Yearbook of Forestry and Wood Economy]*; GUS: Warsaw, Poland, 1981. (In Polish)
69. Mill, W. Dynamic Modelling of Polish Forest Soil Response to Changes in Atmospheric Acid Deposition. *Environ. Prot. Eng.* **2007**, *33*, 39–45.
70. Directorate-General of the State Forests. *Sprawozdanie Finansowo-Gospodarcze za 2020 Rok [Report on Financial and Economic State of State Forests in 2020]*; DGLP: Warsaw, Poland, 2021; p. 56. (In Polish)
71. Augusto, L.; Ranger, J.; Binkley, D.; Rothe, A. Impact of Several Common Tree Species of European Temperate Forests on Soil Fertility. *Ann. For. Sci.* **2002**, *59*, 233–253. [CrossRef]
72. Kowalska, A.; Astel, A.; Boczoń, A.; Polkowska, Ż. Atmospheric Deposition in Coniferous and Deciduous Tree Stands in Poland. *Atmos. Environ.* **2016**, *133*, 145–155. [CrossRef]
73. Sewerniak, P. Survey of Some Attributes of Post-Agricultural Lands in Polish State Forests. *EQ* **2015**, *22*, 9–16. [CrossRef]
74. Zajączkowski, S. Charakterystyka lasów polskich [Characteristics of Polish forests]. In *Lata Powojenne i Współczesność [The Post-War Years and the Present Day]*; Z dziejów Lasów Państwowych i leśnictwa polskiego 1924–2004 [The history of the State Forests and the Polish Forestry 1924–2004]; Bernadzki, E., Ed.; CILP: Warsaw, Poland, 2006; Volume 3.1, pp. 62–96, ISBN 83-88478-94-X. (In Polish)
75. Rozwałka, Z. *Leśna Statystyka 1997–2007 [Forestry Statistics 1997–2007]*; CILP: Warsaw, Poland, 2010; ISBN 978-83-61633-24-2. (In Polish)
76. Blicharska, M.; Angelstam, P.; Elbakidze, M.; Axelsson, R.; Skorupski, M.; Węgiel, A. The Polish Promotional Forest Complexes: Objectives, Implementation and Outcomes towards Sustainable Forest Management? *For. Policy Econ.* **2012**, *23*, 28–39. [CrossRef]
77. Kliczkowska, A.; Zielony, R.; Czępińska-Kamińska, D.; Kowalkowski, A.; Sikorska, E.; Krzyżanowski, A.; Cieśla, A.; Czerepko, J. (Eds.) *Siedliskowe Podstawy Hodowli Lasu [Habitat Basis of Silviculture]—Załącznik do Zasad Hodowli Lasu [Annex to the Silvicultural Principles]*; ORW LP: Bedoń, Poland, 2004; ISBN 83-913320-6-3.
78. Babiński, S.; Białkiewicz, F.; Krajewski, T. Stan melioracji wodnych w lasach [State of water reclamation in forests]. *Zesz. Probl. Postępu Nauk Rol.* **1989**, *375*, 127–138.
79. Zabrocka-Kostrubiec, U. Mała Retencja w Lasach Państwowych—Stan i Perspektywy [Small Retention in State Forests—Present Condition and Future Prospects]. *Stud. I Mater. Cent. Edukac. Przyr.-Leśnej* **2008**, *18*, 55–63.
80. Czerniak, A.; Grajewski, S.; Krysztofiak-Kaniewska, A.; Kurowska, E.E.; Okoński, B.; Górna, M.; Borkowski, R. Engineering Methods of Forest Environment Protection against Meteorological Drought in Poland. *Forests* **2020**, *11*, 614. [CrossRef]

81. Miler, A.T. Mała retencja wodna w polskich lasach nizinnych [Small Water Retention in Polish Lowland Forest]. *Infrastrukt. I Ekol. Teren. Wiej./Infrastruct. Ecol. Rural Areas* **2015**, *IV/1*, 979–992. [CrossRef]
82. Więcko, E. (Ed.) Słownik Encyklopedyczny: Leśnictwa, Drzewnictwa, Ochrony Srodowiska, Łowiectwa Oraz Dziedzin Pokrewnych [Encyclopaedic Dictionary: Forestry, Lumbering, Environmental Protection, Hunting and Related Fields]. SGGW: Warszawa, Poland, 1996; ISBN 978-83-00-02825-2.
83. Burton, J.I.; Zenner, E.K.; Frelich, L.E.; Cornett, M.W. Patterns of Plant Community Structure within and among Primary and Second-Growth Northern Hardwood Forest Stands. *For. Ecol. Manag.* **2009**, *258*, 2556–2568. [CrossRef]
84. Durak, T. Changes in Diversity of the Mountain Beech Forest Herb Layer as a Function of the Forest Management Method. *For. Ecol. Manag.* **2012**, *276*, 154–164. [CrossRef]
85. Muzika, R.M. Opportunities for Silviculture in Management and Restoration of Forests Affected by Invasive Species. *Biol. Invasions* **2017**, *19*, 3419–3435. [CrossRef]
86. Podlaski, R. Can Forest Structural Diversity Be a Response to Anthropogenic Stress? A Case Study in Old-Growth Fir *Abies Alba* Mill. Stands. *Ann. For. Sci.* **2018**, *75*, 99. [CrossRef]
87. Decocq, G.; Aubert, M.; Dupont, F.; Bardat, J.; Wattez-Franger, A.; Saguez, R.; de Foucault, B.; Alard, D.; Delelis-Dusollier, A. Silviculture-Driven Vegetation Change in a European Temperate Deciduous Forest. *Ann. For. Sci.* **2005**, *62*, 313–323. [CrossRef]
88. Haze, M. (Ed.) *Zasady Hodowli Lasu [Silvicultural Principles]*, 1st ed.; CILP: Warsaw, Poland, 2012; ISBN 978-83-61633-65-5. (In Polish)
89. *Rocznik Statystyczny Leśnictwa i Przemysłu Drzewnego 1964 [Statistical Yearbook of Forestry and Wood Industry 1964]*; PWRiL: Warsaw, Poland, 1966. (In Polish)
90. Barzdajn, W. Znaczenie hodowli lasu dla ochrony przyrody [Importance of silviculture for nature conservation]. In *Gospodarka Leśna a Ochrona Przyrody [Forest Management and Nature Conservation]*; Gwiazdowicz, D.J., Ed.; Ornatus Wydawnictwo PTL: Poznań, Poland, 2006; pp. 31–50, ISBN 978-83-921460-7-0.
91. Węgiel, A.; Grzywiński, W.; Ciechanowski, M.; Jaros, R.; Kmiecik, A.; Kmiecik, P.; Węgiel, J. Aktywność żerowiskowa nietoperzy w różnych fazach rozwojowych drzewostanów sosny zwyczajnej [Foraging activity of bats in Scots pine stands in different growth stages]. *Sylwan* **2016**, *160*, 767–776. [CrossRef]
92. Main-Knorn, M.; Hostert, P.; Kozak, J.; Kuemmerle, T. How Pollution Legacies and Land Use Histories Shape Post-Communist Forest Cover Trends in the Western Carpathians. *For. Ecol. Manag.* **2009**, *258*, 60–70. [CrossRef]
93. Broda, J. Etapy rozwoju gospodarstwa leśnego w Polsce Ludowej [Stages of development of the forest economy in People's Poland]. *Sylwan* **1985**, *129*, 1–18.
94. Broda, J. (Ed.) *Lasy Państwowe w Polsce w Latach 1944–1990 [State Forests in Poland 1944–1990]*, 1st ed.; PWN: Warsaw-Poznań, Poland, 1997; ISBN 83-01-12359-1.
95. Schütz, J.-P. *Der Plenterwald und Weitere Formen Strukturierter und Gemischter Wälder*; Parey Buchverlag: Berlin, Germany, 2001; ISBN 978-3-8263-3347-7.
96. Halpern, C.B.; Spies, T.A. Plant Species Diversity in Natural and Managed Forests of the Pacific Northwest. *Ecol. Appl.* **1995**, *5*, 913–934. [CrossRef]
97. Decocq, G.; Aubert, M.; Dupont, F.; Alard, D.; Saguez, R.; Wattez-Franger, A.; Foucault, B.D.; Delelis-Dusollier, A.; Bardat, J. Plant Diversity in a Managed Temperate Deciduous Forest: Understorey Response to Two Silvicultural Systems. *J. Appl. Ecol.* **2004**, *41*, 1065–1079. [CrossRef]
98. Referowska-Chodak, E. Potrzeby społeczne w zakresie ochrony przyrody w Lasach Państwowych [Public needs for nature conservation in the State Forests]. In *Gospodarka i Ochrona Przyrody w Lasach w Oczekiwaniach Społecznych [Management and Nature Conservation in Forests in Public Expectations]*; Grzywacz, A., Ed.; PTL: Gniezno, Poland, 2017; pp. 93–105, ISBN 978-83-941444-5-6.
99. Bender, M.J.; Castleberry, S.B.; Miller, D.A.; Bently Wigley, T. Site Occupancy of Foraging Bats on Landscapes of Managed Pine Forest. *For. Ecol. Manag.* **2015**, *336*, 1–10. [CrossRef]
100. Łuczak, M.; Paschalis-Jakubowicz, P. Uwarunkowania rynku drzewnego w Lasach Państwowych w latach 1918—2008 [Determinants of the wood market of the State Forests in the years 1918—2008]. *Sylwan* **2013**, *157*, 506–515. [CrossRef]
101. Jonsson, M.; Bengtsson, J.; Moen, J.; Gamfeldt, L.; Snäll, T. Stand Age and Climate Influence Forest Ecosystem Service Delivery and Multifunctionality. *Environ. Res. Lett.* **2020**, *15*, 0940a8. [CrossRef]
102. Nijnik, M. To an Economist's Perception on Sustainability in Forestry-in-Transition. *For. Policy Econ.* **2004**, *6*, 403–413. [CrossRef]
103. Kozak, J. Forest Cover Changes and Their Drivers in the Polish Carpathian Mountains since 1800. In *Reforesting Landscapes: Linking Pattern and Process*; Landscape series; Nagendra, H., Southworth, J., Eds.; Springer: Dordrecht, The Netherlands; London, UK; New York, NY, USA, 2010; pp. 253–273, ISBN 978-1-4020-9655-6.
104. Kuemmerle, T.; Hostert, P.; Radeloff, V.C.; Perzanowski, K.; Kruhlov, I. Post-Socialist Forest Disturbance in the Carpathian Border Region of Poland, Slovakia, and Ukraine. *Ecol. Appl.* **2007**, *17*, 1279–1295. [CrossRef]
105. Referowska-Chodak, E. *Ochrona Przyrody w Lasach Państwowych—Potrzeby i Oczekiwania Różnych Grup Społecznych Oraz Ich Konsekwencje [Nature Protection in the State Forests—Needs and Expectations of Various Social Groups and Their Consequences]*, 1st ed.; SGGW: Warsaw, Poland, 2020; ISBN 978-83-7583-976-0.
106. Humphrey, J.W. Benefits to Biodiversity from Developing Old-Growth Conditions in British Upland Spruce Plantations: A Review and Recommendations. *Forestry* **2005**, *78*, 33–53. [CrossRef]

107. BULiGL. *Wyniki Aktualizacji Stanu Powierzchni Leśnej i Zasobów Drzewnych w Lasach Poza Zarządem PGL Lasy Państwowe Na Dzień 1 Stycznia 2021 r. [Results of the Update of Forest Area and Timber Resources in Forests Outside the State Forests National Forest Holding as of 1 January 2021] 2022*; BULiGL: Raszyn, Poland, 2022.
108. Vilén, T.; Gunia, K.; Verkerk, P.J.; Seidl, R.; Schelhaas, M.-J.; Lindner, M.; Bellassen, V. Reconstructed Forest Age Structure in Europe 1950–2010. *For. Ecol. Manag.* **2012**, *286*, 203–218. [CrossRef]
109. Stępniewska, M.; Zwierzchowska, I.; Mizgajski, A. Capability of the Polish Legal System to Introduce the Ecosystem Services Approach into Environmental Management. *Ecosyst. Serv.* **2018**, *29*, 271–281. [CrossRef]
110. Kaliszewski, A.; Gil, W. Cele i priorytety "Polityki leśnej państwa" w świetle porozumień procesu Forest Europe (dawniej MCPFE) [Goals and priorities of the 'National Forest Policy' in the light of the Forest Europe (formerly MCPFE) commitments]. *Sylwan* **2017**, *161*, 648–658. [CrossRef]
111. Edwards, D.; Jay, M.; Jensen, F.S.; Lucas, B.; Marzano, M.; Montagné, C.; Peace, A.; Weiss, G. Public preferences for structural attributes of forests: Towards a pan-European perspective. *For. Policy Econ.* **2012**, *19*, 12–19. [CrossRef]
112. Act on Access to Information about the Environment and Its Protection, Public Participation in Environmental Protection and Environmental Impact Assessments, 2008 [Dz. U. 2008.199.1227, in Polish]. Available online: http://isap.sejm.gov.pl/isap.nsf/download.xsp/WDU20081991227/U/D20081227Lj.pdf (accessed on 15 May 2018).
113. Erickson, J.L.; West, S.D. Associations of Bats with Local Structure and Landscape Features of Forested Stands in Western Oregon and Washington. *Biol. Conserv.* **2003**, *109*, 95–102. [CrossRef]
114. Miles, A.C.; Castleberry, S.B.; Miller, D.A.; Conner, L.M. Multi-Scale Roost-Site Selection by Evening Bats on Pine-Dominated Landscapes in Southwest Georgia. *J. Wildl. Manag.* **2006**, *70*, 1191–1199. [CrossRef]
115. Kepel, A. Oczekiwania środowisk przyrodniczych wobec gospodarki leśnej [Naturalists' expectations towards forest management]. In *Wielofunkcyjna Gospodarka Leśna Wobec Oczekiwań Przemysłu Drzewnego i Ochrony Przyrody [Multifunctional Forest Management in Relation to the Expectations of the Wood Industry and Nature Protection]*; Szabla, K., Ed.; PTL: Darłówko, Poland, 2019; pp. 167–187, ISBN 978-83-954196-0-7.
116. Hogberg, L.K.; Patriquin, K.J.; Barclay, R.M.R. Use by Bats of Patches of Residual Trees in Logged Areas of the Boreal Forest. *Am. Midl. Nat.* **2002**, *148*, 282. [CrossRef]
117. Brooks, R.T. Habitat-Associated and Temporal Patterns of Bat Activity in a Diverse Forest Landscape of Southern New England, USA. *Biodivers. Conserv.* **2009**, *18*, 529–545. [CrossRef]
118. Schütz, J.-P.; Götz, M.; Schmid, W.; Mandallaz, D. Vulnerability of Spruce (*Picea Abies*) and Beech (*Fagus Sylvatica*) Forest Stands to Storms and Consequences for Silviculture. *Eur. J. For. Res.* **2006**, *125*, 291–302. [CrossRef]
119. Bałazy, R.; Zasada, M.; Ciesielski, M.; Waraksa, P.; Zawiła-Niedźwiecki, T. Forest Dieback Processes in the Central European Mountains in the Context of Terrain Topography and Selected Stand Attributes. *For. Ecol. Manag.* **2019**, *435*, 106–119. [CrossRef]
120. Seidl, R.; Schelhaas, M.-J.; Lexer, M.J. Unraveling the Drivers of Intensifying Forest Disturbance Regimes in Europe. *Glob. Change Biol.* **2011**, *17*, 2842–2852. [CrossRef]
121. Modrzyński, J. Defoliation of Older Norway Spruce (*Picea abies* /L./ Karst.) Stands in the Polish Sudety and Carpathian Mountains. *For. Ecol. Manag.* **2003**, *181*, 289–299. [CrossRef]
122. Płotkowski, L.; Zając, S.; Wysocka-Fijorek, E.; Gruchała, A.; Piekutin, J.; Parzych, S. Economic Optimization of the Rotation Age of Stands. *Folia For. Pol.* **2016**, *58*, 188–197. [CrossRef]
123. Parzych, S.; Mandziuk, A.; Wysocka-Fijorek, E. Wpływ zasobności drzewostanów sosnowych na ustalanie ekonomicznego wieku dojrzałości rębnej [Impact of Scots pine stand growing stock on determining the optimal economic rotation age]. *Sylwan* **2018**, *162*, 671–678. [CrossRef]
124. Keith, S.A.; Newton, A.C.; Morecroft, M.D.; Bealey, C.E.; Bullock, J.M. Taxonomic Homogenization of Woodland Plant Communities over 70 Years. *Proc. R. Soc. B* **2009**, *276*, 3539–3544. [CrossRef]
125. Rawlik, K.; Jagodziński, A.M. Ekologiczne znaczenie roślin runa leśnego [Ecological significance of undergrowth plants]. *ACADEMIA PAN* **2019**, *3–4*, 50–53.
126. Reinecke, J.; Klemm, G.; Heinken, T. Vegetation Change and Homogenization of Species Composition in Temperate Nutrient Deficient Scots Pine Forests after 45 Yr. *J. Veg. Sci.* **2014**, *25*, 113–121. [CrossRef]
127. Lindner, M.; Maroschek, M.; Netherer, S.; Kremer, A.; Barbati, A.; Garcia-Gonzalo, J.; Seidl, R.; Delzon, S.; Corona, P.; Kolström, M.; et al. Climate Change Impacts, Adaptive Capacity, and Vulnerability of European Forest Ecosystems. *For. Ecol. Manag.* **2010**, *259*, 698–709. [CrossRef]
128. Bartosz, R.; Bukowska, M.; Chylarecki, P.; Ignatowicz, A.; Puzio, A.; Wilińska, A. *Ocena Wpływu Zmian Klimatu Na Różnorodność Biologiczną Oraz Wynikające z Niej Wytyczne Dla Działań Administracji Ochrony Przyrody Do Roku 2030 [Assessment of the Impact of Climate Change on Biodiversity and the Resulting Guidelines for Conservation Administration Action up to 2030]*; GDOŚ: Warsaw, Poland, 2012. (In Polish)
129. Keča, N.; Koufakis, I.; Dietershagen, J.; Nowakowska, J.A.; Oszako, T. European Oak Decline Phenomenon in Relation to Climatic Changes. *Folia For. Pol.* **2016**, *58*, 170–177. [CrossRef]
130. Grajewski, S. Effectiveness of Forest Fire Security Systems in Poland. *Infrastrukt. I Ekol. Teren. Wiej./Infrastruct. Ecol. Rural Areas* **2017**, *4*, 1563–1576. [CrossRef]
131. Forest Research Institute (Poland) Mapa zagrożenia pożarowego lasu w Polsce [Map of forest fire danger in Poland]. Available online: http://bazapozarow.ibles.pl/zagrozenie (accessed on 25 June 2023).

132. Schütz, J.-P. Opportunities and Strategies of Transforming Regular Forests to Irregular Forests. *For. Ecol. Manag.* **2001**, *151*, 87–94. [CrossRef]
133. Bernadzki, E. Postępy w zagospodarowaniu lasów [Progress in forest management]. In *Lata Powojenne i Współczesność [The Post-War Years and the Present Day]*; Z dziejów Lasów Państwowych i leśnictwa polskiego 1924–2004 [The history of the State Forests and the Polish Forestry 1924–2004]; Bernadzki, E., Ed.; CILP: Warsaw, Poland, 2006; Volume 3.1, pp. 123–172, ISBN 83-88478-94-X. (In Polish)
134. Zielony, R.; Kliczkowska, A. *Regionalizacja Przyrodniczo-Leśna Polski 2010 [Natural-Forest Regionalization of Poland 2010]*; CILP: Warsaw, Poland, 2012; ISBN 978-83-61633-62-4. (In Polish)
135. Jasnowski, M. Aktualny stan i program ochrony torfowisk w Polsce [Current status and conservation programme for peatlands in Poland]. *Chrońmy Przyr. Ojczystą* **1977**, *33*, 18–29.
136. Directorate-General of the State Forests. *Sprawozdanie Finansowo-Gospodarcze za 2016 Rok [Report on Financial and Economic State of State Forests in 2016]*; DGLP: Warsaw, Poland, 2017; p. 43.
137. United Nations SDG 2015. Available online: https://sdgs.un.org/2030agenda (accessed on 15 January 2023).
138. Directorate-General of the State Forests. *Sprawozdanie Finansowo-Gospodarcze za 2019 Rok [Report on Financial and Economic State of State Forests in 2019]*; DGLP: Warsaw, Poland, 2020; p. 56. (In Polish)
139. Rutkowski, A. Lekko w górę [Slightly upwards]. *Głos Lasu* **2019**, *2*, 6–9.
140. Rykowski, K. Badania naukowe w leśnych kompleksach promocyjnych [Research in promotional forest complexes]. In *Mat. Konf. "10 Lat Leśnych Kompleksów Promocyjnych" [Conference Materials "10 Years of Promotional Forest Complexes"]. Rogów, 15–16.11.2004*; CILP: Warsaw, Poland, 2004; pp. 51–54. (In Polish)
141. European Commission, Directorate-General for Environment. *EU Biodiversity Strategy for 2030: Bringing Nature Back into Our Lives*; Publications Office of the European Union: Luxembourg, 2021.

MDPI AG
Grosspeteranlage 5
4052 Basel
Switzerland
Tel.: +41 61 683 77 34

Forests Editorial Office
E-mail: forests@mdpi.com
www.mdpi.com/journal/forests

www.ingramcontent.com/pod-product-compliance
Lightning Source LLC
Chambersburg PA
CBHW040744250926
51850CB00007BB/1404
* 9 7 8 3 7 2 5 8 2 2 4 0 9 *